LONDON MATHEMATICAL SOCIETY
MONOGRAPHS NEW SERIES

LONDON MATHEMATICAL SOCIETY MONOGRAPHS
NEW SERIES

Previous volumes of the LMS monographs were published by Academic Press to whom all enquiries should be addressed. Volumes in the New Series will be published by Oxford University Press throughout the world

NEW SERIES

1. *Diophantine inequalities* R. C. Baker
2. *The Schur multiplier* Gregory Karpilovsky
3. *Existentially closed groups* Graham Higman and Elizabeth Scott
4. *The asymptotic solution of linear differential systems* M. S. P. Eastham

The Asymptotic Solution of Linear Differential Systems

Applications of the Levinson Theorem

M. S. P. Eastham

Mathematics Department,
University of Bahrain

CLARENDON PRESS · OXFORD
1989

Oxford University Press, Walton Street, Oxford OX2 6DP

Oxford New York Toronto
Delhi Bombay Calcutta Madras Karachi
Petaling Jaya Singapore Hong Kong Tokyo
Nairobi Dar es Salaam Cape Town
Melbourne Auckland

and associated companies in
Berlin Ibadan

Oxford is a trade mark of Oxford University Press

Published in the United States
by Oxford University Press, New York

British Library Cataloguing in Publication Data
Eastham, M. S. P. (Michael Stephen Patrick)
The asymptotic solution of linear
differential systems.
1. Linear differential equations
I. Title II. Series
515.3′54
ISBN 0-19-853299-7

Library of Congress Cataloging in Publication Data
Eastham, M. S. P. (Michael Stephen Patrick)
The asymptotic solution of linear
differential systems.
(London Mathematical Society monographs; new ser.)
Bibliography: p. Includes Index.
1. Differential equations, Linear—Asymptotic theory.
I. Title II. Series
QA372.M187 1989 515.3′54 88-25382
ISBN 0-19-853299-7

Typeset by The Universities Press (Belfast) Ltd
Printed in Great Britain
at the University Printing House, Oxford
by David Stanford
Printer to the University

Preface

The asymptotic theorem which provides the basis for the theory in this book was proved by Levinson and published in 1948. At that time, the theorem itself represented a considerable advance on existing related results but, in contrast, initial applications of the theorem were only of a very straightforward kind. It was not until some 25 years later that a real appreciation of the significance and range of applications of the theorem started to become evident through the work of Harris and Lutz (1974). There were of course many asymptotic results for differential equations and systems in the intervening years and, in the main, these results may be categorized roughly as follows. First, there are those whose proof involved some kind of fixed-point argument. In the light of later developments covered in this book, such arguments can be regarded as in effect a re-proof of the Levinson theorem for the particular situation under consideration. Second, there are results obtained directly from the Levinson theorem but where the deductive procedures are particularly complicated and, in the absence of clearly defined techniques, no development of such procedures followed. In the final category, there are specialized results which are proved by quite separate methods, a notable example being the Prüfer transformation method of Atkinson (1954) for second-order equations.

Meanwhile, in the Russian literature, the asymptotic theory of differential systems was also being developed by Rapoport (1954) and others. Once again, although some direct applications were made, the important paper of Fedorjuk (1966*b*) for example, falls into the first category mentioned above.

Two papers which did make direct and significant use of Levinson's theorem in this period were those of Hinton (1968) and Devinatz (1971). To these can be added the more recent papers of Cassell (1981, 1982) Grudniewicz (1980*b*) and Read (1979) which appeared around 1980 and established certain important methods in the application of Levinson's theorem. The way was then open for further developments, of which the paper of mine (Eastham 1985) may be mentioned, and it became clear that there was scope for a book which would give a coherent account of this whole area of asymptotic theory of differential systems.

The purpose of this book is to show that many of the asymptotic results from 1948 onwards can be deduced directly from the Levinson theorem by means of certain well-defined techniques. These techniques involve transformations of the solution vectors of the differential systems, and our general theme is to show how the various transformations are used both individually and in combinations. In Chapter 1, the main types of transformation are identified, with the applications following in the later chapters. We do not go to the extent of applying every possible transformation to every differential system of interest, but we intend to cover, or at least indicate how to cover, the situations of most significance. The book is by no means intended to be the last word on the subject and it is my hope that, with the existing theory presented in an organized manner, further research may be stimulated.

Two particular topics have motivated much of the work on asymptotic theory over the years. The first is the deficiency index problem in which, as part of the spectral theory of self-adjoint differential operators, the problem is to determine the number of linearly independent solutions of the associated differential equation which are $L^2(a, \infty)$. Clearly, if the asymptotic form of the solutions at infinity can be determined sufficiently accurately, the L^2-nature of the solutions follows. Equations of self-adjoint type will be covered in Chapter 3 with the deficiency index application in §§ 3.10 and 3.11. The second topic is the phenomenon of resonance for perturbations of the equation of simple harmonic motion. Again there are applications to spectral theory and the location of eigenvalues embedded in the continuous spectrum. These matters will be developed in Chapter 4.

Asymptotic results of Poincaré–Perron type, that is, results of the form $y'/y \to l$ (a finite or infinite limit), do not fit into our scheme and are in any case a much less precise and less sensitive type of result than those obtained from Levinson's theorem. Nor do we cover in any detail systems with a general perturbed Jordan matrix. Such systems at present appear to have a rather limited range of applications in relation to the technical requirements of the theory. The case of a single elementary Jordan block, however, is important and is fully covered in § 1.10 and subsequently.

Some of the material in the book is new, some is a re-presentation of existing theory in line with the main theme of the book, and much of it has appeared only recently in the research literature. The notes at the end of each chapter are necessarily somewhat copious in view of the nature of the subject, and they contain references to the original papers as well as some historical remarks and comments on other matters of detail which are best separated from the main text.

It is a pleasure to thank a number of friends and colleagues who have kindly read much of the book and made comments: Dr J. S. Cassell (City of London Polytechnic), Prof. F. R. Dias-Agudo (Academy of Sciences, Lisbon), Prof. H. Kalf (University of Munich), Dr J. B. McLeod (Mathematical Institute, Oxford) and Prof. A. D. Wood (N.I.H.E., Dublin). To them and to many others who have provided me with the opportunity to lecture on, and discuss, the material in this book over the past ten years, I would like to express my gratitude.

King's College
University of London
May 1988

M.S.P.E.

Contents

1. Asymptotically diagonal systems **1**

1.1 Introduction 1
1.2 Notation and basic theory 3
1.3 The Levinson theorem 8
1.4 Proof of the Levinson theorem 11
1.5 The Hartman–Wintner theorem 16
1.6 A pointwise condition on $R(x)$ 21
1.7 Conditions on higher derivatives of $R(x)$ 28
1.8 Asymptotically constant systems 34
1.9 Higher-order differential equations 36
1.10 Coefficient matrices of Jordan type 41
1.11 Integral conditions with non-absolute convergence 46
Notes and references 51

2. Two-term differential equations **55**

2.1 The second-order equation 55
2.2 The Liouville–Green asymptotic formulae 57
2.3 Repeated diagonalization 60
2.4 Extended Liouville–Green asymptotic formulae 63
2.5 The Liouville–Green transformation 68
2.6 Equations of Euler type 74
2.7 Subdominant coefficient q 77
2.8 Application of the Hartman–Wintner theorem 84
2.9 Higher-order equations 90
2.10 Higher-order equations of Euler type 93
2.11 Subdominant coefficient q 95
Notes and references 101

3. Equations of self-adjoint type **105**

3.1 Introduction 105
3.2 Eigenvalues of the same magnitude 108
3.3 Eigenvalues of differing magnitudes 113
3.4 Small eigenvalues 121
3.5 The fourth-order equation 126
3.6 Odd-order equations 134
3.7 Equations of generalized hypergeometric type 138
3.8 Integration of asymptotic formulae 143
3.9 Estimation of error terms 147

3.10 The deficiency index problem 152
3.11 Evaluation of deficiency indices 155
Notes and references 162

4. Resonance and non-resonance **166**

4.1 Introduction 166
4.2 Perturbations of harmonic oscillation 170
4.3 Resonance and embedded eigenvalues 174
4.4 Two examples 178
4.5 Higher-order equations 180
4.6 Non-resonance for systems 183
4.7 The matrices Λ_1, Λ_2, Λ_3 188
4.8 The form of P 196
4.9 Values of λ in the resonant set 199
4.10 Resonance for $M \geqslant 2$ 203
4.11 Perturbations without periodicity 211
4.12 Systems with $\Lambda_0 = 0$ 217
4.13 Slowly decaying oscillatory coefficient 221
Notes and references 226

Bibliography **231**

Index **241**

1
Asymptotically diagonal systems

1.1 INTRODUCTION

The basic problem posed by a differential system

$$Y'(x) = A(x)Y(x) \tag{1.1.1}$$

is that the solution vectors $Y(x)$ cannot normally be written as explicit expressions involving the entries of the given square matrix $A(x)$. It is of course this difficulty which creates the challenge and interest in developing a wide range of techniques for investigating the properties of solutions, where (1.1.1) alone is taken as the starting point. The general aim of the asymptotic theory of (1.1.1) is to identify those types of system whose solutions can nevertheless be approximated by such explicit expressions as x approaches some limiting value, the various types being specified by some given property of $A(x)$ in the neighbourhood of the limiting value. Throughout this book we consider the asymptotic situation where $x \to \infty$, and therefore we take (1.1.1) to be defined on some x-interval $[a, \infty)$.

An obvious starting point for the asymptotic theory of (1.1.1) is to examine the case where (1.1.1), though not itself soluble explicitly, does for large x approximate to a simple system

$$Y'(x) = A_0(x)Y(x) \tag{1.1.2}$$

which can be solved explicitly. The question then is whether the solutions of (1.1.1) approximate to the known solutions of (1.1.2) for large x. As we shall see, the answer to this question is not always straighforward and unexpected features do arise.

We now introduce two simple systems (1.1.2) which form the basis for the theory in this chapter.

(i) The constant coefficient system

Here $A_0(x)$ is a constant matrix C, and (1.1.2) is

$$Y'(x) = CY(x). \tag{1.1.3}$$

Let λ be an eigenvalue of C with corresponding eigenvector u. Then a solution of (1.1.3) is

$$Y(x) = u e^{\lambda x} \tag{1.1.4}$$

since here

$$Y'(x) = \lambda u e^{\lambda x} = C u e^{\lambda x} = CY(x).$$

We regard (1.1.4) as an explicit expression for a solution of (1.1.3) in terms of the matrix C. If C is an $n \times n$ matrix and if there are n linearly independent eigenvectors u_k corresponding to eigenvalues λ_k $(1 \leqslant k \leqslant n)$, (1.1.4) gives n linearly independent solutions

$$Y_k(x) = u_k e^{\lambda_k x} \tag{1.1.5}$$

of (1.1.3).

(ii) The diagonal system

Here $A_0(x)$ is a diagonal matrix $\Lambda(x)$, and (1.1.2) is

$$Y'(x) = \Lambda(x) Y(x). \tag{1.1.6}$$

We write

$$\Lambda(x) = \mathrm{dg}(\lambda_1(x), \ldots, \lambda_n(x)).$$

Then, denoting the components of $Y(x)$ by $y_j(x)$ $(1 \leqslant j \leqslant n)$, we can split up (1.1.6) into the first-order equations

$$y_j'(x) = \lambda_j(x) y_j(x)$$

for each $y_j(x)$ separately. Hence

$$y_j(x) = c_j \exp\left(\int_a^x \lambda_j(t)\, dt\right),$$

where the c_j are arbitrary constants. If we now take any integer k $(1 \leqslant k \leqslant n)$ and choose $c_k = 1$ and the other c_j zero, we obtain the solution

$$Y_k(x) = e_k \exp\left(\int_a^x \lambda_k(t)\, dt\right) \tag{1.1.7}$$

of (1.1.6), where e_k is the coordinate vector whose kth component is unity and other components zero. Again, as k varies from 1 to n, we have explicit expressions (1.1.7) for n linearly independent solutions of (1.1.6).

Moving on now to systems (1.1.1) which approximate to (1.1.3) or (1.1.6) when x is large, we consider

$$Y'(x) = \{C + R(x)\} Y(x) \tag{1.1.8}$$

or

$$Y'(x) = \{\Lambda(x) + R(x)\} Y(x), \tag{1.1.9}$$

where $R(x)$ is an $n \times n$ matrix which is small in some sense as $x \to \infty$. We describe these systems as asymptotically constant and asymptotically diagonal respectively. Of the two, (1.1.9) is the more fundamental because, later in the chapter, we obtain the results for (1.1.8) from those for (1.1.9) by means of a simple transformation.

There are various ways in which $R(x)$ can be thought of as small when $x \to \infty$, the most obvious being

$$R(x) \to 0 \tag{1.1.10}$$

as $x \to \infty$. This condition on $R(x)$ is not, however, a suitable one to work with at the outset because, as we shall see, it does allow a considerable variety in the asymptotic behaviour of solutions of (1.1.9), with more detailed properties of $R(x)$ also influencing the results.

In the theorem on which the asymptotic theory of (1.1.9) rests, the condition on $R(x)$ which arises naturally is

$$\int_a^\infty |R(x)|\, dx < \infty, \tag{1.1.11}$$

by which we mean that each entry in $R(x)$ has an absolutely convergent infinite integral. This theorem was proved by Levinson (1948) and it states that, subject to (1.1.11) and a condition on $\Lambda(x)$ which will be introduced into the present account in § 1.3, (1.1.9) has solutions $Y_k(x)$ $(1 \leq k \leq n)$ such that

$$Y_k(x) = \{e_k + o(1)\} \exp\left(\int_a^x \lambda_k(t)\, dt\right) \tag{1.1.12}$$

as $x \to \infty$. Thus, comparing (1.1.12) with the right-hand side in (1.1.7), we see that the Levinson theorem gives conditions on R and Λ under which the solutions of (1.1.9) do approximate to those of (1.1.6) for large x. This result provides the starting point for the asymptotic theory of (1.1.1) that we mentioned towards the beginning of this section.

We refer to a system (1.1.9) in which (1.1.11) holds as having the *Levinson form,* and we prove the Levinson theorem itself in §§ 1.3–1.4. Then, in the later sections of this chapter, we give a number of consequences of the main theorem which deal with the solutions not only of (1.1.8), but also of other systems (1.1.9) where (1.1.11) is replaced by some alternative smallness condition on $R(x)$. These results, together with those on quite different types of system (1.1.1) in subsequent chapters, are all proved by means of transformations which take the system under consideration into a system of the Levinson form.

1.2 NOTATION AND BASIC THEORY

There are several basic facts concerning linear algebra and linear differential systems which will be used on various occasions, and it is convenient to have these results available here for reference. We also introduce some notation that will be used throughout.

Notation For an $m \times n$ matrix $A = (a_{ij})$, we define the modulus $|A|$ by

$$|A| = \sum_{i=1}^{m} \sum_{j=1}^{n} |a_{ij}|. \tag{1.2.1}$$

Then, for two matrices whose product is defined, we have

$$|AB| \leq |A|\,|B|.$$

For a square matrix A, we denote by either $\mathrm{dg}\, A$ or, more fully,

$$\mathrm{dg}\,(a_{11}, \ldots, a_{nn})$$

the diagonal matrix formed by the diagonal entries a_{ii} in A. We also denote by A^{t} and A^* the transposed and adjoint matrices defined by $A^{\mathrm{t}} = (a_{ji})$ and $A^* = (\bar{a}_{ji})$.

For matrices whose entries are differentiable functions of x, we define

$$A'(x) = (a'_{ij}(x)),$$

and similarly for the integral of $A(x)$ over an interval (a, b). If $b > a$, we have immediately from (1.2.1)

$$\left| \int_a^b A(x)\,\mathrm{d}x \right| \leq \int_a^b |A(x)|\,\mathrm{d}x.$$

Statements such as $A(x) \in L^P(a, \infty)$ and $A(x) = O(x^{-1})$ $(x \to \infty)$ mean simply that every entry $a_{ij}(x)$ has the relevant property.

Linear algebra We shall often use the result that an $n \times n$ matrix A with n distinct eigenvalues λ_k can be expressed in the diagonal form

$$T^{-1}AT = \Lambda, \tag{1.2.2}$$

where

$$\Lambda = \mathrm{dg}\,(\lambda_1, \ldots, \lambda_n)$$

and T has the corresponding eigenvectors v_k as its columns. The computation of T^{-1} is generally complicated but, in certain cases, the form of T^{-1} can be obtained directly from the form of T. The simplest case is where A is hermitian and then, with normalized eigenvectors, we have $T^{-1} = T^*$. The following lemma gives a useful result of the same nature for the case where A and its transpose are related by a similarity transformation.

Lemma 1.2.1 *Let the $n \times n$ matrix A have n distinct eigenvalues λ_k with corresponding eigenvectors v_k, and let T be as in* (1.2.2). *Let there be a non-singular matrix F such that*

$$A = F^{-1}A^{\mathrm{t}}F \tag{1.2.3}$$

and define the scalar

$$M_k = (Fv_k)^{\mathrm{t}} v_k. \tag{1.2.4}$$

Then $M_k \neq 0$ *and* T^{-1} *has the rows*

$$(Fv_k)^{\mathrm{t}}/M_k.$$

Proof By (1.2.3), $A^{\mathrm{t}}F^{\mathrm{t}} = F^{\mathrm{t}}A$. Then, using

$$Av_j = \lambda_j v_j, \qquad Av_k = \lambda_k v_k,$$

we have

$$\begin{aligned}\lambda_k (Fv_k)^{\mathrm{t}} v_j &= \lambda_k v_k^{\mathrm{t}} F^{\mathrm{t}} v_j = v_k^{\mathrm{t}} A^{\mathrm{t}} F^{\mathrm{t}} v_j \\ &= v_k^{\mathrm{t}} F^{\mathrm{t}} A v_j = \lambda_j (Fv_k)^{\mathrm{t}} v_j.\end{aligned}$$

Hence, when $j \neq k$,

$$(Fv_k)^{\mathrm{t}} v_j = 0.$$

Now let S denote the matrix with rows $(Fv_k)^{\mathrm{t}}$. By what we have just proved,

$$ST = \mathrm{dg}(M_1, \ldots, M_n),$$

where the M_k are defined by (1.2.4). Since S and T are non-singular, we have $M_k \neq 0$ and the lemma follows. □

The expression (1.2.4) for M_k can be written in a more useful form if A has additional properties and, in the next lemma, we express M_k in terms of the characteristic polynomial of A. We denote by E the $n \times n$ matrix whose entries are zero except on the secondary diagonal, where

$$e_{ij} = 1 \qquad (i + j = n + 1).$$

Lemma 1.2.2 *Let* A *have the form*

$$A = EB, \tag{1.2.5}$$

where B *is a symmetric matrix with*

$$b_{ij} = 0 \qquad (n + 3 \leqslant i + j \leqslant 2n)$$

and b_{ij} $(i + j = n + 2)$ *all non-zero. Let* M_k *be defined by* (1.2.4), *where* $F = E$ *and the first component of each* v_k *is taken to be unity. Then*

$$M_k = \left[\frac{\mathrm{d}}{\mathrm{d}\lambda} \det(\lambda I - A)\right]_{\lambda = \lambda_k} \Big/ \prod_{i+j=n+2} b_{ij},$$

where λ_k *is the eigenvalue of* A *corresponding to* v_k.

We refer to the literature (Eastham 1984, § 3) for the proof of the lemma.

Finally, we mention the Jordan normal form of A which replaces (1.2.2) when A does not necessarily have n linearly independent eigenvectors. A $j \times j$ matrix of the type

$$J(\lambda) = \begin{bmatrix} \lambda & 1 & & 0 \\ & \ddots & \ddots & \\ & & \ddots & 1 \\ 0 & & & \lambda \end{bmatrix} \tag{1.2.6}$$

is called an elementary Jordan matrix. We then have the result that, for any $n \times n$ matrix A, there is a non-singular matrix T such that

$$T^{-1}AT = J_1(\lambda_1) \oplus \cdots \oplus J_m(\lambda_m), \tag{1.2.7}$$

where $\lambda_1, \ldots, \lambda_m$ are eigenvalues of A and $J_i(\lambda_i)$ is an elementary Jordan matrix of some order j_i. The λ_i are not necessarily distinct. The right-hand side of (1.2.7) is a block diagonal matrix formed by the $J_i(\lambda_i)$ and it is known as the Jordan normal form of A.

Linear differential systems The basic existence theorem for (1.1.1) states that solutions $Y(x)$ exist throughout an interval I provided that $A(x)$ is locally Lebesgue integrable in I. The solutions are then locally absolutely continuous in I, and (1.1.1) itself holds almost everywhere in I. If a is a given point in I and b is a given vector, there is a unique solution which satisfies the initial condition $Y(a) = b$. Virtually the whole book, however, can be understood if $A(x)$ is assumed to be continuous or piecewise continuous, and solutions are taken in the familiar sense of continuous (or piecewise continuous) differentiability.

A solution matrix for (1.1.1) is an $n \times n$ matrix $\Phi(x)$ whose columns are solutions $Y_i(x)$ $(1 \leqslant i \leqslant n)$ of (1.1.1). Then $\Phi(x)$ satisfies the corresponding matrix differential equation

$$\Phi'(x) = A(x)\Phi(x). \tag{1.2.8}$$

Further, $\Phi(x)$ is either singular for all x in I or non-singular for all x in I, according as the $Y_i(x)$ are linearly dependent or linearly independent in I. In the latter case, $\Phi(x)$ is said to be a fundamental matrix for (1.1.1).

One of the basic properties of a fundamental matrix $\Phi(x)$ is that any solution of (1.1.1) can be expressed as

$$Y(x) = \Phi(x)c,$$

where c is a constant vector. If we also consider a second differential system

$$Z'(x) = \{A(x) + B(x)\}Z(x) \tag{1.2.9}$$

along with (1.1.1), any solution $Z(x)$ can be related to given linearly independent solutions $Y_i(x)$ of (1.1.1) by the formula

$$Z(x) = \Phi(x)c + \Phi(x)\int_a^x \Phi^{-1}(t)B(t)Z(t)\,\mathrm{d}t, \tag{1.2.10}$$

where again a is a given point in I and c is a suitable constant vector. This formula is often known as the variation of parameters formula. A useful generalization of (1.2.10) is

$$Z(x) = \Phi(x)c + \Phi_1(x)\int_a^x \Phi^{-1}(t)B(t)Z(t)\,\mathrm{d}t$$

$$- \Phi_2(x)\int_x^b \Phi^{-1}(t)B(t)Z(t)\,\mathrm{d}t, \qquad (1.2.11)$$

where a and b are given points in I, and the fundamental matrix $\Phi(x)$ is split up into any two separate solution matrices for (1.1.1) in the form

$$\Phi(x) = \Phi_1(x) + \Phi_2(x).$$

We also note the effect of making the transformation

$$Z(x) = \Phi(x)W(x) \qquad (1.2.12)$$

in (1.2.9). The corresponding W-system is given by

$$\Phi W' + \Phi' W = (A + B)\Phi W.$$

Using (1.2.8), we obtain

$$W' = \Phi^{-1}B\Phi W.$$

Thus the effect of (1.2.12) on (1.2.9) is to remove the matrix $A(x)$.

The determinant of Φ satisfies the Liouville–Jacobi formula

$$\det \Phi(x) = \det \Phi(a) \exp\left(\int_a^x \operatorname{tr} A(t)\,\mathrm{d}t\right), \qquad (1.2.13)$$

where $\operatorname{tr} A$ denotes the trace of A defined by

$$\operatorname{tr} A = \sum_{i=1}^{n} a_{ii},$$

and two simple consequences of this formula will be required later. First, let Ψ be a given non-singular matrix and define

$$A = \Psi' \Psi^{-1}. \qquad (1.2.14)$$

Then $\Psi' = A\Psi$ and we can take $\Phi = \Psi$ in (1.2.13). This gives

$$\frac{\mathrm{d}}{\mathrm{d}x} \log \det \Psi = \operatorname{tr} \Psi' \Psi^{-1} = -\operatorname{tr} \Psi(\Psi^{-1})'. \qquad (1.2.15)$$

Second, let C be a given matrix, and then introduce x as a parameter and define $\Psi(x) = \exp(xC)$. Then (1.2.14) gives $A = C$, and it follows from (1.2.13) that

$$\det \exp(xC) = \exp(x \operatorname{tr} C).$$

Taking $x = 1$, we obtain

$$\det \exp C = \exp \operatorname{tr} C \tag{1.2.16}$$

for any matrix C.

1.3 THE LEVINSON THEOREM

In this section we give the statement of the Levinson theorem on asymptotically diagonal systems together with a number of explanatory comments on the conditions that are imposed in the theorem. The actual proof will follow in § 1.4.

Theorem 1.3.1 (The Levinson theorem) *Let $\Lambda(x)$ be an $n \times n$ diagonal matrix,*

$$\Lambda(x) = \operatorname{dg}(\lambda_1(x), \ldots, \lambda_n(x)),$$

which satisfies the following two-part condition L.

L. *For each pair of integers i and j in $[1, n]$ $(i \neq j)$ and for all x and t such that $a \leqslant t \leqslant x < \infty$, either*

(a)
$$\int_t^x \operatorname{re}\{\lambda_i(s) - \lambda_j(s)\}\, ds \leqslant K_1 \tag{1.3.1}$$

or

(b)
$$\int_t^x \operatorname{re}\{\lambda_i(s) - \lambda_j(s)\}\, ds \geqslant K_2, \tag{1.3.2}$$

where K_1 and K_2 are constants.

Let the $n \times n$ matrix $R(x)$ satisfy

$$\int_a^\infty |R(x)|\, dx < \infty. \tag{1.3.3}$$

Then, as $x \to \infty$, the system

$$Y'(x) = \{\Lambda(x) + R(x)\} Y(x) \tag{1.3.4}$$

has solutions $Y_k(x)$ $(1 \leqslant k \leqslant n)$ with the asymptotic form

$$Y_k(x) = \{e_k + o(1)\} \exp\left(\int_a^x \lambda_k(t)\, dt\right). \tag{1.3.5}$$

To gain some idea of the nature of Condition L, let us note a particular situation which the condition definitely excludes. This situation is where

$$\liminf \int_a^x \operatorname{re}\{\lambda_i(s) - \lambda_j(s)\}\, ds = -\infty$$

and

$$\limsup \int_a^x \operatorname{re}\{\lambda_i(s) - \lambda_j(s)\}\, ds = \infty$$

both occur for one and the same pair (i, j). In terms of (1.3.5) therefore, Condition L implies that we are dealing with systems where the ratio

$$|Y_i(x)|/|Y_j(x)| \tag{1.3.6}$$

of two given solutions does not satisfy both

$$\liminf |Y_i(x)|/|Y_j(x)| = 0$$

and

$$\limsup |Y_i(x)|/|Y_j(x)| = \infty,$$

that is, the ratio does not oscillate between large and small values as $x \to \infty$. The condition therfore represents a regularity situation for the relative sizes of any two solutions as $x \to \infty$.

As they stand, (1.3.1) and (1.3.2) are not mutually exclusive, but they can be re-written in a way which does exhibit this feature if (1.3.1), for example, is sub-divided. Let

$$I_{ij}(x) = \int_a^x \operatorname{re}\{\lambda_i(s) - \lambda_j(s)\}\, ds.$$

Then (1.3.1) is

$$I_{ij}(x) - I_{ij}(t) \leqslant K_1,$$

and it follows immediately that, either $I_{ij}(x) \to -\infty$ as $x \to \infty$, or $I_{ij}(x) \geqslant$ (const.) for all $x \geqslant a$. Putting the latter alternative together with (1.3.2), we can write Condition L in the following equivalent form.

L′. For each pair of integers i and j in $[1, n]$ $(i \neq j)$ and all x and t such that $a \leqslant t \leqslant x < \infty$, either

(a′) $$\int_a^x \operatorname{re}\{\lambda_i(s) - \lambda_j(s)\}\, ds \to -\infty \qquad (x \to \infty) \tag{1.3.7}$$

and

$$\int_t^x \operatorname{re}\{\lambda_i(s) - \lambda_j(s)\}\, ds \leqslant K_1 \tag{1.3.8}$$

or

(b′) $$\int_t^x \operatorname{re}\{\lambda_i(s) - \lambda_j(s)\}\, ds \geqslant K_2. \tag{1.3.9}$$

With a given j, Condition L′ divides the integers i in $[1, n]$ into two classes according as λ_i falls under part (a′) or part (b′). Correspondingly, in terms of (1.3.6), the solutions $Y_i(x)$ are divided into those which are small compared to $Y_j(x)$ as $x \to \infty$ and those which are not. For this reason, Condition L′ is known as the *dichotomy condition*, and we shall also use the same terminology for Condition L. It is possible to subdivide (1.3.9) further into the two cases where $I_{ij}(x) \to \infty$ and $I_{ij}(x) \leqslant$ (const.), and it is in the resulting three-part form that the dichotomy condition (though not under that name of course) was originally introduced by Levinson (1948).

As the theory develops, we shall on occasion meet λ_j of the form

$$\lambda_j = \rho_j'/\rho_j. \tag{1.3.10}$$

In this case, the integration in (1.3.1) and (1.3.2) can be carried out, and the dichotomy condition then states that either

$$|\rho_i(x)/\rho_j(x)| \leqslant K' \, |\rho_i(t)/\rho_j(t)| \tag{1.3.11}$$

or

$$|\rho_i(x)/\rho_j(x)| \geqslant K'' \, |\rho_i(t)/\rho_j(t)| \tag{1.3.12}$$

for each i and j, and $a \leqslant t \leqslant x < \infty$, where K' and K'' (>0) are constants.

Next we mention a stronger form of the dichotomy condition which is often adequate in applications as well as being easy to verify.

L″. For each pair (i, j), let

$$\text{re}\,\{\lambda_i(x) - \lambda_j(x)\} = f(x) + g(x), \tag{1.3.13}$$

where $g(x)$ is $L(a, \infty)$ and either $f(x) \geqslant 0$ in $[a, \infty)$ or $f(x) \leqslant 0$ in $[a, \infty)$. It is clear that either (1.3.1) or (1.3.2) holds when (1.3.13) does.

The necessity of the dichotomy condition in Theorem 1.3.1 can be illustrated by the following example in which $n = 2$, (1.3.3) holds, Condition L is not satisfied, and the asymptotic formula (1.3.5) does not hold. We define

$$\rho(x) = x^2(1 - \sin x) + 1,$$
$$\Lambda(x) = \text{dg}\,(0, \rho'(x)/\rho(x)),$$

and

$$R(x) = \begin{bmatrix} 0 & x^{-2} \\ 0 & 0 \end{bmatrix}$$

in $[a, \infty)$. With $Y = \begin{pmatrix} u \\ v \end{pmatrix}$, the system (1.3.4) is now

$$u' = x^{-2}v, \qquad v' = (\rho'/\rho)v.$$

Hence

$$v(x) = c\rho(x),$$

where c is a constant, and then $u'(x) = c(1 - \sin x) + cx^{-2}$, giving

$$u(x) = c(x + \cos x) - cx^{-1} + d,$$

where d is a further constant. Taking $c = 0$ and $d = 1$, we obtain the solution $Y_1(x) = e_1$, which does of course have the form (1.3.5) since

$\lambda_1(x)=0$ here. However, turning to a solution corresponding to $\lambda_2(x)=\rho'(x)/\rho(x)$, we take $d=0$ and $c=\rho(a)$ to obtain the solution

$$Y_2(x)=\left\{e_2+\begin{bmatrix} w(x) \\ 0 \end{bmatrix}\right\}\rho(x),$$

where

$$w(x)=(x+\cos x-x^{-1})/\{x^2(1-\sin x)+1\}.$$

It is clear that $w(x)\neq o(1)$ as $x\to\infty$ and hence there is no solution $Y_2(x)$ of the form (1.3.5). Condition L fails to hold in this example because here

$$\int_t^x \mathrm{re}\,\{\lambda_2(s)-\lambda_1(s)\}\,\mathrm{d}s=\log\{\rho(x)/\rho(t)\}$$

$$=\log[\{x^2(1-\sin x)+1\}/\{t^2(1-\sin t)+1\}],$$

and the choices $x=(2N+1)\pi$, $t=(2N+\frac{1}{2})\pi$ and $x=(2N+\frac{1}{2})\pi$, $t=2N\pi$ (large N) show that neither (1.3.1) nor (1.3.2) is satisfied.

1.4 PROOF OF THE LEVINSON THEOREM

We consider a particular k in (1.3.5) and begin by making the change of variable

$$Y(x)=Z(x)\exp\left(\int_a^x \lambda_k(t)\,\mathrm{d}t\right)$$

in (1.3.4). Then

$$Z'(x)=\{\Lambda_1(x)+R(x)\}Z(x) \tag{1.4.1}$$

where

$$\Lambda_1(x)=\Lambda(x)-\lambda_k(x)I,$$

and the theorem will be proved if we show that there is a solution $Z(x)$ of (1.4.1) such that

$$Z(x)=e_k+o(1) \tag{1.4.2}$$

as $x\to\infty$.

Working now with (1.4.1), we write

$$\Lambda_1(x)=\mathrm{dg}(\mu_1(x),\ \ldots\ ,\ \mu_n(x)),$$

where

$$\mu_i(x)=\lambda_i(x)-\lambda_k(x)$$

and, in particular,

$$\mu_k(x)=0. \tag{1.4.3}$$

Referring to the dichotomy condition in the form (1.3.7)–(1.3.9), we take $j=k$ and, with a re-ordering if necessary, we arrange that part (a') holds for $1 \leqslant i \leqslant N-1$ and part (b') for $N \leqslant i \leqslant n$, with

$$\mu_N(x) = 0, \tag{1.4.4}$$

corresponding to (1.4.3). The values $N=1$ and $N=n+1$ are allowed with the obvious meanings. If we then define

$$v_i(x) = \exp\left(\int_a^x \mu_i(s)\,\mathrm{d}s\right), \tag{1.4.5}$$

the dichotomy condition L′ implies that

$$v_i(x) \to 0 \qquad (x \to \infty,\ 1 \leqslant i \leqslant N-1), \tag{1.4.6}$$

$$|v_i(x)/v_i(t)| \leqslant K' \qquad (a \leqslant t \leqslant x < \infty,\ 1 \leqslant i \leqslant N-1), \tag{1.4.7}$$

$$|v_i(t)/v_i(x)| \leqslant K'' \qquad (a \leqslant t \leqslant x < \infty,\ N \leqslant i \leqslant n), \tag{1.4.8}$$

where K' and K'' are constants.

We complete the initial stages of the proof by introducing the fundamental matrix $\Phi(x)$ of the diagonal system

$$Z'(x) = \Lambda_1(x)Z(x) \tag{1.4.9}$$

defined by

$$\Phi(x) = \mathrm{dg}(v_1(x), \ldots, v_n(x)).$$

We write

$$\Phi(x) = \Phi_1(x) + \Phi_2(x),$$

where

$$\Phi_1(x) = \mathrm{dg}(v_1(x), \ldots, v_{N-1}(x), 0, \ldots, 0) \tag{1.4.10}$$

and

$$\Phi_2(x) = \mathrm{dg}\,(0, \ldots, 0, v_N(x), \ldots, v_n(x)).$$

Then (1.4.7) implies that

$$|\Phi_1(x)\Phi^{-1}(t)| \leqslant c_1 \qquad (a \leqslant t \leqslant x < \infty) \tag{1.4.11}$$

while (1.4.8), with x and t interchanged, implies that

$$|\Phi_2(x)\Phi^{-1}(t)| \leqslant c_2 \qquad (a \leqslant x \leqslant t < \infty), \tag{1.4.12}$$

where c_1 and c_2 are constants.

Moving on to the main part of the proof, the aim is to show that the integral equation

$$Z(x) = e_N + \Phi_1(x)\int_a^x \Phi^{-1}(t)R(t)Z(t)\,\mathrm{d}t$$
$$- \int_x^\infty \Phi_2(x)\Phi^{-1}(t)R(t)Z(t)\,\mathrm{d}t \tag{1.4.13}$$

has a solution $Z(x)$ and then that $Z(x)$ satisfies (1.4.2) with $k=N$. It is immediately clear, by differentiation, that a solution $Z(x)$ of (1.4.13) satisfies (1.4.1) and, in fact, (1.4.13) is suggested by the standard formula (1.2.11) as applied to (1.4.1) and (1.4.9). The important difference between (1.2.11) and (1.4.13) is that here we have the upper limit ∞ in the second integral with, as a consequence, a convergence question to be considered.

Applying the method of successive approximations to (1.4.13), we define the sequence $\{Z_m(x)\}$ $(m = 1, 2, \ldots)$ by $Z_1(x) = e_N$ and

$$Z_{m+1}(x) = e_N + \Phi_1(x)\int_a^x \Phi^{-1}(t)R(t)Z_m(t)\,\mathrm{d}t - \int_x^\infty \Phi_2(x)\Phi^{-1}(t)R(t)Z_m(t)\,\mathrm{d}t \tag{1.4.14}$$

for $m \geqslant 1$. First we prove that each $Z_m(x)$ is bounded in $[a, \infty)$ and this, together with (1.3.3) and (1.4.12), ensures that the infinite integral in (1.4.14) converges. Proceeding by induction on m, we assume that $|Z_m(x)| \leqslant C_m$ for some constant C_m. Then, by (1.4.11)–(1.4.14), we have

$$\begin{aligned}|Z_{m+1}(x)| &\leqslant 1 + c_1\int_a^x |R(t)|\,|Z_m(t)|\,\mathrm{d}t + c_2\int_x^\infty |R(t)|\,|Z_m(t)|\,\mathrm{d}t \\ &\leqslant 1 + (c_1+c_2)C_m\int_a^\infty |R(t)|\,\mathrm{d}t.\end{aligned}$$

Denoting this last expression by C_{m+1}, we have $|Z_{m+1}(x)| \leqslant C_{m+1}$ and, since $Z_1(x) = e_N$, the boundedness of $Z_m(x)$ for each m follows.

Next we prove that $Z_m(x)$ converges uniformly on $[a, \infty)$ to a limiting function $Z(x)$. By (1.4.11)–(1.4.14) again, we have

$$|Z_{m+2}(x) - Z_{m+1}(x)| \leqslant (c_1+c_2)\int_a^\infty |R(t)|\,|Z_{m+1}(t) - Z_m(t)|\,\mathrm{d}t.$$

Hence, by induction again,

$$|Z_{m+1}(x) - Z_m(x)| \leqslant \left((c_1+c_2)\int_a^\infty |R(t)|\,\mathrm{d}t\right)^m. \tag{1.4.15}$$

By making a suitable choice for the point a in (1.4.14), we can arrange that

$$(c_1+c_2)\int_a^\infty |R(t)|\,\mathrm{d}t < 1. \tag{1.4.16}$$

Then, by comparison with a geometric series, $\Sigma\,\{Z_{m+1}(x) - Z_m(x)\}$

converges uniformly in $[a, \infty)$. Hence we can define

$$Z(x) = Z_1(x) + \sum_{l=1}^{\infty} \{Z_{l+1}(x) - Z_l(x)\} = \lim_{m \to \infty} Z_m(x), \qquad (1.4.17)$$

and it follows from (1.4.15)–(1.4.16) that $Z(x)$ is bounded in $[a, \infty)$. Thus

$$|Z(x)| \leqslant K \qquad (1.4.18)$$

for some constant K. On letting $m \to \infty$ in (1.4.14), we obtain (1.4.13) with $Z(x)$ as just defined.

To prove that $Z(x)$ satisfies (1.4.2) with $k = N$, we write (1.4.13) as

$$Z(x) = e_N + \Phi_1(x) \int_a^X \Phi^{-1}(t) R(t) Z(t)\, dt + \Psi(x), \qquad (1.4.19)$$

where X will be chosen suitably in $[a, x]$ and

$$\Psi(x) = \Phi_1(x) \int_X^x \Phi^{-1}(t) R(t) Z(t)\, dt - \int_x^{\infty} \Phi_2(x) \Phi^{-1}(t) R(t) Z(t)\, dt.$$

As before, using (1.4.11), (1.4.12), and (1.4.18), we have

$$|\Psi(x)| \leqslant (c_1 + c_2) K \int_X^{\infty} |R(t)|\, dt.$$

Hence, given ϵ (>0), we can choose X so that $|\Psi(x)| < \epsilon$. Further, by (1.4.6) and (1.4.10), we have $\Phi_1(x) \to 0$ as $x \to \infty$. Hence, with X already chosen, (1.4.19) gives

$$|Z(x) - e_N| < 2\epsilon$$

for all sufficiently large x. This completes the proof of (1.4.2) and so of (1.3.5). □

We end this section by making three further comments on the conditions in Theorem 1.3.1. The first is that, if the dichotomy condition L holds, not for all i and j, but for $1 \leqslant i \leqslant n$ and only a particular value of j (say $j = k$), then (1.3.5) continues to hold for that value k. Indeed, the above proof is written out for any one such k.

The second comment concerns the necessity for some condition on $R(x)$ such as (1.3.3). Certainly (1.3.3) holds when $R(x) = O(x^{-\alpha})$ for some $\alpha > 1$. On the other hand, the condition $R(x) = O(x^{-1})$ is not sufficient for (1.3.5), and a simple example illustrates this point. Let us take $n = 1$, $\Lambda(x) = 0$, and $R(x) = 1/x$. Then (1.3.2) has the general solution $Y(x) = (\text{const.})x$ and no solution satisfies (1.3.5). In the following sections we discuss further smallness conditions on $R(x)$ under which modifications of (1.3.5) hold.

The final comment concerns the $o(1)$ term in (1.3.5). A more explicit estimate for this term can be obtained from (1.4.13), with the result that the accuracy of (1.3.5) can be improved subject to a strengthening of the condition (1.3.3) on R. To see this, we define

$$\xi_i(x) = \mathrm{re}\,\{\lambda_i(x) - \lambda_k(x)\} \tag{1.4.20}$$

with $1 \leqslant i \leqslant n$ and a chosen k. We consider separately those i for which

$$\int_a^\infty \xi_i(s)\,\mathrm{d}s = -\infty, \tag{1.4.21}$$

as in (1.3.7), and those for which

$$\int_t^x \xi_i(s)\,\mathrm{d}s > K_2 \tag{1.4.22}$$

as in (1.3.9). In (1.4.2) and (1.4.13) (with $N = k$), we write

$$Z = e_k + u, \tag{1.4.23}$$

where u has components u_i, and we take the ith component of (1.4.13). If i is as in (1.4.21), the ith diagonal entry of Φ_2 is zero, and (1.4.13) gives

$$|u_i(x)| \leqslant (\mathrm{const.}) \int_a^x \left\{\exp\left(\int_t^x \xi_i(s)\,\mathrm{d}s\right)\right\} |R(t)|\,\mathrm{d}t, \tag{1.4.24}$$

where we use the fact that Z is bounded in $[a, \infty)$. If, however, i is as in (1.4.22), the ith diagonal entry of Φ_1 is zero, and then (1.4.13) gives

$$|u_i(x)| \leqslant (\mathrm{const.}) \int_x^\infty \left\{\exp\left(-\int_x^t \xi_i(s)\,\mathrm{d}s\right)\right\} |R(t)|\,\mathrm{d}t. \tag{1.4.25}$$

Since $u(x)$ is the $o(1)$ term which appears in (1.3.5), we now have a more explicit estimate for this term.

We note two special cases of (1.4.24) and (1.4.25).

(i) $\xi_i(x) = \text{constant}$. Here (1.4.21) and (1.4.24) apply when $\xi_i(x) = -c$ with $c > 0$ and, on splitting the range of integration (a, x) in (1.4.24) into $(a, \delta x)$ and $(\delta x, x)$ with a fixed $\delta < 1$, we obtain

$$u_i(x) = O(\mathrm{e}^{-c(1-\delta)x}) + O\left(\int_{\delta x}^x |R(t)|\,\mathrm{d}t\right). \tag{1.4.26}$$

Also, (1.4.22) and (1.4.25) apply when $\xi_i(x) = c$ *with* $c \geqslant 0$, and then (1.4.25) gives immediately

$$u_i(x) = O\left(\int_x^\infty |R(t)|\,\mathrm{d}t\right). \tag{1.4.27}$$

Thus if, for example,

$$\int_x^\infty |R(t)|\,\mathrm{d}t = O(x^{-\alpha}) \tag{1.4.28}$$

for some $\alpha > 0$, we obtain $u(x) = O(x^{-\alpha})$ from both (1.4.26) and (1.4.27). Alternatively, if

$$\int_x^\infty |R(t)|\, dt = O(e^{-\alpha x})$$

for some $\alpha > 0$, the choice $\delta = c/(\alpha + c)$ in (1.4.26) gives

$$u_i(x) = O(e^{-\alpha c x/(\alpha + c)})$$

while (1.4.27) enjoys the better estimate

$$u_i(x) = O(e^{-\alpha x}).$$

(ii) $\xi_i(x) = (\text{const.})\rho'(x)/\rho(x)$, where $\rho(x) \to \infty$. This time (1.4.21) and (1.4.24) apply when $\xi_i(x) = -c\rho'(x)/\rho(x)$ $(c > 0)$, and we obtain

$$u_i(x) = O\left(\rho^{-c}(x) \int_a^x \rho^c(t)\, |R(t)|\, dt\right). \tag{1.4.29}$$

When $\xi_i(x) = c\rho'(x)/\rho(x)$ $(c \geqslant 0)$, (1.4.25) once again gives (1.4.27), since

$$\begin{aligned} u_i(x) &= O\left(\int_x^\infty \{\rho(x)/\rho(t)\}^c\, |R(t)|\, dt\right) \\ &= O\left(\int_x^\infty |R(t)|\, dt\right), \end{aligned} \tag{1.4.30}$$

by (1.4.22) with x and t interchanged.

1.5 THE HARTMAN–WINTNER THEOREM

In this section we consider again the system

$$Y'(x) = \{\Lambda(x) + R(x)\}Y(x) \tag{1.5.1}$$

and derive the asymptotic form of solutions under conditions on Λ and R which, although broadly similar to those in Theorem 1.3.1, also differ from the previous conditions in important respects. The theorem which follows is due to Hartman and Wintner (1955). The original proof is lengthy, but there is now a simpler proof which is due to Harris and Lutz (1977). The idea in this simpler proof is to make a transformation

$$Y(x) = \{I + Q(x)\}Z(x), \tag{1.5.2}$$

where I is the identity matrix and $Q(x)$ is a matrix chosen in such a way that $Q(x) \to 0$ as $x \to \infty$ while the corresponding Z-system has the Levinson form considered in § 1.3. The use of transformations of the type (1.5.2) is one of the basic techniques in asymptotic theory, and several such transformations will be developed in other situations in subsequent sections.

Theorem 1.5.1 (The Hartman–Wintner theorem) *Let $\Lambda(x)$ be an $n \times n$ diagonal matrix,*

$$\Lambda(x) = \mathrm{dg}\,(\lambda_1(x), \ldots, \lambda_n(x)),$$

and let there be a constant δ (>0) such that

$$|\mathrm{re}\,\{\lambda_i(x) - \lambda_j(x)\}| \geqslant \delta \tag{1.5.3}$$

in some interval $[a, \infty)$ for each unequal pair of integers i and j in $[1, n]$. Let the $n \times n$ matrix $R(x)$ satisfy

$$\int_a^\infty |R(x)|^p \,\mathrm{d}x < \infty \tag{1.5.4}$$

for some p such that $1 < p \leqslant 2$. Then (1.5.1) has solutions $Y_k(x)$ $(1 \leqslant k \leqslant n)$ with the asymptotic form as $x \to \infty$

$$Y_k(x) = \{e_k + o(1)\} \exp\left(\int_a^x \{\lambda_k(t) + r_{kk}(t)\}\,\mathrm{d}t\right). \tag{1.5.5}$$

Proof We begin by defining

$$\Lambda_1 = \Lambda + \mathrm{dg}\,R = \mathrm{dg}\,(\lambda_i + r_{ii}) \tag{1.5.6}$$

and

$$\tilde{R} = R - \mathrm{dg}\,R. \tag{1.5.7}$$

With Q still to be chosen, we substitute (1.5.2) into (1.5.1) to obtain

$$\begin{aligned}(I+Q)Z' &= \{(\Lambda + R)(I+Q) - Q'\}Z \\ &= \{(I+Q)\Lambda + \Lambda Q - Q\Lambda + R - Q' + RQ\}Z \\ &= \{(I+Q)\Lambda_1 + \Lambda Q - Q\Lambda + \tilde{R} - Q' + RQ - Q\,\mathrm{dg}\,R\}Z,\end{aligned} \tag{1.5.8}$$

by (1.5.6) and (1.5.7). We now choose Q so that

$$Q' = \Lambda Q - Q\Lambda + \tilde{R}, \tag{1.5.9}$$

and then (1.5.8) becomes

$$Z' = (\Lambda_1 + R_1)Z, \tag{1.5.10}$$

where

$$R_1 = (I+Q)^{-1}(RQ - Q\,\mathrm{dg}\,R). \tag{1.5.11}$$

We have now to show that (1.5.9) has a solution Q such that, first,

$$Q(x) \to 0 \quad \text{as} \quad x \to \infty \tag{1.5.12}$$

and second

$$\int_X^\infty |R_1(x)|\,\mathrm{d}x < \infty, \tag{1.5.13}$$

where we note that (1.5.12) guarantees the existence of $(I+Q)^{-1}$ and therefore of R_1 in some interval $[X, \infty)$. We shall then have the Levinson form in (1.5.10), and Theorem 1.3.1 will be applicable to the system (1.5.10).

In terms of the individual entries in the matrices, (1.5.9) is

$$\begin{aligned} q'_{ii} &= 0, \\ q'_{ij} &= (\lambda_i - \lambda_j)q_{ij} + r_{ij} \qquad (i \neq j). \end{aligned} \tag{1.5.14}$$

We dispose of q_{ii} by defining $q_{ii}(x)=0$ for all x. Next, (1.5.14) is a first-order differential equation for q_{ij} and, referring to (1.5.3) and considering first those i and j for which

$$\text{re}\,\{\lambda_i(x) - \lambda_j(x)\} \geq \delta, \tag{1.5.15}$$

we take

$$q_{ij}(x) = -e^{\mu(x)} \int_x^\infty e^{-\mu(t)} r_{ij}(t)\, dt, \tag{1.5.16}$$

where

$$\mu(x) = \int_a^x \{\lambda_i(s) - \lambda_j(s)\}\, ds. \tag{1.5.17}$$

For those i and j where, on the other hand,

$$\text{re}\,\{\lambda_i(x) - \lambda_j(x)\} \leq -\delta, \tag{1.5.18}$$

we take

$$q_{ij}(x) = e^{\mu(x)} \int_a^x e^{-\mu(t)} r_{ij}(t)\, dt. \tag{1.5.19}$$

It follows immediately from (1.5.4) and (1.5.15) that the infinite integral in (1.5.16) converges and, further, that

$$|q_{ij}(x)| \leq \int_x^\infty e^{-\delta(t-x)}\, |r_{ij}(t)|\, dt. \tag{1.5.20}$$

A Hölder inequality then shows that $q_{ij}(x) \to 0$ as $x \to \infty$. In the case of (1.5.18) and (1.5.19), we have

$$\begin{aligned} |q_{ij}(x)| &\leq \int_a^x e^{-\delta(x-t)}\, |r_{ij}(t)|\, dt \\ &\leq e^{-\frac{1}{2}\delta x} \int_a^{\frac{1}{2}x} |r_{ij}(t)|\, dt + \int_{\frac{1}{2}x}^x e^{-\delta(x-t)}\, |r_{ij}(t)|\, dt, \end{aligned} \tag{1.5.21}$$

and again a Hölder inequality shows that $q_{ij}(x) \to 0$ as $x \to \infty$. So far, therefore, we have shown that Q can be chosen to satisfy (1.5.9) and (1.5.12).

Next we establish (1.5.13), where R_1 is defined by (1.5.11), and to do this we require an additional property of Q, that

$$Q \text{ is } L^p(a, \infty). \tag{1.5.22}$$

To prove (1.5.22), we write (1.5.20) as

$$|q_{ij}(x)| \leqslant \int_0^\infty e^{-\delta u} |r_{ij}(x+u)| \, du.$$

Then a Minkowski inequality gives

$$\left(\int_a^\infty |q_{ij}(x)|^p \, dx\right)^{1/p} \leqslant \int_0^\infty e^{-\delta u}\left(\int_a^\infty |r_{ij}(x+u)|^p \, dx\right)^{1/p} du,$$

and the right-hand side is finite by (1.5.4). Hence the q_{ij} in (1.5.20) are $L^p(a, \infty)$. Similarly, in the case of (1.5.21), we have

$$|q_{ij}(x)| \leqslant \int_0^\infty e^{-\delta u} |r_{ij}(x-u)| \, du$$

where, for convenience, we take $r_{ij}(x) = 0$ for $x < a$. Again it follows that these q_{ij} are $L^p(a, \infty)$, and (1.5.22) is proved.

By (1.5.11), (1.5.12) and the Hölder inequality, we have

$$\int_X^\infty |R_1(x)| \, dx \leqslant (\text{const.}) \left(\int_X^\infty |R(x)|^p \, dx\right)^{1/p} \left(\int_X^\infty |Q(x)|^{p'} \, dx\right)^{1/p'}, \tag{1.5.23}$$

where $1/p + 1/p' = 1$. At this point we use the hypothesis that $p \leqslant 2$ to say that $p' \geqslant p$. It then follows from (1.5.12) and (1.5.22) that Q is $L^{p'}(X, \infty)$. The right-hand side of (1.5.23) is therefore finite, and (1.5.13) is now proved.

By (1.5.13), we can apply Theorem 1.3.1 to (1.5.10) provided that Λ_1, as defined in (1.5.6), satisfies the dichotomy condition L in § 1.3. That this is so follows immediately once we note that, if re $\{\lambda_i(x) - \lambda_j(x)\} \geqslant \delta$ in (1.5.3), we have

$$\begin{aligned} &\int_t^x \text{re}\,\{\lambda_i(s) + r_{ii}(s) - \lambda_j(s) - r_{jj}(s)\} \, ds \\ &\geqslant \int_t^x \text{re}\,\{\lambda_i(s) - \lambda_j(s)\} - |r_{ii}(s)| - |r_{jj}(s)|) \, ds \\ &\geqslant \delta(x-t) - (\text{const.})(x-t)^{1/p'} \end{aligned} \tag{1.5.24}$$

for $x > t$, by (1.5.4) and the Hölder inequality. Thus (1.3.2) is satisfied. Similarly, (1.3.1) is satisfied if re $\{\lambda_i(x) - \lambda_j(x)\} \leqslant -\delta$ in (1.5.3). Theorem 1.3.1 can now be applied to (1.5.10) to establish the existence of solutions Z_k of (1.5.10) such that

$$Z_k(x) = \{e_k + o(1)\} \exp\left(\int_a^x \{\lambda_k(t) + r_{kk}(t)\} \, dt\right),$$

and then (1.5.5) follows when we transform back to (1.5.1) by means of (1.5.2). □

There are three comments which can be made on the conditions imposed in the above theorem. The first is that, in comparison with the conditions in Theorem 1.3.1, there are certainly matrices $R(x)$ which satisfy (1.5.4) but not (1.3.3), the most important examples occurring when $R(x)$ is $O(x^{-1})$ as $x \to \infty$. Indeed, (1.5.4) is more general than (1.3.3) as far as those $R(x)$ which are $o(1)$ $(x \to \infty)$ are concerned. On the other hand, the condition (1.5.3) on Λ is more restrictive than the dichotomy condition in § 1.3, as may be seen most clearly on comparing (1.5.3) with (1.3.13).

The second comment is that, if (1.5.3) holds, not for all i and j, but for $1 \leqslant i \leqslant n$ and only a particular value of j (say $j = k$), then (1.5.5) continues to hold for that value k. We made a similar remark in connection with Theorem 1.3.1 in § 1.4.

The third point is the important observation that the condition $p \leqslant 2$ was not used in the proof until (1.5.23). Everything in the proof up to and including (1.5.22) is also valid for $p > 2$. The implication for the matrix R_1 in (1.5.11) is that, by (1.5.4) and (1.5.22), R_1 is $L^{p/2}(X, \infty)$ when $p > 2$. The effect of the transformation defined by (1.5.2) and (1.5.9) is therefore to replace a system (1.5.1), in which R is $L^p(X, \infty)$ $(p > 1)$, by a system (1.5.10) in which R_1 is $L^{p(1)}(X, \infty)$, where

$$p(1) = \max(1, p/2). \tag{1.5.25}$$

If $p(1) > 1$, that is, if $p > 2$, the transformation procedure can be repeated with (1.5.10) as the new starting point. The Levinson form would then be obtained after the second transformation when $2 < p \leqslant 4$. For larger values of p, further iterations of the procedure are required before the Levinson form is obtained.

To give the details of the general asymptotic result that arises from this process, we suppose that (1.5.4) holds with some $p > 1$ and we choose the integer M so that

$$2^{M-1} < p \leqslant 2^M. \tag{1.5.26}$$

Then, iterating the formulae (1.5.6), (1.5.9), and (1.5.11) which connect the Y and Z systems (1.5.1) and (1.5.10), we define

$$\left.\begin{aligned} \Lambda_{m+1} &= \Lambda_m + \operatorname{dg} R_m, \\ Q'_m &= \Lambda_m Q_m - Q_m \Lambda_m + \tilde{R}_m, \\ R_{m+1} &= (I + Q_m)^{-1}(R_m Q_m - Q_m \operatorname{dg} R_m), \end{aligned}\right\} \tag{1.5.27}$$

for $m = 1, 2, \ldots, M-1$, where $\tilde{R}_m = R_m - \operatorname{dg} R_m$, $\operatorname{dg} Q_m = 0$, and $Q_m = o(1)$ as $x \to \infty$. The iterated transformation

$$Y = (I + Q)(I + Q_1)(\cdots)(I + Q_{M-1})Z_M \tag{1.5.28}$$

then gives a system

$$Z'_M = (\Lambda_M + R_M)Z_M \tag{1.5.29}$$

in which R_M is $L^{p(M)}(X, \infty)$ where, corresponding to (1.5.25),

$$p(M) = \max(1, p/2^M) = 1,$$

by (1.5.26). Also, as in the case of (1.5.24), (1.5.3) implies that Λ_M satisfies the dichotomy condition of § 1.3 We can therefore apply the Levinson theorem to (1.5.29) and, since (1.5.28) is simply of the form $Y = \{I + o(1)\}Z_M$, we obtain the following result for (1.5.1).

Theorem 1.5.2 *Let the diagonal matrix* $\Lambda(x)$ *satisfy* (1.5.3) *and let* $R(x)$ *satisfy* (1.5.4) *with some* $p > 1$. *Let the integer* M *be as in* (1.5.26) *and let* Λ_M *be defined by* (1.5.27), *where* Λ_1 *and* R_1 *are as in* (1.5.6) *and* (1.5.11). *Then, with*

$$\Lambda_M(x) = \mathrm{dg}\,(\lambda_1^{(M)}(x), \dots, \lambda_n^{(M)}(x)),$$

(1.5.1) *has solutions* $Y_k(x)$ $(1 \leq k \leq n)$ *such that*

$$Y_k(x) = \{e_k + o(1)\} \exp\left(\int_a^x \lambda_k^{(M)}(t)\,\mathrm{d}t\right).$$

The computation of Λ_M in any particular example of (1.5.1) may of course be a lengthy matter, but we shall give the details in § 2.8 for a system with $n = 2$ which arises from a second-order differential equation. The most obvious matrices $R(x)$ covered by Theorem 1.5.2 are those where $R(x) = O(x^{-\alpha})$ as $x \to \infty$ for some $\alpha > 0$.

1.6 A POINTWISE CONDITION ON $R(x)$

In the Levinson and Hartman–Wintner theorems, a condition involving integration is imposed on $R(x)$, and the convergence in (1.3.3) and (1.5.4) states that $R(x)$ is small in a certain average sense as $x \to \infty$. The theorem in this section gives the asymptotic form of the solutions of

$$Y'(x) = \{\Lambda(x) + R(x)\}Y(x) \tag{1.6.1}$$

subject to a pointwise condition on $R(x)$ as well as an integral condition, with the latter relating to $R'(x)$ rather than $R(x)$. The condition on $R(x)$ is not, however, simply $R(x) \to 0$ as $x \to \infty$ but rather that $R(x)$ should be small compared to $\Lambda(x)$ in the sense specified in (1.6.2). We note that, apart from its pointwise nature, (1.6.2) stands in contrast to (1.3.3) and (1.5.4) in its involvement with $\Lambda(x)$.

Theorem 1.6.1 *Let* $\Lambda(x)$ *be an* $n \times n$ *diagonal matrix,*

$$\Lambda(x) = \mathrm{dg}\,(\lambda_1(x), \dots, \lambda_n(x)),$$

in which $\lambda_i(x) - \lambda_j(x)$ is nowhere zero in some interval $[a, \infty)$ for each pair of unequal integers i and j in $[1, n]$. Let the $n \times n$ matrix $R(x)$ satisfy

(i) $$\{\lambda_i(x) - \lambda_j(x)\}^{-1} R(x) \to 0 \quad \text{as } x \to \infty; \tag{1.6.2}$$

(ii) $$(\{\lambda_i(x) - \lambda_j(x)\}^{-1} R(x))' \in L(a, \infty) \tag{1.6.3}$$

for each unequal i and j in $[1, n]$. Also, let the eigenvalues $\mu_k(x)$ $(1 \leqslant k \leqslant n)$ of $\Lambda(x) + R(x)$ satisfy the dichotomy condition L *in* § 1.3. *Then, as $x \to \infty$,* (1.6.1) *has solutions $Y_k(x)$ with the asymptotic form*

$$Y_k(x) = \{e_k + o(1)\} \exp\left(\int_a^x \mu_k(t)\, \mathrm{d}t\right). \tag{1.6.4}$$

Proof As in the proof of the Hartman–Wintner theorem in § 1.5, we use a transformation

$$Y(x) = \{I + Q(x)\} Z(x), \tag{1.6.5}$$

where $Q(x) \to 0$ as $x \to \infty$, but with a different choice of Q from the one made in (1.5.9). The general idea is to express $\Lambda + R$ in the diagonal form

$$T^{-1}(\Lambda + R)T = \Lambda_1 \tag{1.6.6}$$

in some interval $[X, \infty)$, where

$$\Lambda_1 = \mathrm{dg}\,(\mu_1, \ldots, \mu_n),$$

and in the main part of the proof we show that T can be written as

$$T = I + Q, \tag{1.6.7}$$

where $Q(x)$ has the two properties

$$Q(x) \to 0 \quad \text{as } x \to \infty \tag{1.6.8}$$

and

$$Q'(x) \quad \text{is} \quad L(X, \infty). \tag{1.6.9}$$

Once this is done, we can substitute (1.6.5) into (1.6.1) to obtain

$$(I + Q)Z' = \{(\Lambda + R)(I + Q) - Q'\} Z$$

which, by (1.6.6) and (1.6.7), becomes

$$Z' = (\Lambda_1 + R_1) Z \tag{1.6.10}$$

with

$$R_1 = -(I + Q)^{-1} Q'. \tag{1.6.11}$$

Then (1.6.8) and (1.6.9) imply that (1.6.10) has the Levinson form, and Theorem 1.3.1 can therefore be applied to this system.

To establish the existence of Q with the above properties, we write

$$\Lambda_1 = \Lambda + D \tag{1.6.12}$$

in (1.6.6) and we substitute from (1.6.7) to write (1.6.6) as

$$(\Lambda + R)(I + Q) = (I + Q)(\Lambda + D),$$

that is,

$$Q\Lambda - \Lambda Q + D = R + RQ - QD. \tag{1.6.13}$$

We regard (1.6.13) as an equation to solve for both Q and D, and we proceed by considering the individual entries q_{ij} in Q and d_i in the diagonal matrix D. We begin by taking

$$q_{ii} = 0 \qquad (1 \leqslant i \leqslant n), \tag{1.6.14}$$

this representing no more than a normalization of the columns of T. Then, equating the (i, j) entries on each side of (1.6.13), we obtain

$$d_i = r_{ii} + \sum_{\nu=1}^{n} r_{i\nu} q_{\nu i} \tag{1.6.15}$$

and, for $i \neq j$,

$$(\lambda_j - \lambda_i) q_{ij} = r_{ij} + \sum_{\nu=1}^{n} r_{i\nu} q_{\nu j} - q_{ij} d_j. \tag{1.6.16}$$

On substituting for d_i from (1.6.15) into (1.6.16), we obtain

$$(\lambda_j + r_{jj} - \lambda_i - r_{ii}) q_{ij} = r_{ij} + \sum_{\nu \neq i,j} r_{i\nu} q_{\nu j} - q_{ij} \sum_{\nu \neq j} r_{j\nu} q_{\nu j} \tag{1.6.17}$$

as a set of $n^2 - n$ equations to solve for the q_{ij} $(i \neq j)$. We now define

$$u_{ij\nu} = r_{i\nu} / (\lambda_j + r_{jj} - \lambda_i - r_{ii}) \qquad (i \neq j), \tag{1.6.18}$$

the denominator here being non-zero in some interval $[X, \infty)$ by (1.6.2), and we write (1.6.17) as

$$q_{ij} = u_{ijj} + \sum_{\nu \neq i,j} u_{ij\nu} q_{\nu j} + q_{ij} \sum_{\nu \neq j} u_{ji\nu} q_{\nu j}. \tag{1.6.19}$$

The given conditions (1.6.2) and (1.6.3), when applied to (1.6.18), show that

$$u_{ij\nu}(x) \to 0 \qquad \text{as } x \to \infty \tag{1.6.20}$$

and

$$u'_{ij\nu}(x) \in L(X, \infty), \tag{1.6.21}$$

and these are the two properties that we need in order to solve (1.6.19) for the q_{ij}.

A convenient way of proceeding is to express (1.6.19) in the vector form

$$q = q_0 + f(u, q), \tag{1.6.22}$$

where q and q_0 are the (n^2-n)-component vectors formed by the q_{ij} and u_{ijj} $(i \neq j)$, f is the vector function which arises from the other terms in (1.6.19) and u indicates the dependence of f on the u_{ijv}. We solve (1.6.22) by the method of successive approximation in which we define the vectors q_m $(m=1, 2, \ldots)$ by

$$q_m = q_0 + f(u, q_{m-1}) \tag{1.6.23}$$

and examine the convergence of q_m as $m \to \infty$.

We write

$$U(x) = \max |u_{ijv}(x)|, \tag{1.6.24}$$

the maximum being taken over all i, j, v $(i \neq j)$. By (1.6.20), we can choose X so that

$$nU + 4n^2U^2 \leqslant \tfrac{1}{2} \tag{1.6.25}$$

in $[X, \infty)$. From (1.6.19) and (1.6.23), we have

$$|q_m| \leqslant |q_0| + nU\,|q_{m-1}| + U\,|q_{m-1}|^2.$$

Since

$$|q_0| \leqslant n^2U, \tag{1.6.26}$$

it follows from (1.6.25) and a simple induction argument that

$$|q_m| \leqslant 2\,|q_0| \tag{1.6.27}$$

in $[X, \infty)$ for all m.

From (1.6.19) and (1.6.23) again, we have

$$\begin{aligned} |q_{m+1}-q_m| &\leqslant nU\,|q_m-q_{m-1}| + U(|q_m|+|q_{m-1}|)\,|q_m-q_{m-1}| \\ &\leqslant (nU+4n^2U^2)\,|q_m-q_{m-1}| \\ &\leqslant \tfrac{1}{2}\,|q_m-q_{m-1}|, \end{aligned}$$

on using (1.6.27), (1.6.26) and (1.6.25). Hence

$$|q_{m+1}-q_m| \leqslant |q_0|/2^{m+1} \tag{1.6.28}$$

for $m \geqslant 0$. Then, since

$$q_m = q_0 + \sum_{v=0}^{m-1} (q_{v+1}-q_v),$$

(1.6.28) implies that q_m converges to a limiting vector q uniformly in the interval $[X, \infty)$. From (1.6.27), we have

$$|q| \leqslant 2\,|q_0| \tag{1.6.29}$$

and hence $q(x) \to 0$ as $x \to \infty$. The components q_{ij} of q satisfy (1.6.19) and therefore, recalling (1.6.14), we have the existence of the matrix Q in (1.6.7) and (1.6.13) with the property (1.6.8).

We have now to show that Q' exists and satisfies (1.6.9). Certainly q'_m exists for each m by (1.6.23) and, to examine the convergence of q'_m as $m \to \infty$, we begin by defining

$$V(x) = \max |u'_{ij\nu}(x)|, \tag{1.6.30}$$

the maximum being taken over all i, j, ν $(i \neq j)$. Then

$$V(x) \in L(X, \infty) \tag{1.6.31}$$

by (1.6.21). On differentiating (1.6.23) and using (1.6.19), (1.6.24) and (1.6.30), we obtain

$$\begin{aligned} |q'_m| &\leq |q'_0| + nV\,|q_{m-1}| + nU\,|q'_{m-1}| + V\,|q_{m-1}|^2 + 2U\,|q_{m-1}|\,|q'_{m-1}| \\ &\leq V(n^2 + 2n\,|q_0| + 4\,|q_0|^2) + (nU + 4n^2U^2)\,|q'_{m-1}|, \end{aligned}$$

by (1.6.26) and (1.6.27). From (1.6.25) and a further choice of X, we therefore have

$$|q'_m| \leq 2n^2V + \tfrac{1}{2}\,|q'_{m-1}|$$

in $[X, \infty)$ and hence, by induction on m,

$$|q'_m| \leq 4n^2V. \tag{1.6.32}$$

From the differentiated form of (1.6.23) again, we have

$$\begin{aligned} |q'_{m+1} - q'_m| &\leq nV\,|q_m - q_{m-1}| + nU\,|q'_m - q'_{m-1}| + U\,|q_m|\,|q'_m - q'_{m-1}| \\ &\quad + V(|q_m| + |q_{m-1}|)\,|q_m - q_{m-1}| + U\,|q'_{m-1}|\,|q_m - q_{m-1}| \\ &\leq V(n + 4\,|q_0| + 4n^2U)\,|q_m - q_{m-1}| + (nU + 2n^2U^2)\,|q'_m - q'_{m-1}| \\ &\leq W/2^m + \tfrac{1}{2}\,|q'_m - q'_{m-1}|, \end{aligned} \tag{1.6.33}$$

where

$$W = n^2(n + 8n^2U)UV$$

and we have used (1.6.26), (1.6.27), (1.6.28) and (1.6.25). In addition, the case $m = 1$ of (1.6.23) gives

$$|q'_1 - q'_0| \leq 2W.$$

By induction on m, it now follows from (1.6.33) that

$$|q'_{m+1} - q'_m| \leq (m + 1)W/2^{m-1}$$

for $m \geq 0$. Then, as in the case of (1.6.28), we have the convergence of q'_m almost everywhere in $[X, \infty)$ to a limiting vector p. Further, by (1.6.31) and (1.6.32), we have dominated convergence in $[X, \infty)$ and therefore p is $L(X, \infty)$ with

$$\int_X^x p(t)\,\mathrm{d}t = \lim_{m \to \infty} \int_X^x q'_m(t)\,\mathrm{d}t = q(x) - q(X).$$

Hence q is absolutely continuous with q' being $L(X, \infty)$. The same is therefore true of Q and we note that, in terms of Q, (1.6.32) gives

$$|Q'| \leqslant 4n^2 V. \tag{1.6.34}$$

We have now established both (1.6.8) and (1.6.9), and it follows that the system (1.6.10) has the Levinson form. Then, applying Theorem 1.3.1 to (1.6.10), we obtain solutions $Z_k(x)$ $(1 \leqslant k \leqslant n)$ such that

$$Z_k(x) = \{e_k + o(1)\} \exp\left(\int_a^x \mu_k(t)\, \mathrm{d}t\right)$$

as $x \to \infty$. Finally, (1.6.4) follows on transforming back to (1.6.1) by means of (1.6.5). □

We emphasize that, in the statement of Theorem 1.6.1, it is the $\mu_k(x)$ rather than the $\lambda_k(x)$ which appear in (1.6.4) and which are obliged to satisfy the dichotomy condition. However, in practice, it may not be easy to calculate the $\mu_k(x)$ and therefore it is useful to know that, if the additional condition

$$|\lambda_i(x) - \lambda_j(x)|^{-1} \, |R(x)|^2 \in L(X, \infty) \tag{1.6.35}$$

holds for all $i \neq j$, $\mu_k(x)$ can be replaced by the simple expression $\lambda_k(x) + r_{kk}(x)$ in the statement of the theorem. To see this, we use (1.6.12) and (1.6.15) to write

$$\begin{aligned} \mu_k &= \lambda_k + d_k = \lambda_k + r_{kk} + \sum_{\nu \neq k} r_{k\nu} q_{\nu k} \\ &= \lambda_k + r_{kk} + O\left(\max_{\nu \neq k} |R|^2 / |\lambda_\nu - \lambda_k|\right) \end{aligned} \tag{1.6.36}$$

by (1.6.29), (1.6.26), (1.6.24) and (1.6.18). The O-term here is $L(X, \infty)$ if (1.6.35) holds, and the required simplification follows. For future reference, then, we note that when the additional condition (1.6.35) holds, (1.6.4) can be replaced by

$$Y_k(x) = \{e_k + o(1)\} \exp\left(\int_a^x \{\lambda_k(t) + r_{kk}(t)\}\, \mathrm{d}t\right). \tag{1.6.37}$$

We note that (1.6.37) agrees with the formula (1.5.5) in Theorem 1.5.1 and, in fact, (1.6.35) is implied by the case $p = 2$ of (1.5.3) and (1.5.4). However, although Theorems 1.5.1 and 1.6.1 certainly overlap, it is clear that neither theorem includes the other.

We can take (1.6.36) further. Bearing in mind that $q_{\nu\nu} = 0$ and substituting for $q_{\nu k}$ from (1.6.17), we have

$$\mu_k = \lambda_k + r_{kk} + \sum_{\nu \neq k} r_{k\nu} r_{\nu k} / (\lambda_k + r_{kk} - \lambda_\nu - r_{\nu\nu}) + O\left(\max_{\nu \neq k} |R|^3 / |\lambda_\nu - \lambda_k|^2\right)$$

$$= \lambda_k + r_{kk} + \sum_{\nu \neq k} r_{k\nu} r_{\nu k} / (\lambda_k - \lambda_\nu) + O\left(\max_{\nu \neq k} |R|^3 / |\lambda_\nu - \lambda_k|^2\right) \quad (1.6.38)$$

by (1.6.2) and (1.6.29). Hence, if (1.6.35) is replaced by

$$|\lambda_i(x) - \lambda_j(x)|^{-2} |R(x)|^3 \in L(X, \infty), \quad (1.6.39)$$

the exponential factor in (1.6.4) can be replaced by

$$\exp\left(\int_a^x \{\lambda_k(t) + r_{kk}(t) + \rho_k(t)\} \, dt\right),$$

where ρ_k denotes the summation term in (1.6.38). This process can of course be continued beyond (1.6.39) but the details rapidly become complicated.

Theorem 1.6.1 has a number of important applications which will be covered later in the book, but we mention here the case where Λ is a constant matrix with distinct diagonal entries. In this case (1.6.2) and (1.6.3) become simply

$$R(x) \to 0 \quad \text{as } x \to \infty \quad (1.6.40)$$

and

$$\int_a^\infty |R'(x)| \, dx < \infty. \quad (1.6.41)$$

These conditions are satisfied when, for example, the entries in R have the form (const.)$x^{-\alpha}$ or even (const.)$(\log x)^{-\alpha}$, with $\alpha > 0$ and α can differ amongst the various entries.

For a more specialized example where (1.6.2) and (1.6.3) hold, this time with a non-constant Λ, we take $n = 2$ and

$$\lambda_1(x) = 1 + \sin x^\beta, \qquad \lambda_2(x) = -1 + \cos x^\beta,$$
$$R(x) = (\text{const.})x^{-\alpha},$$

where $0 < \beta < \alpha$.

Finally, we mention that the theory both in this section and in § 1.5 also applies to the more general system

$$Y'(x) = \{\Lambda(x) + R(x) + S(x)\} Y(x), \quad (1.6.42)$$

where $S(x)$ is $L(a, \infty)$. We use the same transformation (1.5.2) or (1.6.5) as when $S(x)=0$. The transformed system then contains the extra term $(I+Q)^{-1}S(I+Q)$ which, being $L(X,\infty)$, contributes to the Levinson form and does not influence the asymptotic formulae (1.5.5) and (1.6.4).

1.7 CONDITIONS ON HIGHER DERIVATIVES OF $R(x)$

The effect of the transformation (1.6.5) is to replace the condition in the Levinson theorem that R should be $L(a,\infty)$ by the corresponding condition (1.6.3) that $\{(\lambda_i-\lambda_j)^{-1}R\}'$ should be $L(a,\infty)$, subject of course to the pointwise condition (1.6.2) as well. If a second, similar, transformation is then applied with (1.6.10) as the starting point, (1.6.3) can in turn be replaced by a corresponding condition involving R'', subject again to a pointwise condition involving R'. This process can be repeated to bring in successively higher derivatives of R, and the Levinson theorem can be applied at a given stage in the process to obtain an asymptotic result which involves conditions on the derivatives of R up to that stage.

The difficulty with the process in practice is that the details rapidly become too cumbersome to handle effectively. There is however a considerable simplification that can be made in the choice of Q in (1.6.5) and (1.6.13) when, in addition to (1.6.2),

$$\int_a^\infty |\lambda_i(x)-\lambda_j(x)|^{-1}\,|R(x)|^2\,dx<\infty, \tag{1.7.1}$$

and we deal with this point first. We mentioned (1.7.1) previously in (1.6.35) but now we take the consequences of this extra condition somewhat further.

It is clear from (1.6.15), (1.6.29) and (1.6.18) that, when (1.7.1) holds, (1.6.13) contains terms which turn out to be $L(a,\infty)$ and, as we have noted in connection with (1.6.42), such terms do not influence the final result (1.6.4). Accordingly, removing these terms from (1.6.13), we define a new Q not by (1.6.13) but by

$$Q\Lambda-\Lambda Q+\operatorname{dg}R=R. \tag{1.7.2}$$

This gives straightaway

$$q_{ij}=r_{ij}/(\lambda_j-\lambda_i)\qquad(i\neq j). \tag{1.7.3}$$

As before, we take

$$q_{ii}=0. \tag{1.7.4}$$

When this new Q is used in (1.6.5), the resulting Z-system (1.6.10) becomes

$$Z'=(\Lambda+\operatorname{dg}R+R_1+S)Z, \tag{1.7.5}$$

where

$$R_1 = -(I+Q)^{-1}Q', \tag{1.7.6}$$

and we use the symbol S throughout this section to denote any matrices which are $L(a, \infty)$. From (1.7.1) and (1.7.3), we see that $(\lambda_i - \lambda_j)^{1/2} q_{ij}$ is $L^2(a, \infty)$. Further, if we define the matrix $R^{[1]}$ by

$$R^{[1]} = -Q' \tag{1.7.7}$$

and assume in addition to (1.7.1) that

$$\int_a^\infty |\lambda_i(x) - \lambda_j(x)|^{-1}\, |R^{[1]}(x)|^2\, dx < \infty, \tag{1.7.8}$$

then $(\lambda_i - \lambda_j)^{-1/2} Q'$ is also $L^2(a, \infty)$ and consequently QQ' is $L(a, \infty)$. We can then write

$$R_1 = -\{I - (I+Q)^{-1}Q\}Q' = -Q' + S = R^{[1]} + S.$$

Thus, subject to both (1.7.1) and (1.7.8), (1.7.5) simplifies to

$$Z' = (\Lambda + \text{dg}\, R + R^{[1]} + S)Z, \tag{1.7.9}$$

where $R^{[1]}$ is given by (1.7.7), (1.7.3), and (1.7.4).

In the proof of the theorem which follows, we apply these ideas based on (1.7.1), (1.7.2), and (1.7.8) to a sequence of transformed systems in which (1.7.9) is the first stage.

Theorem 1.7.1 *Let $\Lambda(x)$ be an $n \times n$ diagonal matrix,*

$$\Lambda(x) = \text{dg}\,(\lambda_1(x), \ldots, \lambda_n(x)),$$

in which $\lambda_i(x) - \lambda_j(x)$ is nowhere zero in some interval $[a, \infty)$ for each pair of unequal integers i and j in $[1, n]$. Let $\Lambda(x)$ and $R(x)$ have locally absolutely continuous derivatives up to order $M-1$ in $[a, \infty)$ for some $M \geq 1$. With the matrices $R^{[m]}$ $(0 \leq m \leq M)$ defined by $R^{[0]} = R$,

$$r_{ij}^{[m]} = \{r_{ij}^{[m-1]}/(\lambda_i - \lambda_j)\}' \qquad (i \neq j,\ m \geq 1), \tag{1.7.10}$$

and

$$\text{dg}\, R^{[m]} = 0 \qquad (m \geq 1), \tag{1.7.11}$$

let the $R^{[m]}$ $(0 \leq m \leq M-1)$ satisfy

(i) $$\{\lambda_k(x) - \lambda_l(x)\}^{-1} R^{[m]}(x) \to 0 \qquad \textit{as } x \to \infty; \tag{1.7.12}$$

(ii) $$\int_a^\infty |\lambda_k(x) - \lambda_l(x)|^{-1}\, |R^{[m]}(x)|^2\, dx < \infty \tag{1.7.13}$$

for each k and l in $[1, n]$ $(k \neq l)$; and let

(iii) $$\int_a^\infty |R^{[M]}(x)|\, dx < \infty. \tag{1.7.14}$$

Also, let the $\lambda_k(x) + r_{kk}(x)$ *satisfy the dichotomy condition* L *in* § 1.3. *Then, as* $x \to \infty$, (1.6.1) *has solutions* $Y_k(x)$ *with the asymptotic form*

$$Y_k(x) = \{e_k + o(1)\} \exp\left(\int_a^x \{\lambda_k(t) + r_{kk}(t)\}\, dt\right). \tag{1.7.15}$$

Proof As indicated by (1.7.3) and (1.7.4), we define the matrices Q_m $(0 \leqslant m \leqslant M-1)$ by

$$(q_m)_{ij} = r_{ij}^{[m]}/(\lambda_j - \lambda_i) \qquad (i \neq j) \tag{1.7.16}$$

and $\text{dg}\, Q_m = 0$. We note that

$$Q_m(x) \to 0 \qquad \text{as } x \to \infty, \tag{1.7.17}$$

by (1.7.12), and that $R^{[m]} = -Q'_{m-1}$ $(m \geqslant 1)$, by (1.7.10) and (1.7.11). The initial transformation

$$Y = (I + Q_0)Z_1 \tag{1.7.18}$$

is simply the one which takes (1.6.1) into

$$Z_1' = (\Lambda + \text{dg}\, R + R^{[1]} + S)Z_1 \tag{1.7.19}$$

as we have already mentioned in (1.7.9). Next, the case $m = 1$ of (1.7.12) and the cases $m = 1$ and $m = 2$ of (1.7.13) show that $R^{[1]}$ satisfies the conditions which allow the ideas involved in (1.7.18) to be repeated for (1.7.19). Hence the transformation

$$Z_1 = (I + Q_1)Z_2$$

takes (1.7.19) into

$$Z_2' = (\Lambda + \text{dg}\, R + R^{[2]} + S)Z_2.$$

We repeat the process until we make the final transformation

$$Z_{M-1} = (I + Q_{M-1})Z_M$$

to obtain

$$Z_M' = (\Lambda + \text{dg}\, R + R_M + S)Z_M \tag{1.7.20}$$

where, as in (1.7.6),

$$R_M = -(I + Q_{M-1})^{-1}Q'_{M-1} = (I + Q_{M-1})^{-1}R^{[M]}.$$

The inverse matrix here exists in some interval $[X, \infty)$, by (1.7.17), and it then follows from (1.7.14) also that

$$\int_X^\infty |R_M(x)|\, dx < \infty.$$

Hence (1.7.20) is in the Levinson form. The entire transformation from

(1.6.1) to (1.7.20) is

$$Y = (I + Q_0)(\cdots)(I + Q_{M-1})Z_M$$
$$= \{I + o(1)\}Z_M, \qquad (1.7.21)$$

by (1.7.17) again. The proof of (1.7.15) is therefore completed by applying the Levinson theorem to (1.7.20) and using (1.7.21) to transform back to Y. □

A useful expression for $r_{ij}^{[m]}$ in terms of r_{ij} can be obtained from (1.7.10) if we define the variable s_{ij} by

$$s_{ij} = \int_a^x \{\lambda_i(t) - \lambda_j(t)\}\,dt \qquad (1.7.22)$$

and write

$$u_{ij} = r_{ij}/(\lambda_i - \lambda_j).$$

Then, in terms of differentiation with respect to s_{ij}, (1.7.10) becomes simply

$$r_{ij}^{[m]} = (\lambda_i - \lambda_j)\,d^m u_{ij}/ds_{ij}^m. \qquad (1.7.23)$$

Example 1.7.1 A straightforward example on Theorem 1.7.1 arises when the $\lambda_i(x)$ all have the form

$$\lambda_i(x) = (\text{const.})\lambda(x) \qquad (1.7.24)$$

with $\lambda(x)$ independent of i. Let us define

$$s = \int_a^x \lambda(t)\,dt \qquad (1.7.25)$$

and write $D = d/ds$. Then (1.7.22) and (1.7.23) give

$$s_{ij} = (\text{const.})s, \qquad r_{ij}^{[m]} = (\text{const.})\lambda D^m(\lambda^{-1}r_{ij}),$$

and the conditions (1.7.12)–(1.7.14) become

$$D^m(\lambda^{-1}R) \to 0 \quad \text{as } x \to \infty, \qquad (1.7.26)$$

$$\lambda^{1/2}D^m(\lambda^{-1}R) \in L^2(a, \infty), \qquad (1.7.27)$$

both for $0 \leqslant m \leqslant M-1$, and

$$\lambda D^M(\lambda^{-1}R) \in L(a, \infty). \qquad (1.7.28)$$

In the simplest case where $\lambda(x)$ is constant, the derivatives in (1.7.26)–(1.7.28) reduce to x-derivatives and the conditions become

$$R^{(m)}(x) \to 0 \qquad \text{as } x \to \infty, \qquad (1.7.29)$$

$$R^{(m)}(x) \in L^2(a, \infty), \qquad (1.7.30)$$

both for $0 \leqslant m \leqslant M-1$, and

$$R^{(M)}(x) \in L(a, \infty). \tag{1.7.31}$$

These conditions on R do of course correspond to those given for $M=1$ in (1.6.40) and (1.6.41), taken together with the simplifying condition (1.6.35) which forms the basis for the theory so far in this section.

Some idea of the nature of (1.7.26)–(1.7.28) when $M>1$ can be gained by first taking

$$R(x) = r(x)C, \tag{1.7.32}$$

where $r(x)$ is a scalar factor and C is a constant matrix, and then considering the special case

$$\lambda(x) = x^{\alpha} f(x^{\gamma}), \qquad r(x) = x^{\beta} F(x^{\gamma}), \tag{1.7.33}$$

where $f(t)$ and $F(t)$ are periodic functions of t with $f(t)$ nowhere zero, and α, β, γ are constants with $\beta \geqslant -1$ and $\gamma \geqslant 0$. Then (1.7.26)–(1.7.28) hold if

$$\alpha > 2\beta + 1, \qquad \alpha > \gamma - 1 \tag{1.7.34}$$

and

$$M(\alpha + 1 - \gamma) > \beta + 1. \tag{1.7.35}$$

If therefore (1.7.34) holds, (1.7.35) shows that Theorem 1.7.1 applies when M is chosen large enough. In other words, the nearer γ is to $\alpha + 1$, the greater the number of transformations that must be made to arrive at a system of the Levinson form. This concludes for the present our discussion of the example (1.7.24).

At the beginning of this section we referred to the difficulties involved in the process of transforming (1.6.1) repeatedly using the ideas of § 1.6. It is, however, possible to carry out the first step in this process as far as (1.6.10) and then apply the simplified transformation process of this section to (1.6.10) rather than (1.6.1). The effect of doing this is to remove the condition (1.7.1) (which is the case $m=0$ of (1.7.13)) from the analysis. Let us therefore consider briefly the idea of applying Theorem 1.7.1 (with $M-1$ in place of M) to (1.6.10), where Q is as in § 1.6, and then making the transformation (1.6.5) back to Y. The conditions (i)–(iii) will appear first as conditions on Q, and they must then be expressed in terms of R by means of the formula (1.6.17) which relates Q and R. The details of these calculations are somewhat lengthy, and we refer to the literature (Eastham 1986) for a full analysis. Here we merely state the asymptotic theorem for (1.6.1) which is obtained in this way.

The theorem contains an extra condition (1.7.38), which implies that QQ' is $L(a, \infty)$, and the consequence is that we can write (1.6.10) in the simpler form

$$Z' = (\Lambda_1 - Q' + S)Z, \tag{1.7.36}$$

as we did in the case of (1.7.5) and (1.7.9). Then Theorem 1.7.1 can be more readily applied to (1.7.36) than to (1.6.10), although the main problem is still caused by the Q in (1.7.36) being the one in (1.6.17), not (1.7.3). We require the following notation, in which $i \neq j$ and $k \neq l$:

$$\Omega(x) = \max |\{\lambda_i(x) - \lambda_j(x)\}/\{\lambda_k(x) - \lambda_l(x)\}|, \qquad (1.7.37)$$

the maximum being taken over all such i, j, k, l in $[1, n]$;

$$D_{ij} = (\lambda_i - \lambda_j)^{-1}\, \mathrm{d}/\mathrm{d}x = d/s_{ij},$$

where s_{ij} is as in (1.7.22);

$$R(k, l)(x) = \{\lambda_k(x) - \lambda_l(x)\}^{-1} R(x).$$

Theorem 1.7.2 *Let $\Lambda(x)$ be an $n \times n$ diagonal matrix,*

$$\Lambda(x) = \mathrm{dg}\,(\lambda_1(x), \ldots, \lambda_n(x)),$$

in which $\lambda_i(x) - \lambda_j(x)$ is nowhere zero in some interval $[a, \infty)$ for each pair of unequal integers i and j in $[1, n]$. Let $\Lambda(x)$ and $R(x)$ have locally absolutely continuous derivatives up to order $M-1$ in $[a, \infty)$ for some $M \geq 2$. For each i, j, k, l in $[1, n]$ $(i \neq j, k \neq l)$ and each m in $[1, M-1]$, let

(i) $$R(k, l)(x) \to 0 \quad \textit{as } x \to \infty;$$

(ii) $$\int_a^\infty |R(k, l)(x)|\, |R'(i, j)(x)|\, \mathrm{d}x < \infty; \qquad (1.7.38)$$

(iii) $$\Omega(x) D_{ij}^m R(k, l)(x) \to 0 \quad \textit{as } x \to \infty;$$

(iv) $$\int_a^\infty \Omega(x)\, |\lambda_i(x) - \lambda_j(x)|\, |D_{ij}^m R(k, l)(x)|^2\, \mathrm{d}x < \infty;$$

(v) $$\int_a^\infty |\lambda_i(x) - \lambda_j(x)|\, |D_{ij}^M R(k, l)(x)|\, \mathrm{d}x < \infty.$$

Also, let the eigenvalues $\mu_k(x)$ $(1 \leq k \leq n)$ of $\Lambda(x) + R(x)$ satisfy the dichotomy condition of § 1.3. *Then, as $x \to \infty$,* (1.6.1) *has solutions $Y_k(x)$ with the asymptotic form*

$$Y_k(x) = \{e_k + o(1)\} \exp\left(\int_a^x \mu_k(t)\, \mathrm{d}t\right). \qquad (1.7.39)$$

Returning to Example 1.7.1, in which (1.7.24) holds, we have $\Omega(x)$ constant in this situation, by (1.7.37), and the conditions (i)–(v) in Theorem 1.7.2 reduce to (1.7.26)–(1.7.28) again except that the case $m = 0$ of (1.7.27) is omitted and replaced by the less restrictive condition that

$$|\lambda^{-1} R|\, |(\lambda^{-1} R)'| \in L(a, \infty).$$

With the choice (1.7.32) and (1.7.33), the condition (1.7.34) is then improved to

$$\gamma < \min\{2(\alpha-\beta), \alpha+1\} \tag{1.7.40}$$

while (1.7.35) is unchanged.

Finally in this section, we mention that it is possible to avoid the imposition of (1.7.38) and work directly with (1.6.10) rather than (1.7.36) (Eastham 1986, § 5(iv)). The consequent change in (1.7.39) is that μ_k must be replaced by $\mu_k - \{(I+Q)^{-1}Q'\}_{kk}$. However, it is not clear just how informative this new expression is in terms of the original Λ and R.

1.8 ASYMPTOTICALLY CONSTANT SYSTEMS

Turning now to the asymptotically constant system

$$Y'(x) = \{C + R(x)\}Y(x) \tag{1.8.1}$$

which was introduced in (1.1.8), we consider the situation where (1.8.1) can be transformed into an asymptotically diagonal system of the type already investigated. The form of the solutions of (1.8.1) as $x \to \infty$ is then obtained by applying the theorems in §§ 1.3–1.7 and transforming back.

We denote the eigenvalues of the constant $n \times n$ matrix C by λ_k and the eigenvectors by u_k. The basic requirement concerning C in this section is that there are n linearly independent eigenvectors u_k. Then C can be expressed in the diagonal form

$$T^{-1}CT = \Lambda, \tag{1.8.2}$$

where

$$\Lambda = \mathrm{dg}(\lambda_1, \ldots, \lambda_n) \tag{1.8.3}$$

with the eigenvalues being repeated according to their multiplicity, and T is the non-singular matrix

$$T = (u_1 \quad \cdots \quad u_n). \tag{1.8.4}$$

The most common situation in which (1.8.2) holds is where C has n distinct eigenvalues, as we noted in (1.2.2).

The first theorem in this section corresponds to Theorem 1.3.1.

Theorem 1.8.1 *Let C be an $n \times n$ constant matrix with n linearly independent eigenvectors u_k. Let the $n \times n$ matrix $R(x)$ satisfy*

$$\int_a^\infty |R(x)|\,\mathrm{d}x < \infty. \tag{1.8.5}$$

Then, as $x \to \infty$, (1.8.1) has solutions $Y_k(x)$ $(1 \le k \le n)$ with the asymptotic form

$$Y_k(x) = \{u_k + o(1)\}\exp(\lambda_k x). \tag{1.8.6}$$

Proof We make the transformation

$$Y = TZ \tag{1.8.7}$$

with T as in (1.8.2) and (1.8.4). Then (1.8.1) becomes

$$Z'(x) = \{\Lambda + R_1(x)\}Z(x), \tag{1.8.8}$$

where Λ is as in (1.8.3) and $R_1 = T^{-1}RT$. Since T is a constant matrix, (1.8.5) implies that

$$\int_a^\infty |R_1(x)| \, \mathrm{d}x < \infty,$$

and therefore (1.8.8) is in the Levinson form. The dichotomy condition (1.3.13) is clearly satisfied since the λ_i are constant here. Then, by Theorem 1.3.1, (1.8.8) has solutions Z_k such that

$$Z_k(x) = \{e_k + o(1)\} \exp(\lambda_k x).$$

On transforming back to Y by means of (1.8.4) and (1.8.7), we obtain the corresponding solutions $Y_k = TZ_k$ of (1.8.1) which satisfy (1.8.6). □

The next two theorems are obtained in a similar way from Theorems 1.5.1 and 1.6.1 by means of the same transformation (1.8.7).

Theorem 1.8.2 *Let C be an $n \times n$ constant matrix and let the eigenvalues λ_k of C have distinct real parts. Let the $n \times n$ matrix $R(x)$ satisfy*

$$\int_a^\infty |R(x)|^p \, \mathrm{d}x < \infty \tag{1.8.9}$$

for some p such that $1 < p \leqslant 2$. Then, as $x \to \infty$, (1.8.1) has solutions $Y_k(x)$ $(1 \leqslant k \leqslant n)$ with the asymptotic form

$$Y_k(x) = \{u_k + o(1)\} \exp\left(\lambda_k x + \int_a^x \rho_k(t) \, \mathrm{d}t\right),$$

where ρ_k denotes the kth diagonal element of $T^{-1}RT$.

Although it may well be possible to work out ρ_k explicitly in terms of C and R in individual cases, there is no helpful expression for ρ_k in general.

Theorem 1.8.3 *Let C be an $n \times n$ constant matrix with n distinct eigenvalues. Let the $n \times n$ matrix $R(x)$ be locally absolutely continuous in $[a, \infty)$, and let*

$$R(x) \to 0 \quad \text{as } x \to \infty \tag{1.8.10}$$

and

$$\int_a^\infty |R'(x)| \, \mathrm{d}x < \infty. \tag{1.8.11}$$

Let the eigenvalues $\mu_k(x)$ *of* $C+R(x)$ *satisfy the dichotomy condition* L *in* § 1.3. *Then, as* $x\to\infty$, (1.8.1) *has solutions* $Y_k(x)$ $(1\leqslant k\leqslant n)$ *with the asymptotic form*

$$Y_k(x)=\{u_k+o(1)\}\exp\left(\int_a^x \mu_k(t)\,\mathrm{d}t\right). \tag{1.8.12}$$

Finally, from the case $\Omega(x)=$ constant of Therorem 1.7.2, we obtain the following theorem, again using (1.8.7) first.

Theorem 1.8.4 *Let* C *be as in Theorem* 1.8.3. *Let the* $n\times n$ *matrix* $R(x)$ *have locally absolutely continuous derivatives up to order* $M-1$ *in* $[a,\infty)$ *for some* $M\geqslant 2$, *and let*

(i) $$R^{(m)}(x)\to 0 \quad \text{as } x\to\infty \quad (0\leqslant m\leqslant M-1);$$

(ii) $$\int_a^\infty |R(x)|\,|R'(x)|\,\mathrm{d}x<\infty;$$

(iii) $$\int_a^\infty |R^{(m)}(x)|^2\,\mathrm{d}x<\infty \quad (1\leqslant m\leqslant M-1);$$

(iv) $$\int_a^\infty |R^{(M)}(x)|\,\mathrm{d}x<\infty.$$

Let the eigenvalues $\mu_k(x)$ *of* $C+R(x)$ *satisfy the dichotomy condition* L *in* § 1.3. *Then, as* $x\to\infty$, (1.8.1) *has solutions* $Y_k(x)$ *with the asymptotic form* (1.8.12).

1.9 HIGHER-ORDER DIFFERENTIAL EQUATIONS

In this section we discuss two separate ways in which a differential equation of order n $(\geqslant 2)$ can be expressed as a first-order system. Subject to suitable conditions on the coefficients in the differential equation, the system can be transformed into the asymptotically diagonal form (1.1.9), and the theorems in §§ 1.3–1.7 can then be applied to yield asymptotic results for the nth-order equation. In this section we concentrate on the relatively simple case where the system initially has the asymptotically constant form (1.8.1), leaving to subsequent chapters the more intricate analysis required for other cases.

We begin with the familiar way of expressing the nth-order equation

$$y^{(n)}(x)+a_1(x)y^{(n-1)}(x)+\cdots+a_n(x)y(x)=0 \tag{1.9.1}$$

as a first-order system in terms of the vector $Y(x)$ whose n components are

$$y^{(i-1)}(x) \quad (1\leqslant i\leqslant n). \tag{1.9.2}$$

It is easy to check that, when $y(x)$ is a solution of (1.9.1), $Y(x)$ satisfies

$$Y'(x) = A(x)Y(x), \tag{1.9.3}$$

where $A(x)$ is the $n \times n$ matrix whose entries $a_{ij}(x)$ are given by

$$\left.\begin{aligned} a_{i,i+1}(x) &= 1 & &(1 \leqslant i \leqslant n-1) \\ a_{n,j}(x) &= -a_{n-j+1}(x) & &(1 \leqslant j \leqslant n) \end{aligned}\right\} \tag{1.9.4}$$

and the other entries are zero. The basic existence and uniqueness theorems for the solutions of (1.9.3) apply under the condition that the a_{n-j+1} $(1 \leqslant j \leqslant n)$ are locally Lebesgue integrable.

Here we deal with the asymptotically constant situation where (1.9.1) is

$$y^{(n)}(x) + \{c_1 + r_1(x)\}y^{(n-1)}(x) + \cdots + \{c_n + r_n(x)\}y(x) = 0, \tag{1.9.5}$$

the c_j being constants and the $r_j(x)$ small in some sense as $x \to \infty$. We concentrate on the asymptotic result which is obtained from Theorem 1.8.1.

Theorem 1.9.1 *Let the constants* c_j $(1 \leqslant j \leqslant n)$ *be such that the polynomial*

$$\lambda^n + c_1\lambda^{n-1} + \cdots + c_n \tag{1.9.6}$$

has n distinct zeros λ_k $(1 \leqslant k \leqslant n)$. *Let the functions* $r_j(x)$ *satisfy*

$$\int_a^\infty |r_j(x)|\, dx < \infty. \tag{1.9.7}$$

Then, as $x \to \infty$, (1.9.5) *has n solutions* $y_k(x)$ *which, together with their derivatives up to order* $n-1$, *have the asymptotic form*

$$y_k^{(i-1)}(x) = \{\lambda_k^{i-1} + o(1)\} \exp(\lambda_k x) \qquad (1 \leqslant i \leqslant n). \tag{1.9.8}$$

Proof In (1.9.3) we now have $A(x) = C + R(x)$ where, by (1.9.4), C has entries zero except for

$$\begin{aligned} c_{i,i+1} &= 1 & &(1 \leqslant i \leqslant n-1), \\ c_{n,j} &= -c_{n-j+1} & &(1 \leqslant j \leqslant n) \end{aligned}$$

and, by (1.9.7), $R(x)$ satisfies (1.8.5). It is straightforward to verify that the characteristic polynomial $\det(\lambda I - C)$ is just (1.9.6) and that an eigenvector u_k of C corresponding to the eigenvalue λ_k has components

$$\lambda_k^{i-1} \qquad (1 \leqslant i \leqslant n). \tag{1.9.9}$$

Then (1.9.8) follows immediately from (1.8.6) and (1.9.2). □

The only points that we mention in connection with the application of Theorems 1.8.2–4 to (1.9.5) concern the modifications which are made to the exponential factor in (1.9.8). In Theorem 1.8.2, there is an extra term denoted by ρ_k and, after a calculation, it follows from (1.8.4) and (1.9.9) that

$$\rho_k = -(r_1\lambda_k^{n-1} + r_2\lambda_k^{n-2} + \cdots + r_n)\Big/\prod_{i\neq k}(\lambda_k - \lambda_i) \tag{1.9.10}$$

in the present situation. Also, in Theorems 1.8.3 and 1.8.4, the μ_k are the eigenvalues of $C + R(x)$ and are therefore now the zeros of the polynomial

$$\lambda^n + \{c_1 + r_1(x)\}\lambda^{n-1} + \cdots + \{c_n + r_n(x)\} = 0 \tag{1.9.11}$$

rather than (1.9.6).

We move on to the second way of expressing higher-order equations as first-order systems, and we start with the even-order equation

$$\{p_0(x)y^{(m)}(x)\}^{(m)} + \{p_1(x)y^{(m-1)}(x)\}^{(m-1)} + \cdots + p_m(x)y(x) = 0. \tag{1.9.12}$$

This equation is of self-adjoint form when the coefficients are real-valued but, for the purposes of formulating the first-order system, the restriction to real-valued coefficients is not required. In place of (1.9.4), the $2m \times 2m$ matrix $A(x)$ is now defined by

$$\left.\begin{aligned} a_{i,i+1}(x) &= 1 \qquad && (1 \leqslant i \leqslant 2m-1,\ i \neq m), \\ a_{m,m+1}(x) &= 1/p_0(x), && \\ a_{2m-j+1,j}(x) &= -p_{m-j+1}(x) && (1 \leqslant j \leqslant m), \end{aligned}\right\} \tag{1.9.13}$$

with the other entries zero. It is then easy to check that a solution $y(x)$ of (1.9.12) gives rise to a vector $Y(x)$ which satisfies

$$Y'(x) = A(x)Y(x), \tag{1.9.14}$$

the first component of $Y(x)$ being $y(x)$. The basic existence and uniqueness theorems for the solutions of (1.9.14) apply under the condition that $1/p_0$ and p_j $(1 \leqslant j \leqslant m)$ are locally Lebesgue integrable. Indeed, under this condition on the coefficients, (1.9.14) can be regarded as a more general formulation of (1.9.12) which avoids the obvious differentiability requirements on the coefficients that are implied by the form (1.9.12). We also make the important observation that (1.9.13) defines a matrix A which has the form (1.2.5). This property of A will be used in Chapter 3.

The components of Y in (1.9.14) are no longer the derivatives of y as they were in (1.9.2). The components now involve the p_j as well as y, and they are known as the *quasi-derivatives* of y. In the following definition, we introduce this terminology for suitable general functions f defined in $[a, \infty)$.

Definition 1.9.1 Let f be locally absolutely continuous in $[a, \infty)$. The quasi-derivatives $f^{[\nu-1]}$ $(1 \leqslant \nu \leqslant 2m+1)$ are defined by

$$
\begin{aligned}
f^{[i-1]}(x) &= f^{(i-1)}(x) && (1 \leqslant i \leqslant m),\\
f^{[m]}(x) &= p_0(x) f^{(m)}(x), &&\\
f^{[m+i]}(x) &= p_i(x) f^{(m-i)}(x) + \frac{\mathrm{d}}{\mathrm{d}x}(f^{[m+i-1]}(x)) && (1 \leqslant i \leqslant m).
\end{aligned}
$$

The definition proceeds inductively through to $\nu = 2m+1$ provided that, at each stage, the quasi-derivatives already defined are locally absolutely continuous. If Y is a solution of (1.9.14) with first component y, it follows that the components of Y are

$$y^{[\nu-1]} \qquad (1 \leqslant \nu \leqslant 2m) \tag{1.9.15}$$

and that (1.9.12) is simply

$$y^{[2m]}(x) = 0.$$

The system (1.9.14) is known as the quasi-derivative formulation of (1.9.12).

The asymptotic result which corresponds to Theorem 1.9.1 in the case of the equation (1.9.12) is as follows.

Theorem 1.9.2 *Let the constants c_j $(0 \leqslant j \leqslant m)$ be such that the polynomial*

$$c_0\lambda^{2m} + c_1\lambda^{2m-2} + \cdots + c_{m-1}\lambda^2 + c_m \tag{1.9.16}$$

has $2m$ distinct zeros λ_k $(1 \leqslant k \leqslant 2m)$. Let

$$1/p_0(x) = c_0^{-1} + r_0(x), \quad p_j(x) = c_j + r_j(x) \qquad (1 \leqslant j \leqslant m), \tag{1.9.17}$$

where

$$\int_a^\infty |r_j(x)|\, \mathrm{d}x < \infty \qquad (0 \leqslant j \leqslant m).$$

Then, as $x \to \infty$, (1.9.12) has solutions $y_k(x)$ which have the asymptotic form

$$y_k(x) \sim \exp(\lambda_k x). \tag{1.9.18}$$

Proof In (1.9.14), we have $A(x) = C + R(x)$, where $R(x)$ satisfies (1.8.5) and C is a constant matrix whose characteristic polynomial $\det(\lambda I - C)$ is (1.9.16). An eigenvector of C corresponding to an eigenvalue λ is easily seen to have components

$$\lambda^{i-1} \qquad (1 \leqslant i \leqslant m), \quad c_0\lambda^m,$$

and

$$c_0\lambda^{m+i} + c_1\lambda^{m+i-2} + \cdots + c_i\lambda^{m-i} \qquad (1 \leqslant i \leqslant m-1).$$

From (1.8.6) and (1.9.15) we then obtain the asymptotic form of $y_k^{[\nu-1]}(x)$ $(1 \leq \nu \leq 2m)$ as $x \to \infty$, and (1.9.18) is the result for the first component, $\nu = 1$. □

There are also similar remarks to those made in (1.9.10) and (1.9.11) concerning the application of Theorems 1.8.2–4 to (1.9.12) and (1.9.17). In place of (1.9.10), for example, we now have

$$\rho_k = \{c_0^2 r_0 \lambda_k^{2m} - r_1 \lambda_k^{2m-2} - \cdots - r_{m-1}\lambda_k^2 - r_m\}/c_0 \prod_{i \neq k} (\lambda_k - \lambda_i),$$

where Lemma 1.2.2 is used to calculate the inverse matrix T^{-1} which occurs in (1.8.2) and Theorem 1.8.2.

So far we have been dealing with (1.9.12). Let us now augment the differential equation by introducing the extra terms

$$\tfrac{1}{2} \sum_{j=0}^{m-1} [\{q_{m-j}(x) y^{(j)}(x)\}^{(j+1)} + \{q_{m-j}(x) y^{(j+1)}(x)\}^{(j)}] \qquad (1.9.19)$$

into the left-hand side of (1.9.12). The new equation is still of order $2m$, and the factor $\frac{1}{2}$ has been inserted for convenience in applications to be made later. We note that the differential expression (1.9.19) is self-adjoint when the $q_{m-j}(x)$ are all pure imaginary but, once again, we do not need to restrict ourselves to this case. In the quasi-derivative formulation (1.9.14) of the new equation, $A(x)$ has the same entries as in (1.9.13) except that we add

$$a_{2m-j,j}(x) = a_{2m-j+1,j+1}(x) = -\tfrac{1}{2} q_{m-j+1}(x)$$

$$(1 \leq j \leq m-1), \qquad (1.9.20)$$

$$a_{mm}(x) = a_{m+1,m+1}(x) = -\tfrac{1}{2} q_1(x)/p_0(x),$$

and make the change

$$a_{m+1,m}(x) = -p_1(x) + \tfrac{1}{4} q_1^2(x)/p_0(x). \qquad (1.9.21)$$

The new system (1.9.14) is then equivalent to (1.9.12), augmented by (1.9.19), with the first component of Y being y. There is, for example, an obvious result corresponding to Theorem 1.9.2, in which we add to (1.9.17)

$$q_j(x) = d_j + s_j(x) \qquad (1 \leq j \leq m),$$

where $s_j(x)$ is $L(a, \infty)$, and replace (1.9.16) by

$$c_0 \lambda^{2m} + d_1 \lambda^{2m-1} + c_1 \lambda^{2m-2} + \cdots + c_{m-1}\lambda^2 + d_m \lambda + c_m.$$

Finally we mention the equation

$$\sum_{j=0}^{m} [\tfrac{1}{2}\{(q_{m-j} y^{(j)})^{(j+1)} + (q_{m-j} y^{(j+1)})^{(j)}\} + (p_{m-j} y^{(j)})^{(j)}] = 0, \qquad (1.9.22)$$

which again consists of the terms in (1.9.12) and (1.9.19) but is now of odd order $2m+1$. In the corresponding quasi-derivative formulation (1.9.14), $A(x)$ is now the $(2m+1)\times(2m+1)$ matrix whose non-zero entries are given by

$$
\begin{aligned}
a_{i,i+1} &= \begin{cases} 1 & (1 \leqslant i \leqslant 2m,\ i \neq m,\ m+1) \\ q_0^{-1/2} & (i = m,\ m+1) \end{cases} \\
a_{2m-j+2,j} &= \begin{cases} -p_{m-j+1} & (1 \leqslant j \leqslant m) \\ -p_0 q_0^{-1} & (j = m+1) \end{cases} \\
a_{2m-j+1,j} = a_{2m-j+2,j+1} &= \begin{cases} -\tfrac{1}{2} q_{m-j+1} & (1 \leqslant j \leqslant m-1) \\ -\tfrac{1}{2} q_1 q_0^{-1/2} & (j = m). \end{cases}
\end{aligned}
\tag{1.9.23}
$$

As in the case of (1.9.13), the matrices A in (1.9.20), (1.9.21), and (1.9.23) have the form (1.2.5) which was considered in Lemma 1.2.2.

Example 1.9.1 The familiar equation

$$y''(x) + \{c^2 + r(x)\}y(x) = 0, \tag{1.9.24}$$

in which c is a non-zero constant, is a simple example of (1.9.5), and we collect together here the asymptotic results that are obtained by applying Theorems 1.8.1–3 to (1.9.24). We state only the asymptotic forms of the solutions themselves, the forms for the derivatives being the obvious ones.

(i) Let $r(x)$ be $L(a, \infty)$. Then there are solutions

$$y(x) \sim \exp(\pm icx).$$

This follows from (1.9.8) with $\lambda_1 = ic$, $\lambda_2 = -ic$.

(ii) Let $r(x)$ be $L^p(a, \infty)$ for some p in $(1, 2]$ and let c be non-real. Then there are solutions

$$y(x) \sim \exp\left\{\pm i\left(cx + \frac{1}{2c}\int_a^x r(t)\,dt\right)\right\}.$$

Here the modified exponential term follows from (1.9.10).

(iii) Let $r(x) \to 0$, let $r'(x)$ be $L(a, \infty)$, and let $\operatorname{im}\{c^2 + r(x)\}^{1/2}$ have one sign in $[a, \infty)$. Then there are solutions

$$y(x) \sim \exp\left(\pm i\int_a^x \{c^2 + r(t)\}^{1/2}\,dt\right).$$

Here the exponential term follows from (1.9.11).

1.10 COEFFICIENT MATRICES OF JORDAN TYPE

In § 1.2 we introduced the elementary Jordan matrices, in terms of which the block diagonal form of a matrix A is expressed in cases where the pure diagonal form does not apply. These Jordan matrices arise in differential systems (1.1.1) when either the coefficient matrix $A(x)$ is itself a Jordan matrix or the block diagonal form of $A(x)$ occurs after a transformation $Y = TZ$.

For differential systems, a slight generalization of (1.2.6) is appropriate, and we refer to a matrix

$$J = \begin{bmatrix} \lambda & \mu & & 0 \\ & \ddots & \ddots & \\ & & \ddots & \mu \\ 0 & & & \lambda \end{bmatrix} \tag{1.10.1}$$

as being of Jordan type when $\mu \neq 0$. We write

$$J = \lambda I + \mu E$$

with the obvious definition of the matrix E. In this section, we examine the asymptotic form of the solutions of the system

$$Y'(x) = \{J(x) + R(x)\}Y(x) \tag{1.10.2}$$

in which a matrix of Jordan type replaces the diagonal matrix $\Lambda(x)$ previously considered. Once again, the results are obtained by a transformation into the Levinson form.

Theorem 1.10.1 *Let $J(x)$ be an $n \times n$ matrix of Jordan type,*

$$J(x) = \lambda(x)I + \mu(x)E, \tag{1.10.3}$$

in an interval $[a, \infty)$. With $\sigma(x)$ defined by

$$\sigma(x) = \int_a^x \mu(s)\,ds, \tag{1.10.4}$$

let $\sigma(x)$ be nowhere zero in some interval $[X, \infty)$, and let

$$|\sigma(x)/\sigma(t)| \geq K \qquad (X \leq t \leq x < \infty), \tag{1.10.5}$$

where K (>0) is a constant. Let the $n \times n$ matrix $R(x)$ satisfy

$$DRD^{-1} \in L(X, \infty), \tag{1.10.6}$$

where

$$D = \mathrm{dg}\,(1, \sigma, \ldots, \sigma^{n-2}, \sigma^{n-1}). \tag{1.10.7}$$

Then, as $x \to \infty$, (1.10.2) has solutions $Y_k(x)$ $(1 \leq k \leq n)$ with the *asymptotic form*

$$Y_k(x) = \begin{bmatrix} \sigma^{k-1}(x)/(k-1)! + o(\sigma^{k-1}(x)) \\ \sigma^{k-2}(x)/(k-2)! + o(\sigma^{k-2}(x)) \\ \vdots \\ \sigma(x) + o(\sigma(x)) \\ 1 + o(1) \\ o(\sigma^{-1}(x)) \\ \vdots \\ o(\sigma^{-n+k}(x)) \end{bmatrix} \exp\left(\int_a^x \lambda(t)\,\mathrm{d}t\right). \quad (1.10.8)$$

Proof We define the $n \times n$ matrix Φ by

$$\Phi = \begin{bmatrix} 1 & \sigma & \sigma^2/2! & \cdots & \sigma^{n-1}/(n-1)! \\ & 1 & \sigma & \cdots & \sigma^{n-2}/(n-2)! \\ & & \ddots & \ddots & \vdots \\ & & & \ddots & \sigma \\ 0 & & & & 1 \end{bmatrix} \quad (1.10.9)$$

where the entries below the main diagonal are zero. Then Φ is a solution of the matrix equation

$$\Phi' = \mu E \Phi, \quad (1.10.10)$$

and we also note that Φ can be expressed in the form

$$\Phi = D^{-1} B D, \quad (1.10.11)$$

where D is given by (1.10.7) and B is the constant matrix

$$B = \begin{bmatrix} 1 & 1 & 1/2! & \cdots & 1/(n-1)! \\ & 1 & 1 & \cdots & 1/(n-2)! \\ & & \ddots & \ddots & \vdots \\ & & & \ddots & 1 \\ 0 & & & & 1 \end{bmatrix}. \quad (1.10.12)$$

By (1.2.12) and (1.10.10), the transformation

$$Y = \Phi Z \quad (1.10.13)$$

removes μE from (1.10.2) and takes (1.10.2) into

$$Z' = (\lambda I + \Phi^{-1} R \Phi) Z.$$

By (1.10.11), the further transformation

$$Z = D^{-1} W \quad (1.10.14)$$

gives

$$W' = (\Lambda + B^{-1}DRD^{-1}B)W, \tag{1.10.15}$$

where

$$\Lambda = \lambda I + D'D^{-1}. \tag{1.10.16}$$

Since B is constant, (1.10.6) is the condition under which the Levinson theorem, Theorem 1.3.1, can be applied to (1.10.15), provided that Λ satisfies the dichotomy condition L in § 1.3. In the case (1.10.16) and (1.10.7) which we have here,

$$\int_t^x \operatorname{re}\{\lambda_i(s) - \lambda_j(s)\}\, \mathrm{d}s = c \log|\sigma(x)/\sigma(t)|,$$

where c is a constant, and therefore (1.10.5) is the dichotomy condition in the form (1.3.1) or (1.3.2) according as $c<0$ or $c>0$.

It now follows from (1.10.16) and (1.10.7) that the W-system (1.10.15) has solutions

$$W_k(x) = \{e_k + o(1)\} \exp\left(\int_X^x \{\lambda(t) + (k-1)\sigma'(t)/\sigma(t)\}\, \mathrm{d}t\right).$$

By (1.10.11), (1.10.13) and (1.10.14), the transformation back to Y takes the simple form

$$Y = D^{-1}BW, \tag{1.10.17}$$

and (1.10.8) follows from (1.10.7) and (1.10.12) when we adjust the solutions by constant multiples. □

The meaning of (1.10.5) becomes clearer if we introduce

$$\sigma_- = \liminf|\sigma(x)|, \qquad \sigma_+ = \limsup|\sigma(x)|,$$

both as $x \to \infty$. It then follows from (1.10.5) that, if σ_- is finite, σ_+ must also be finite. We therefore have two alternatives: either $\sigma(x)$ is bounded as $x \to \infty$ or $|\sigma(x)| \to \infty$ as $x \to \infty$. The interest of the theorem lies mainly in the second alternative. If, however, $\sigma(x)$ is not only bounded but converges to a finite limit as $x \to \infty$ we can go further and replace (1.10.4) by

$$\sigma(x) = -\int_x^\infty \mu(s)\, \mathrm{d}s.$$

Then (1.10.5) is replaced by $|\sigma(x)/\sigma(t)| \leqslant K$, and (1.10.8) follows as before with the new $\sigma(x)$.

It is useful to note that (1.10.9) and (1.10.12) can be written

$$\Phi = \exp \sigma E, \qquad B = \exp E.$$

It follows that

$$B^{-1} = \exp(-E) \tag{1.10.18}$$

and therefore that B^{-1} is simply (1.10.9) with σ replaced by -1. Of course, the form of B^{-1} can easily be checked directly once it is known.

It is also possible to apply Theorems 1.5.1 and 1.6.1 to the system (1.10.15) to obtain further asymptotic results for (1.10.2). By (1.10.15) and (1.10.16), the relevant conditions in the case of Theorem 1.5.1 are

$$|\mathrm{re}\, \sigma'/\sigma| \geq \delta > 0 \tag{1.10.19}$$

and

$$DRD^{-1} \in L^p(X, \infty)$$

for some p in $(1, 2]$ while, in the case of Theorem 1.6.1, the conditions are

$$(\sigma/\sigma')DRD^{-1} \to 0 \tag{1.10.20}$$

and

$$\{(\sigma/\sigma')DRD^{-1}\}' \in L(X, \infty). \tag{1.10.21}$$

In both cases, the exponential factor on the right-hand side of (1.10.8) is modified in accordance with (1.5.5) and (1.6.4).

To give an application of Theorem 1.10.1, we return to the nth order equation (1.9.1) and consider the case where all the coefficients $a_j(x)$ are small in a suitable sense as $x \to \infty$. This corresponds to the situation where all the c_j are zero in (1.9.5) and is not covered by Theorem 1.9.1.

Theorem 1.10.2 *In* (1.9.1), *let*

$$\int_a^\infty x^{j-1}\, |a_j(x)|\, \mathrm{d}x < \infty \qquad (1 \leq j \leq n). \tag{1.10.22}$$

Then, as $x \to \infty$, (1.9.1) *has n solutions* $y_k(x)$ *which, together with their derivatives up to order* $n-1$, *have the asymptotic form*

$$\left.\begin{aligned} y_k^{(i-1)}(x) &\sim x^{k-i}/(k-i)! \qquad && (1 \leq i \leq k), \\ y_k^{(i-1)}(x) &= o(x^{k-i}) && (k+1 \leq i \leq n). \end{aligned}\right\} \tag{1.10.23}$$

Proof We express the system (1.9.3) in the form (1.10.2) by writing

$$A(x) = J(x) + R(x), \tag{1.10.24}$$

where $J(x) = E$ and $R(x)$ has (n, j) entries $-a_{n-j+1}(x)$ $(1 \leq j \leq n)$ with zeros elsewhere. Thus we have the case $\lambda(x) = 0$, $\mu(x) = 1$ of (1.10.3), and we can take $\sigma(x) = x$. Clearly (1.10.5) is satisfied, and (1.10.6) is simply (1.10.22). Then (1.10.23) follows immediately from (1.10.8) since the components of Y_k are $y_k^{(i-1)}$, by (1.9.2). □

We also give the alternative result for (1.9.1) which is obtained when the conditions (1.10.20) and (1.10.21) are imposed. These conditions arise from Theorem 1.6.1, which we use here in the simplified form given by (1.6.35) and (1.6.37).

Theorem 1.10.3 *In* (1.9.1), *let*

(i) $x^j a_j(x) \to 0 \qquad as\ x \to \infty$ (1.10.25)

(ii) $\{x^j a_j(x)\}' \in L(a, \infty)$ (1.10.26)

(iii) $x^{2j-1} a_j^2(x) \in L(a, \infty)$. (1.10.27)

Then (1.10.23) *continues to hold except that the right-hand side throughout is multiplied by*

$$\exp\left\{\{(-1)^{n-k+1}/(n-k)!\}\int_a^x \left(\sum_{j=n-k+1}^{n} t^{j-1} a_j(t)/(j-n+k-1)!\right) \mathrm{d}t\right\}. \tag{1.10.28}$$

Proof We start with (1.10.24) again and then apply Theorem 1.6.1 to the corresponding W-system (1.10.15). The conditions (1.10.25) and (1.10.26) are just (1.10.20) and (1.10.21). Also, (1.10.27) is the simplifying condition (1.6.35) which allows us to apply (1.6.37) to (1.10.15). We therefore require the (k, k) diagonal entry in $B^{-1}DRD^{-1}B$ and, after a simple calculation using (1.10.12) and (1.10.18), we find that the extra factor $\exp\left(\int_a^x r_{kk}(t)\,\mathrm{d}t\right)$ which appears in (1.6.37) is now the expression (1.10.28). □

The example

$$a_j(x) = (\text{const.})x^{-j}(\log x)^{-\alpha_j}$$

is covered by Theorem 1.10.2 if $\alpha_j > 1$ and by Theorem 1.10.3 if $\alpha_j > \frac{1}{2}$. This last inequality arises only from the simplifying condition (1.10.27), the other two conditions being satisfied when $\alpha_j > 0$.

1.11 INTEGRAL CONDITIONS WITH NON-ABSOLUTE CONVERGENCE

Finally in this chapter, we consider a further $I + Q$ type of transformation which has more specialized implications than the previous types. We introduce the transformation in terms of the asymptotically diagonal system

$$Y'(x) = \{\Lambda(x) + R(x)\}Y(x) \tag{1.11.1}$$

and, in the following theorem, the main feature is that the smallness conditions on $R(x)$ involve an infinite integral in which the convergence may not be absolute.

Theorem 1.11.1 *Let* $\Lambda(x)$ *be an* $n \times n$ *diagonal matrix,*

$$\Lambda(x) = \mathrm{dg}\,(\lambda_1(x), \ldots, \lambda_n(x)),$$

which satisfies the dichotomy condition in § 1.3. *Let the* $n \times n$ *matrix* $R(x)$ *satisfy*

(i) $$\int_a^\infty R(t)\,\mathrm{d}t \quad \textit{converges}; \tag{1.11.2}$$

(ii) $$\Lambda(x)\int_x^\infty R(t)\,\mathrm{d}t - \left(\int_x^\infty R(t)\,\mathrm{d}t\right)\Lambda(x) \in L(a, \infty); \tag{1.11.3}$$

(iii) $$R(x)\int_x^\infty R(t)\,\mathrm{d}t \in L(a, \infty). \tag{1.11.4}$$

Then, as $x \to \infty$, (1.11.1) *has solutions* $Y_k(x)$ $(1 \leqslant k \leqslant n)$ *such that*

$$Y_k(x) = \{e_k + o(1)\} \exp\left(\int_a^x \lambda_k(t)\,\mathrm{d}t\right). \tag{1.11.5}$$

Proof The transformation $Y = (I + Q)Z$ takes (1.11.1) into

$$\begin{aligned}(I+Q)Z' &= \{(\Lambda + R)(I+Q) - Q'\}Z \\ &= \{(I+Q)\Lambda + \Lambda Q - Q\Lambda + R - Q' + RQ\}Z. \end{aligned} \tag{1.11.6}$$

With the choice

$$Q(x) = -\int_x^\infty R(t)\,\mathrm{d}t, \tag{1.11.7}$$

we have $Q(x) \to 0$ as $x \to \infty$, by (1.11.2), and in (1.11.6) we have $Q' = R$. Then, by (1.11.3) and (1.11.4), (1.11.6) becomes

$$(I+Q)Z' = \{(I+Q)\Lambda + S\}Z$$

where, as on previous occasions, S denotes any $L(a, \infty)$ matrix. Hence, in some interval $[X, \infty)$, we have

$$Z' = (\Lambda + S)Z,$$

and the Levinson theorem is applicable. Then (1.11.5) follows when we transform back to Y. □

As a simple example on the theorem, we consider

$$R(x) = (x^{-\alpha} \sin x)C$$

where C is a constant matrix. When $\alpha \leqslant 1$, this $R(x)$ is not covered by Theorem 1.3.1. However, (1.11.2) and (1.11.4) do hold when $\alpha > \frac{1}{2}$. Then (1.11.3) is a smallness condition on $\Lambda(x)$ which is satisfied when, for example, $\Lambda(x) = O(x^{-\beta})$ with $\alpha + \beta > 1$. We also note that this $\Lambda(x)$ is not covered by Theorem 1.5.1 and that Theorem 1.6.1 is not applicable either.

In §§ 1.5 and 1.7, the previous two $I + Q$ transformations were applied repeatedly to extend their range of usefulness, and the same can be done with the transformation based on (1.11.7). However, we defer a detailed analysis to Chapter 4 where we examine the main class of systems for which this idea produces significant results.

The choice (1.11.7) can also be made in the case of the system

$$Y'(x) = \{J(x) + R(x)\}Y(x), \tag{1.11.8}$$

which was the subject of § 1.10. However, unlike the case of Theorem 1.11.1, the conditions for (1.11.8) involve $(I+Q)^{-1}$ and are not easy to verify in general. An exception occurs when $Q^2 = 0$ since then $(I+Q)^{-1} = I - Q$, and progress can be made. We deal with an example of this situation in (1.11.16) below.

As applications of the choice (1.11.7), we prove two further theorems for the nth order equation (1.9.1), in which the main conditions on the coefficients involve only non-absolutely convergent integrals. These results include error terms of the type $o(x^{-\alpha})$ and they also draw on some of the earlier theory in the chapter.

Theorem 1.11.2 *In* (1.9.1), *let*

$$\int_x^\infty |a_1(t)|\, dt = o(x^{-\alpha}) \tag{1.11.9}$$

and

$$\int_x^\infty a_j(t)\, dt = o(x^{-j+1-\alpha}) \qquad (2 \leq j \leq n) \tag{1.11.10}$$

for some constant α *with* $0 < \alpha < 1$. *Then, as* $x \to \infty$, (1.9.1) *has solutions* $y_k(x)$ $(1 \leq k \leq n)$ *such that*

$$\left.\begin{aligned} y_k^{(i-1)}(x) &= \{x^{k-i}/(k-i)!\}\{1 + o(x^{-\alpha})\} && (1 \leq i \leq k) \\ y_k^{(i-1)}(x) &= o(x^{k-i-\alpha}) && (k+1 \leq i \leq n). \end{aligned}\right\} \tag{1.11.11}$$

Proof As in the proof of Theorem 1.10.2, we start with the system (1.10.2) in which $\lambda(x) = 0$, $\mu(x) = 1$, and $R(x)$ has (n, j) entries $-a_{n-j+1}(x)$ $(1 \leq j \leq n)$. We separate from $R(x)$ the term $-a_1(x)$ by writing

$$R = R_1 + R_2, \tag{1.11.12}$$

where $R_1(x)$ has the (n, n) entry $-a_1(x)$ and zeros elsewhere. In (1.10.2) we then make the transformation

$$Y = (I + Q)Z, \tag{1.11.13}$$

where

$$Q(x) = -\int_x^\infty R_2(t)\, dt. \tag{1.11.14}$$

Corresponding to (1.11.6), we have

$$(I+Q)Z' = \{E + EQ + R_2 - Q' + R_2Q + R_1(I+Q)\}Z. \tag{1.11.15}$$

It follows from the decomposition (1.11.12) that

$$Q^2 = 0, \qquad R_2 Q = 0. \tag{1.11.16}$$

Then also $(I+Q)^{-1} = I - Q$ and, by (1.11.14) and (1.11.16), the system (1.11.15) becomes

$$Z' = \{E + EQ - QE - QEQ + (I-Q)R_1(I+Q)\}Z. \tag{1.11.17}$$

Now we proceed to the Levinson form as in (1.10.17) by making the transformation

$$Z = D^{-1}BW, \tag{1.11.18}$$

where

$$D = \mathrm{dg}\,(1, x, \ldots, x^{n-1}). \tag{1.11.19}$$

Then (1.11.17) becomes

$$W' = (D'D^{-1} + S_1 + S_2)W \tag{1.11.20}$$

where, as in (1.10.15),

$$S_1 = B^{-1}D(I-Q)R_1(I+Q)D^{-1}B$$

and

$$S_2 = B^{-1}D(EQ - QE - QEQ)D^{-1}B.$$

It follows from (1.11.10), (1.11.14), and (1.11.19) that

$$S_1(x) = O\{|a_1(x)|\}, \qquad S_2(x) = o(x^{-1-\alpha}).$$

Hence, by (1.11.9), the system (1.11.20) has the form

$$W' = (x^{-1}\Lambda_0 + S)W, \tag{1.11.21}$$

where

$$\Lambda_0 = \mathrm{dg}\,(0, 1, \ldots, n-1) \tag{1.11.22}$$

and

$$\int_x^\infty |S(t)|\,\mathrm{d}t = o(x^{-\alpha}). \tag{1.11.23}$$

We can now apply Theorem 1.3.1 to the W-system (1.11.21), with the error estimates (1.4.29) and (1.4.30). By (1.11.22), we have $c \geqslant 1$ in (1.4.29) and this gives $u_i(x) = o(x^{-\alpha})$ since $\alpha < 1$ by hypothesis. In (1.4.30), we have the same estimate for $u_i(x)$, valid for any $\alpha > 0$ by (1.11.23). Hence (1.11.21) has solutions

$$W_k(x) = x^{k-1}\{e_k + o(x^{-\alpha})\}.$$

The $o(x^{-\alpha})$ term is preserved when we transform back to Y via (1.11.18) and (1.11.13) since $QD^{-1} = o(x^{-n+1-\alpha})$, by (1.11.10) and (1.11.14). Then (1.11.11) follows in the same way as (1.10.23) but with $o(x^{-\alpha})$ in place of $o(1)$. □

The next theorem gives a corresponding result for the case where (1.9.1) has the form (1.9.5).

Theorem 1.11.3 *Let the constants* c_j $(1 \leqslant j \leqslant n)$ *be such that the polynomial* (1.9.6) *has n zeros* λ_k *with distinct real parts. Let*

$$\int_x^\infty |r_1(t)|\,\mathrm{d}t = o(x^{-\alpha}) \tag{1.11.24}$$

and

$$\int_x^\infty r_j(t)\,\mathrm{d}t = o(x^{-\alpha}) \qquad (2 \leqslant j \leqslant n) \tag{1.11.25}$$

for some constant $\alpha > \frac{1}{2}$. *Then, as* $x \to \infty$, (1.9.5) *has n solutions* $y_k(x)$ *such that*

$$y_k^{(i-1)}(x) = \{\lambda_k^{i-1} + o(x^{-\beta})\} \exp(\lambda_k x) \qquad (1 \leqslant i \leqslant n), \tag{1.11.26}$$

where

$$\beta = \min(\alpha, 2\alpha - 1). \tag{1.11.27}$$

Proof We begin as in the proof of Theorem 1.9.1 by writing $A(x) = C + R(x)$ in (1.9.3) and then expressing C in its diagonal form

$$T^{-1}CT = \Lambda,$$

where

$$\Lambda = \mathrm{dg}(\lambda_1, \ldots, \lambda_n).$$

The transformation $Y = TZ$ gives

$$Z' = (\Lambda + T^{-1}RT)Z.$$

We again use the decomposition (1.11.12) where, in the present notation, $R_1(x)$ has the (n, n) entry $-r_1(x)$ and zeros elsewhere. Next, we make the transformation

$$Z = (I + Q)W, \tag{1.11.28}$$

where

$$Q(x) = -T^{-1}\left(\int_x^\infty R_2(t)\,\mathrm{d}t\right)T, \tag{1.11.29}$$

corresponding to (1.11.14). Then, as in (1.11.16), we have

$$Q^2 = 0, \qquad T^{-1}R_2TQ = 0, \qquad Q\Lambda Q = 0$$

and, corresponding to (1.11.17), we obtain

$$W' = (\Lambda + S + S_1)W,$$

where

$$S = \Lambda Q - Q\Lambda \tag{1.11.30}$$

and

$$S_1 = (I - Q)T^{-1}R_1T(I + Q).$$

It follows from (1.11.24) that

$$\int_x^\infty |S_1(t)|\,\mathrm{d}t = o(x^{-\alpha}). \tag{1.11.31}$$

Also, by (1.11.25) and (1.11.29),

$$Q(x) = o(x^{-\alpha}) \tag{1.11.32}$$

as $x \to \infty$, and hence also

$$S(x) = o(x^{-\alpha}). \tag{1.11.33}$$

Since $\alpha > \frac{1}{2}$ by hypothesis, we have $S \in L^2(a, \infty)$ and, leaving aside S_1 for the moment, we can make the Hartman–Wintner transformation

$$W = (I + Q_1)V \tag{1.11.34}$$

of the type used to obtain (1.5.10). By (1.11.33), (1.5.20), and (1.5.21), we have $Q_1 = o(x^{-\alpha})$ and hence the transformed system (1.5.10) has the form

$$V' = \{\Lambda + o(x^{-2\alpha}) + S_1\}V,$$

with a modified S_1 which still satisfies (1.11.31). Also, we retain Λ rather than Λ_1 because $\mathrm{dg}\, S = 0$, by (1.11.30). Then, with β as in (1.11.27), it follows from (1.11.31) and (1.4.28) that the V-system has solutions

$$V_k(x) = \{e_k + o(x^{-\beta})\} \exp(\lambda_k x). \tag{1.11.35}$$

Finally, by (1.11.28), (1.11.32), and (1.11.34), the transformation back to Y has the form

$$Y = T\{I + o(x^{-\alpha})\},$$

and (1.11.26) follows. □

It is also possible to say something when $0 < \alpha \leqslant \frac{1}{2}$. Everything down to (1.11.33) continues to hold, but (1.11.34) would be replaced by the iterated transformation of the type (1.5.28). By (1.5.27), the exponential factor in (1.11.35) would then be replaced by

$$\exp\left\{\lambda_k x + O\left(\int_a^x |S(t)|^2\,\mathrm{d}t\right)\right\}$$
$$= \begin{cases} \exp\{\lambda_k x + o(\log x)\} & (\alpha = \frac{1}{2}) \\ \exp\{\lambda_k x + o(x^{1-2\alpha})\} & (0 < \alpha < \frac{1}{2}). \end{cases}$$

NOTES AND REFERENCES

1.1 It is not our purpose to describe the historical development of the asymptotic theory of differential equations but, of the early work, we mention Poincaré (1885), Kneser (1896/9, 1897), Horn (1897), Bôcher (1900), Dunkel (1902/3), Birkhoff (1909), Perron (1913, 1929, 1960), and Sternberg (1919). Surveys within the wider context of stability theory are given by Cesari (1971, Chapter II, §§ 3.1–3.7) and Coppel (1965). We also refer to McHugh (1971) and Schlissel (1977) for other relevant historical surveys.

The investigation of asymptotically constant differential equations and systems goes back to Poincaré (1885), but the more fundamental concept of an asymptotically diagonal system is due to Cesari (1940, 1971, Chapter II, § 3.5) and Perron (1929).

1.2 A property like (1.2.3), but involving A^*, is also known in other contexts; see Pease (1965, Chapter 9) and Yakubovich and Starzhinskii (1975, pp. 156–7). A particular case of Lemma 1.2.2 was given by Fedorjuk (1966*b*, Lemma 2.1), and this case will be considered in Chapter 3. Lemmas 1.2.1 and 1.2.2 are due to Eastham (1984). A recent account of the Jordan normal form (1.2.7) is given by Väliaho (1986). We refer to Coddington and Levinson (1955, p. 28), for example, for the Liouville–Jacobi formula (1.2.13).

1.3 The original statement of the Levinson theorem (Levinson 1948) differed in two respects from the one given here. First, it was expressed in terms of the asymptotically constant system (1.1.8). The actual proof, however, was based on the asymptotically diagonal system (1.1.9). The same presentation was repeated in the books of Bellman (1953, pp. 50–5) and Coddington and Levinson (1955, pp. 92–7). In Coppel (1965, pp. 88–9) and subsequently, it is recognized that the fundamental statement of the theorem is in terms of (1.1.9).

Second, the original statement (Levinson 1948) contained a combination of the two theorems given separately here as Theorems 1.8.1 and 1.8.3. Again, it is now recognized that the two distinct ideas involved in the original proof are best separated, and that the result based on (1.3.3) (or (1.8.5)) is the fundamental one.

Of the earlier results leading up to the Levinson theorem, we mention those of Levinson (1946) (and the references therein) and Weyl (1946). There is also the result of Wintner (1946*b*) for the case $\Lambda(x)=0$ of (1.3.4) which states that, subject to (1.3.3), all solutions $Y(x)$ converge to a finite limit as $x \to \infty$; see also (Bellman 1947, § 7).

The example indicating the necessity of the dichotomy condition appears to be new here.

1.4 The estimates for the $o(1)$ terms in (1.4.24) and (1.4.25) can be improved further under stronger conditions on R and in more specialized situations. We refer to the work of Ràb (1958, 1966*a*, *b*, 1969, 1972), Olver (1961, 1974, Chapter 6), Taylor (1978, 1982), and Smith (1986) for details. For exponentially small estimates, we refer to Trench (1976) and Fubini (1937). In the case of asymptotically constant systems, a comprehensive asymptotic theory, including $o(x^{-\alpha})$ error terms is given by Hartman (1982, pp. 304–21).

1.5 The extension of the Hartman–Wintner theorem to the range $p>2$, given in Theorem 1.5.2, is due to Harris and Lutz (1977, § 3).

1.6 Theorem 1.6.1 is due to Eastham (1985). A similar result under more restrictive conditions was obtained by Gingold (1985b). The ideas in § 1.6 up to (1.6.17) are in essence due to Harris and Lutz (1974, § 2) in the case of constant Λ. Other methods for constructing the transformation (1.6.5), and the connection with 'L-diagonal' theory, are discussed by Cesari (1971, Chapter II, § 3.5), Coppel (1965, pp. 111–2), Harris and Lutz (1974, § 2), and Rapoport (1954).

1.7 In the case of constant Λ, Harris and Lutz (1974, Lemma 2) discuss the simplified transformation based on (1.7.1)–(1.7.3), and they introduce the idea of repeatedly applying transformations of this type in § 3 of their paper. Theorems 1.7.1 and 1.7.2 are due to Eastham (1986). Ben-Artzi and Devinatz (1979, § 3) consider an example of (1.7.33) with $\lambda(x)$ constant and $F(t)=\sin t$; then (1.7.40) partially covers Lemma 3.1 of their paper.

1.8 Theorems 1.8.1 and 1.8.3 are due to Levinson (1948) and, as mentioned in the notes on § 1.3, these two theorems were combined in Levinson's original statement of his theorem; see also the books of Naimark (1968, pp. 145–70) and Coppel (1965, pp. 114–5) and the paper of Rapoport (1951). Theorem 1.8.4 is due to Devinatz (1971) in the case $M=2$ and to Harris and Lutz (1974) for general M.

A generalization of Theorem 1.8.3, due to Fedorjuk (1966*c*, Theorem 1) deals with the system $Y'=\phi\Phi(C+R)\Phi^{-1}Y$, where Φ is a diagonal matrix whose entries are powers of the complex-valued function ϕ. This system is also covered by Theorem 1.6.1 (Eastham 1985, § 3).

1.9 Theorem 1.9.1 is due to Faedo (1946, p. 116). The formula (1.9.10), together with further detailed results for (1.9.5), are given by Hartman (1982, pp. 314–21).

A full account of quasi-derivatives, including the system formulation (1.9.14) for (1.9.12), (1.9.19), and (1.9.22), is given by Everitt and Zettl (1979); see also Naimark (1968, pp. 55–6). Theorem 1.9.2 is given by Naimark (1968, p. 175), but with the unnecessary restriction that the λ_k have distinct real parts.

Of the results (i)–(iii) in Example 1.9.1, (i) goes back to Kneser (1897), Weyl (1910), and (after a change of variable) Bôcher (1900); see also Wintner (1948, § 10). Next, (ii) is due to Wintner (1947*a*, § 2, 1948,

§ 10) in the case $p = 2$. Wintner imposed an extra condition that $r' \in L(a, \infty)$, and this condition was removed by Hartman (1948), who also extended the range of p to $1 < p \leq 2$; see also Bellman (1950, § 2, 1953, pp. 130–3) and, for improvements to the earlier work, Ascoli (1953*a*) and Hartman and Wintner (1955, p. 79). The third result (iii) is due to Ascoli (1941, 1947) and Wintner (1947*a*, Appendix, 1948, § 4).

1.10 In the case $\sigma = 1$, Theorem 1.10.1 is given by Coppel (1965, p. 91); see also Faedo (1947). The statement with general σ appears to be new here.

The condition (1.10.22) was used by Haupt (1942) to prove the less precise result that $x^{i-n}y^{(i-1)}(x)$ converges to a finite limit as $x \to \infty$ for all solutions $y(x)$. For other early papers with this condition, see Bellman (1947, § 8) and Hartman (1982, p. 321). Theorem 1.10.2 itself is due to Faedo (1946) and Ghizzetti (1949), but it goes back to Bôcher (1900) in the case $n = 2$. The theorem can also be obtained from Theorem 1.9.1 by the change of variable $x = \exp t$, and this fact was observed by Faedo (1946, p. 116).

There are also asymptotic results for the general situation where A has the Jordan normal form consisting of diagonal blocks of the type (1.10.1). We refer to Devinatz (1965), Devinatz and Kaplan (1972), Medina and Pinto (1985) and, for earlier results, Levinson (1948, Theorem 3), Coppel (1965, pp. 92–6), and Hartman (1982, pp. 304–5).

1.11 Theorem 1.11.1 is due to Harris and Lutz (1975, Theorem 3.1). In the case $\Lambda(x) = 0$, both the theorem and its proof are due to Wintner (1954), and the conditions on $R(x)$ are then simply (1.11.2) and (1.11.4); see also Harris and Lutz (1977, § 2(ii)). The earlier paper of Wintner (1949*b*) contained more restrictive conditions on $R(x)$. The case where $\Lambda = \Omega'\Omega^{-1}$, with Ω diagonal, is discussed by Hinton (1984, pp. 293–300) with a proof of Theorem 1.11.1 from first principles. An iterated version of Theorem 1.11.1 is given by Medina and Pinto (1985).

Theorem 1.11.2 is due to Trench (1984) but without the restriction to $\alpha < 1$; see also Trench (1976) and Hartman (1982, pp. 315–6) for earlier results of this type. Theorem 1.11.3 is an extension of a result due to Šimša (1985, Theorem 1). Both Trench (1976, 1984) and Šimša (1982, 1984, 1985) use complicated fixed-point arguments, and it is not clear whether the more refined results in these papers can also be deduced by the methods in this chapter.

2
Two-term differential equations

2.1 THE SECOND-ORDER EQUATION

The main theorems in Chapter 1 apply to asymptotically diagonal systems (1.1.9). The aim now is to consider other types of system which can nevertheless be transformed into the asymptotically diagonal cases covered in §§ 1.3–1.7, 1.10, and 1.11. The asymptotic form of the solutions for these wider classes of system is then obtained by applying the theorems in Chapter 1 and transforming back. We have already carried out this procedure in the simple case of asymptotically constant systems in § 1.8, but nothing has been said so far about systems (1.1.1) where, for example, $A(x)$ has off-diagonal entries which are large as $x \to \infty$. An important feature of the transformations to be introduced is that, like those in Chapter 1, they involve only the solution vector of the system and not the independent variable x.

We begin this development of the theory in the previous chapter by discussing the familiar second-order equation

$$\{p(x)y'(x)\}' - q(x)y(x) = 0. \tag{2.1.1}$$

Although the equation is written here in the usual Sturm–Liouville form, we emphasize that there is no assumption that p and q are necessarily real-valued. The most obvious way of expressing (2.1.1) as a first-order system

$$Y' = AY \tag{2.1.2}$$

is to take

$$Y = \begin{bmatrix} y \\ py' \end{bmatrix}, \qquad A = \begin{bmatrix} 0 & 1/p \\ q & 0 \end{bmatrix}, \tag{2.1.3}$$

and we note that (2.1.1) and (2.1.3) are just the case $m = 1$ of (1.9.12) and (1.9.13). The basic conditions on p and q at this stage are that $1/p$ and q are locally Lebesgue integrable in an interval $[a, \infty)$.

There are then three main methods of transforming (2.1.2) into an asymptotically diagonal system

$$Z' = (\Lambda + R)Z \tag{2.1.4}$$

with a suitably chosen transformation

$$Y = TZ. \tag{2.1.5}$$

One method is appropriate when q is in a certain sense large compared to p as $x \to \infty$, and the others when q is subdominant. The three transformations are as follows.

(i) Let A be expressed in its diagonal form

$$T^{-1}AT = \Lambda, \tag{2.1.6}$$

where

$$\Lambda = \mathrm{dg}\,\{(q/p)^{1/2}, -(q/p)^{1/2}\} \tag{2.1.7}$$

and

$$T = \begin{bmatrix} 1 & 1 \\ (pq)^{1/2} & -(pq)^{1/2} \end{bmatrix}, \tag{2.1.8}$$

with any fixed and consistent choice of the square roots. Then, in (2.1.4), we have

$$R = -T^{-1}T' = -\tfrac{1}{4}(pq)'/pq \begin{bmatrix} 1 & -1 \\ -1 & 1 \end{bmatrix}. \tag{2.1.9}$$

These formulae are valid if p and q are locally absolutely continuous and nowhere zero in $[a, \infty)$. The nature of (2.1.4) depends on the relative sizes of Λ and R as $x \to \infty$, but it can be described as asymptotically diagonal if q is dominant in the sense that $(q/p)^{1/2}$ is large compared to $(pq)'/pq$ as $x \to \infty$.

(ii) In the second transformation we write

$$Y = DZ, \tag{2.1.10}$$

where D is a non-singular diagonal matrix. Then (2.1.2) becomes

$$Z' = (-D^{-1}D' + D^{-1}AD)Z, \tag{2.1.11}$$

and we have (2.1.4) with

$$\Lambda = -D^{-1}D', \qquad R = D^{-1}AD.$$

Various choices can be made for D, but this time (2.1.11) may be described as asymptotically diagonal if A, and in particular the entry q, is subdominant as $x \to \infty$.

(iii) The theory of § 1.10 provides a further effective method of dealing with (2.1.1) in the case where q is subdominant. Here the matrix A in (2.1.3) is regarded as a perturbation of a matrix of Jordan type, the entry q being the perturbing term. This method will be taken up in § 2.7.

In the later sections of the chapter, the ideas just outlined will be developed further to cover a class of higher-order two-term differential equations, of which (2.1.1) is the simplest example.

2.2 THE LIOUVILLE–GREEN ASYMPTOTIC FORMULAE

We now consider the system (2.1.4) in greater detail, where Λ and R are defined by (2.1.7) and (2.1.9). The main asymptotic theorems in Chapter 1 can all be applied to (2.1.4) to give the form of solutions of (2.1.1) under various conditions on p and q. We start with the most important of these results, which is obtained from Theorem 1.6.1.

Theorem 2.2.1 *Let p and q be nowhere zero and have locally absolutely continuous first derivatives in $[a, \infty)$. Let*

$$(pq)'/pq = o\{(q/p)^{1/2}\} \qquad (x \to \infty) \tag{2.2.1}$$

and let

$$\{p^{-1/2}q^{-3/2}(pq)'\}' \in L(a, \infty). \tag{2.2.2}$$

Let

$$\text{re}\,(q/p + r^2)^{1/2} \quad \textit{have one sign in } [a, \infty), \tag{2.2.3}$$

where

$$r = \tfrac{1}{4}(pq)'/pq. \tag{2.2.4}$$

Then (2.1.1) *has solutions y_1 and y_2 such that*

$$y_1 \sim (pq)^{-1/4} \exp\left(\int_a^x (q/p + r^2)^{1/2}\,\mathrm{d}t\right), \tag{2.2.5}$$

$$py_1' \sim (pq)^{1/4} \exp\left(\int_a^x (q/p + r^2)^{1/2}\,\mathrm{d}t\right), \tag{2.2.6}$$

with similar formulae for y_2 containing $-(\cdots)^{1/2}$ in the exponential term.

Proof We apply Theorem 1.6.1 to (2.1.4). By (2.1.7) and (2.1.9), the conditions (2.2.1) and (2.2.2) are just (1.6.2) and (1.6.3) in the present situation. Also, the eigenvalues of $\Lambda + R$ are

$$\mu_1 = -r + (q/p + r^2)^{1/2}, \qquad \mu_2 = -r - (q/p + r^2)^{1/2}. \tag{2.2.7}$$

Hence

$$\text{re}\,(\mu_1 - \mu_2) = 2\,\text{re}\,(q/p + r^2)^{1/2},$$

and (2.2.3) is then the dichotomy condition for μ_1 and μ_2 in the form (1.3.13). The conditions of Theorem 1.6.1 are therefore satisfied by (2.1.4), and it follows that there are solutions $Z_k(x)$ $(k = 1, 2)$ such that

$$Z_k(x) = \{e_k + o(1)\} \exp\left(\int_a^x \mu_k(t)\,\mathrm{d}t\right).$$

We then substitute from (2.2.7) and write

$$\int_a^x r(t)\,\mathrm{d}t = \frac{1}{4}\int_a^x (pq)'/pq\,\mathrm{d}t = \log\,(pq)^{1/4} + \text{const.}$$

Finally, on transforming back to y via (2.1.8), (2.1.5), and (2.1.3), we obtain (2.2.5) and (2.2.6) after adjusting y_1 by a constant multiple, and similarly for y_2. □

In (1.6.35) and (1.6.37), we discussed a simplified version of Theorem 1.6.1. The same remarks apply to Theorem 2.2.1. However, since we are working only with 2×2 matrices here, rather than quote § 1.6 again, we can deal with the simplification directly as follows.

Corollary 2.2.1 *Let p and q be nowhere zero and have locally absolutely continuous first derivatives in $[a, \infty)$. Let (2.2.1) and (2.2.2) hold and, in addition, let*

$$p^{-3/2}q^{-5/2}(pq)'^2 \in L(a, \infty). \tag{2.2.8}$$

Let

$$\mathrm{re}\,(q/p)^{1/2} \quad \text{have one sign in } [a, \infty). \tag{2.2.9}$$

Then (2.1.1) has solutions y_1 and y_2 such that

$$y_1 \sim (pq)^{-1/4} \exp\left(\int_a^x (q/p)^{1/2}\,dt\right), \tag{2.2.10}$$

$$py_1' \sim (pq)^{1/4} \exp\left(\int_a^x (q/p)^{1/2}\,dt\right), \tag{2.2.11}$$

with similar formulae for y_2 containing $-(q/p)^{1/2}$ in the exponential term.

Proof By (2.2.1) and (2.2.4), we have $r = o\{(q/p)^{1/2}\}$ *and therefore, in* (2.2.7), *we can write*

$$(q/p + r^2)^{1/2} = (q/p)^{1/2} + O\{r^2(p/q)^{1/2}\}.$$

The O-term here is $L(a, \infty)$ by (2.2.8), and then it follows from (2.2.9) that μ_1 and μ_2 again satisfy the dichotomy condition in the form (1.3.13). Also, (2.2.10) and (2.2.11) follow from (2.2.5) and (2.2.6) when y_1 is adjusted by a constant multiple. □

The formulae (2.2.10) and (2.2.11), together with the corresponding ones for y_2, are known as the *Liouville–Green asymptotic formulae.* This appellation stems from a long-standing alternative proof of the formulae which is based on a certain transformation of both x and y in (2.1.1). This transformation was introduced independently by Liouville and Green in a quite different context in 1837, well ante-dating the statement of Corollary 2.2.1. We discuss the alternative proof in § 2.5.

There are a number of remarks to be made concerning the nature of the conditions imposed on p and q in Theorem 2.2.1 and Corollary 2.2.1.

(a) Dealing first with (2.2.3) and (2.2.9), these conditions are clearly satisfied when

$$p \text{ and } q \quad \text{are real-valued in } [a, \infty). \tag{2.2.12}$$

Moving to complex-valued coefficients, (2.2.9) is also clearly satisfied when

$$-\pi + \omega \leqslant \arg(q/p) \leqslant \pi - \omega \tag{2.2.13}$$

in $[a, \infty)$, where $\omega\,(>0)$ is fixed. Again, since (2.2.1) states that $r^2 = o(q/p)$, (2.2.13) also implies that (2.2.3) holds after a possible adjustment of the point $x = a$.

(b) Turning to particular coefficients p and q, the simplest example is

$$p(x) = (\text{const.})x^{\alpha}, \qquad q(x) = (\text{const.})x^{\beta}, \tag{2.2.14}$$

where α and β are constants. Then (2.2.1), (2.2.2), and (2.2.8) all reduce to the same condition

$$\alpha - \beta < 2. \tag{2.2.15}$$

A more general example is

$$p(x) = x^{\alpha} f(x^{\gamma}), \qquad q(x) = x^{\beta} g(x^{\eta}), \tag{2.2.16}$$

where $f(t)$ and $g(t)$ are periodic functions of t which are nowhere zero, and γ and η are constants. In this case, (2.2.15) is replaced by

$$\alpha - \beta + 4\Gamma < 2, \tag{2.2.17}$$

where $\Gamma = \max(\gamma, \eta, 0)$.

A different example is

$$p(x) = (\text{const.}) \exp x^{\alpha}, \qquad q(x) = (\text{const.}) \exp x^{\beta}. \tag{2.2.18}$$

This time (2.2.1), (2.2.2), and (2.2.8) are all satisfied when $\alpha < \beta$ but, when $\alpha = \beta$, (2.2.1) and (2.2.2) hold if $\alpha\,(=\beta) < 1$, while (2.2.8) only holds if $\alpha\,(=\beta) < \frac{1}{2}$.

(c) The general nature of (2.2.1), (2.2.2), and (2.2.8) is that these conditions represent a restriction on the absolute size of the first two derivatives of p and q in terms of the sizes of p and q themselves. There is therefore a limitation on the oscillatory nature of p and q. This limitation is seen in the example (2.2.16), where (2.2.17) restricts the size of γ and η for given α and β.

(d) The question arises whether (2.2.8) is already implied by (2.2.1) and (2.2.2). That it is not is shown by the example (2.2.18) and the case $\alpha = \beta$ already mentioned. In general terms, the connection between the three conditions can be brought out if we define

$$h = (pq)'/p^{1/2}q^{3/2}.$$

In terms of h, (2.2.1) and (2.2.2) are

$$h = o(1), \qquad h' \in L(a, \infty), \tag{2.2.19}$$

while (2.2.8) is

$$q^{1/2}h^2/p^{1/2} \in L(a, \infty). \tag{2.2.20}$$

Clearly the two conditions (2.2.19) can be satisfied in circumstances when (2.2.20) is not.

2.3 REPEATED DIAGONALIZATION

Theorem 2.2.1 is the first in a sequence of results that involve successively higher derivatives of p and q. These results are obtained by using the technique introduced at the beginning of § 1.7 where the idea is to apply repeatedly the diagonalizing transformation (1.6.5) on which the proof of Theorem 2.2.1 is based. In the general case of $n \times n$ matrices, which was considered in § 1.7, the computations which occur in the repeated diagonalization become too cumbersome to handle effectively, and this difficulty was circumvented in § 1.7 with the use of a simplified repeated transformation subject to additional conditions on R of the type (1.7.1). In this section, however, we have only 2×2 matrices and repeated diagonalization is feasible without the need of additional simplifying conditions. We deal with the algebraic details first and leave to § 2.4 the statement of the asymptotic result for (2.1.1) which follows from the method.

To discuss the process of repeated diagonalization as applied to (2.1.2), we examine a typical step in the process. We start with (2.1.4), where Λ and R are given by (2.1.7) and (2.1.9), and with a slight change of notation we write the system as

$$Z_1' = (\Lambda_1 + R_1)Z_1 \tag{2.3.1}$$

where

$$\Lambda_1 = \Lambda - rI, \qquad R_1 = r\begin{bmatrix} 0 & 1 \\ 1 & 0 \end{bmatrix}, \tag{2.3.2}$$

and

$$r = \tfrac{1}{4}(pq)'/pq. \tag{2.3.3}$$

Let us now suppose that we have a system

$$Z_m' = (\Lambda_m + R_m)Z_m \tag{2.3.4}$$

in which

$$\Lambda_m = \mathrm{dg}\,(q_m - \rho_m, -q_m - \rho_m) \tag{2.3.5}$$

and

$$R_m = r_m\begin{bmatrix} 0 & 1 \\ (-1)^{m-1} & 0 \end{bmatrix}. \tag{2.3.6}$$

We express $\Lambda_m + R_m$ in the diagonal form

$$T_m^{-1}(\Lambda_m + R_m)T_m = \mathrm{dg}\,(q_{m+1} - \rho_m,\ -q_{m+1} - \rho_m)$$

by defining

$$T_m = \begin{bmatrix} 1 & -a_m \\ (-1)^{m-1}a_m & 1 \end{bmatrix} \tag{2.3.7}$$

and

$$q_{m+1} = \{q_m^2 + (-1)^{m-1}r_m^2\}^{1/2}, \tag{2.3.8}$$

where

$$a_m = r_m/[q_m + \surd\{q_m^2 + (-1)^{m-1}r_m^2\}]. \tag{2.3.9}$$

The transformation

$$Z_m = T_m Z_{m+1} \tag{2.3.10}$$

then takes (2.3.4) into

$$Z'_{m+1} = (\Lambda_{m+1} + R_{m+1})Z_{m+1}, \tag{2.3.11}$$

in which

$$\Lambda_{m+1} = \mathrm{dg}\,(q_{m+1} - \rho_{m+1},\ -q_{m+1} - \rho_{m+1}) \tag{2.3.12}$$

and

$$R_{m+1} = r_{m+1}\begin{bmatrix} 0 & 1 \\ (-1)^m & 0 \end{bmatrix} \tag{2.3.13}$$

with

$$\rho_{m+1} = \rho_m + (-1)^{m-1}a_m a_m'/\{1 + (-1)^{m-1}a_m^2\}. \tag{2.3.14}$$

and

$$r_{m+1} = a_m'/\{1 + (-1)^{m-1}a_m^2\}. \tag{2.3.15}$$

The formulae (2.3.11)–(2.3.13) have the same form as (2.3.4)–(2.3.6) with $m+1$ in place of m. This analysis is valid in some interval $[X, \infty)$ provided that T_m is non-singular there and, by (2.3.7), this is so if

$$a_m = o(1) \qquad (x \to \infty). \tag{2.3.16}$$

It follows from (2.3.14), (2.3.15) and (2.3.9) that Λ_{m+1} and R_{m+1} contain first derivatives of terms in Λ_m and R_m. Starting therefore with (2.3.1) and (2.3.2), which involve the first derivatives of p and q, the above process can be repeated until we reach

$$Z'_{M+1} = (\Lambda_{M+1} + R_{M+1})Z_{M+1}, \tag{2.3.17}$$

provided that p and q have locally absolutely continuous derivatives up to order M in $[a, \infty)$ for some $M \geqslant 1$.

We have next to determine the nature of Λ_{M+1} and R_{M+1} in some detail in terms of p and q. Since $\rho_1 = r$, (2.3.14) and (2.3.3) give

$$\rho_{M+1} = \frac{\mathrm{d}}{\mathrm{d}x}\left\{\log\left((pq)^{1/4}\prod_{m=1}^{M}\{1 + (-1)^{m-1}a_m^2\}^{1/2}\right)\right\}. \tag{2.3.18}$$

Also, if we define

$$u_m = r_m/q_m, \tag{2.3.19}$$

(2.3.8) gives

$$q_{M+1} = (q/p)^{1/2} \prod_{m=1}^{M} \{1+(-1)^{m-1}u_m^2\}^{1/2}, \tag{2.3.20}$$

where we have used the fact that, by (2.1.7) and (2.3.2), $q_1 = (q/p)^{1/2}$. Once we have analysed the nature of u_m, (2.3.16), (2.3.18), (2.3.20) and (2.3.12) give the form of Λ_{M+1} sufficiently for our purposes.

We can obtain a recurrence formula for the u_m if, by (2.3.9) and (2.3.19), we write

$$a_m = u_m/[1+\surd\{1+(-1)^{m-1}u_m^2\}]. \tag{2.3.21}$$

Then (2.3.15) gives

$$r_{m+1} = \tfrac{1}{2}u_m'/\{1+(-1)^{m-1}u_m^2\}, \tag{2.3.22}$$

and hence

$$\begin{aligned} u_{m+1} &= r_{m+1}/q_{m+1} \\ &= \tfrac{1}{2}(p/q)^{1/2}(1+u_1^2)^{-1/2}(\cdots)\{1+(-1)^{m-2}u_{m-1}^2\}^{-1/2} \\ &\quad \times \{1+(-1)^{m-1}u_m^2\}^{-3/2}u_m', \end{aligned}$$

by (2.3.19) and (2.3.20). If we now define

$$t = \int_X^x \{q(s)/p(s)\}^{1/2}\,\mathrm{d}s \tag{2.3.23}$$

and write $D = \mathrm{d}/\mathrm{d}t$, we obtain

$$\begin{aligned} u_{m+1} &= \frac{1}{2}\left(\prod_{\nu=1}^{m-1} \{1+(-1)^{\nu-1}u_\nu^2\}\right)^{-1/2} \\ &\quad \times \{1+(-1)^{m-1}u_m^2\}^{-3/2}Du_m. \end{aligned} \tag{2.3.24}$$

From (2.1.7), (2.3.2), and (2.3.3), we have

$$u_1 = \tfrac{1}{4}(pq)'/p^{1/2}q^{3/2}, \tag{2.3.25}$$

and then (2.3.24) determines the u_m $(m>1)$ in terms of u_1 and its D-derivatives. Then also (2.3.22) gives the nature of R_{M+1} for use in (2.3.17).

The Levinson theorem, Theorem 1.3.1, can be applied to (2.3.17) if three conditions are satisfied. First we need

$$u_m = o(1) \qquad (x\to\infty,\ 1\leqslant m\leqslant M), \tag{2.3.26}$$

this condition, by (2.3.21), implying that (2.3.16) holds, and therefore that the analysis leading to (2.3.17) is valid. Second, by (2.3.12), we require that

$$\operatorname{re} q_{M+1} \quad \text{has one sign in } [X, \infty), \tag{2.3.27}$$

this being the dichotomy condition for (2.3.17) in the form (1.3.13). Finally, we require that

$$u'_M \in L(X, \infty) \tag{2.3.28}$$

since this, by (2.3.22), (2.3.26), and (2.3.13), implies that the condition (1.3.3) is satisfied by R_{M+1}. In the next section we state and prove the asymptotic theorem that is obtained when these conditions are put in terms of u_1 and therefore, by (2.3.25), in terms of p and q.

2.4 EXTENDED LIOUVILLE–GREEN ASYMPTOTIC FORMULAE

Theorem 2.4.1 *Let p and q be nowhere zero and have locally absolutely continuous derivatives up to order M in $[a, \infty)$ for some $M \geq 1$. With u_m $(1 \leq m \leq M)$ and q_{M+1} defined in some interval $[X, \infty)$ by (2.3.24), (2.3.25) and (2.3.20), let*

$$\operatorname{re} q_{M+1} \quad \textit{have one sign in } [X, \infty). \tag{2.4.1}$$

With t defined by (2.3.23) and $D = \mathrm{d}/\mathrm{d}t$, let

(i) $$D^{m-1} u_1 = o(1) \qquad (x \to \infty,\ 1 \leq m \leq M) \tag{2.4.2}$$

(ii) $$(q/p)^{1/2}(D^{k_1}u_1)(\cdots)(D^{k_j}u_1) \in L(X, \infty) \tag{2.4.3}$$

for all integers k_i in $[0, M]$ (not necessarily distinct) such that

(a) $$j \textit{ is odd and } k_1 + \cdots + k_j - M \textit{ is even} \tag{2.4.4}$$

(b) $$M \leq k_1 + \cdots + k_j \leq d_{M+1}, \tag{2.4.5}$$

where the sequence d_n is defined by

$$d_{m+1} = 2d_m + d_{m-1} \quad (m \geq 2), \quad d_1 = 0, \quad d_2 = 1. \tag{2.4.6}$$

(c) *the number j_0 of the k_j which are zero satisfies*

$$j_0 \leq d_M. \tag{2.4.7}$$

Then (2.1.1) has solutions y_1 and y_2 such that

$$\left.\begin{aligned} y_1 &\sim (pq)^{-1/4} \exp\left(\int_X^x q_{M+1}(s)\,\mathrm{d}s\right), \\ y_2 &\sim (pq)^{-1/4} \exp\left(-\int_X^x q_{M+1}(s)\,\mathrm{d}s\right). \end{aligned}\right\} \tag{2.4.8}$$

Proof We have to check that (2.3.26)–(2.3.28) hold. It is clear from (2.3.24) and a simple induction argument that u_m is a finite sum of terms of the form

$$\Pi_k P_k \qquad (k = 1, 2, \ldots), \tag{2.4.9}$$

where each Π_k is a product

$$\Pi_k = \prod_{\nu=1}^{m-1} \{1 + (-1)^{\nu-1} u_\nu^2\}^{-\gamma(\nu)} \tag{2.4.10}$$

with some $\gamma(\nu) > 0$ and P_k is a polynomial in the D-derivatives of u_1 up to order $m - 1$. It then follows immediately from (2.4.2) that (2.3.26) holds. We also note in passing that (2.3.24) and (2.4.2) give

$$u_m = D^{m-1} u_1 + v_{m-1} \qquad (1 \leqslant m \leqslant M),$$

where v_{m-1} only involves derivatives of u_1 up to order $m - 2$ and $v_{m-1} = o(1)$ as $x \to \infty$.

Next, (2.3.27) is simply re-stated in (2.4.1), and we shall return to this condition for further comment later on.

To deal with (2.3.28) we have to examine the P_k in (2.4.9) in greater detail. Proceeding by induction on m, we suppose that u_m only contains terms of the type

$$\Pi_k (D^{k_1} u_1)(\cdots)(D^{k_j} u_1), \tag{2.4.11}$$

where each k_i is in $[0, m - 1]$ and, as indicated by (2.4.4) and (2.4.5) but with $m - 1$ in place of M,

$$j \text{ is odd}, \qquad k_1 + \cdots + k_j - m \text{ is odd}, \tag{2.4.12}$$

$$m - 1 \leqslant k_1 + \cdots + k_j \leqslant d_m. \tag{2.4.13}$$

We suppose that this is true for $m = 1, 2, \ldots$ up to a general value of m. We then apply the operator D to (2.4.11) and substitute into (2.3.24) to show that u_{m+1} has the required form given by (2.4.11)–(2.4.13) with $m + 1$ in place of m. There is no difficulty as far as terms

$$\Pi_k D\{(D^{k_1} u_1)(\cdots)(D^{k_j} u_1)\}$$

are concerned, and it remains to deal with

$$(D\Pi_k)(D^{k_1} u_1)(\cdots)(D^{k_j} u_1).$$

By (2.4.10) and (2.3.24), a typical term here involves

$$\tilde{\Pi}_k u_r u_{r+1} (D^{k_1} u_1)(\cdots)(D^{k_j} u_1) \qquad (1 \leqslant r \leqslant m - 1) \tag{2.4.14}$$

with a new $\tilde{\Pi}_k$. We now apply the induction hypothesis (2.4.11)–(2.4.13) both as it stands and to u_r and u_{r+1}. Then, apart from the Π_k factor,

(2.4.14) has the form

$$(D^{K_1}u_1)(\cdots)(D^{K_J}u_1),$$

where J is odd, $K_1 + \cdots + K_J - (m+1)$ is odd, and

$$m \leqslant K_1 + \cdots + K_J \leqslant d_m + (d_{m-1} + d_m) = d_{m+1}, \qquad (2.4.15)$$

by (2.4.6). Thus all the terms in u_{m+1} have the form (2.4.11)–(2.4.13) with $m+1$ in place of m (and new Π_k factors). The induction hypothesis is clearly true for $m = 1$ and for $m = 2$ since (2.3.24) gives

$$u_2 = \tfrac{1}{2}(1 + u_1^2)^{-3/2} Du_1. \qquad (2.4.16)$$

This completes the proof by induction that u_m contains only terms of the form (2.4.11)–(2.4.13).

We have also to examine the factor u_1^k which arises from those k_i which are zero in (2.4.11). We define

$$\delta_m = \max k, \qquad (2.4.17)$$

the maximum being taken over all the terms (2.4.11) which occur in u_m. The same argument that led to (2.4.15) also gives

$$\delta_{m+1} = 2\delta_m + \delta_{m-1} \qquad (m \geqslant 2),$$

as in the definition (2.4.6) of d_m. There is a difference however in that $\delta_1 = 1$ and, by (2.4.16), $\delta_2 = 0$. Further, $\delta_3 = 1$ by (2.3.24) and (2.4.16). From this follows the information that we need:

$$\delta_m = d_{m-1} \qquad (m \geqslant 2). \qquad (2.4.18)$$

We can now deal with (2.3.28), which we write as

$$(q/p)^{1/2} Du_M \in L(X, \infty). \qquad (2.4.19)$$

Starting at (2.4.11)–(2.4.13) with $m = M$, the working which led to (2.4.15) shows that Du_M contains only u_1-derivatives of the form indicated in (2.4.3)–(2.4.5), and therefore (2.4.19) follows from (2.4.3). Also, (2.4.7) follows in the same way from the case $m = M+1$ of (2.4.17) and (2.4.18).

The Levinson theorem, Theorem 1.3.1, can now be applied to show that (2.3.17) has two solutions with the asymptotic forms

$$\{e_j + o(1)\} \exp\left(\int_X^x \{\pm q_{M+1}(s) - \rho_{M+1}(s)\}\, ds\right), \qquad (2.4.20)$$

where e_j $(j = 1, 2)$ denote the usual unit coordinate vectors, and we have used the form of Λ_{M+1} given by (2.3.12). Then, by (2.3.16) and (2.3.18), we can write (2.4.20) as

$$(pq)^{-1/4}\{e_j + o(1)\} \exp\left(\pm \int_X^x q_{M+1}(s)\, ds\right). \qquad (2.4.21)$$

Finally, we transform back to (2.1.2). The transformation of Y into Z_{M+1} is simply of the form

$$Y = \{I + o(1)\}Z_{M+1},$$

by (2.3.7), (2.3.10) and (2.3.16), and then (2.4.8) follows immediately from (2.4.21). □

The formulae (2.4.8) are somewhat intricate for general M, based as they are on the complicated recurrence relation (2.3.24) which defines the u_m. We now examine low values of M and work out in greater detail both the condition (2.4.3) and the final result (2.4.8). The case $M = 1$ has already been covered in Theorem 2.2.1 and we therefore move on to $M = 2$ and $M = 3$.

(i) $M = 2$. Here (2.4.2) and (2.4.3) are

$$u_1 = o(1), \quad Du_1 = o(1) \qquad (x \to \infty) \tag{2.4.22}$$

and

$$(q/p)^{1/2}D^2u_1 \in L(X, \infty), \qquad (q/p)^{1/2}u_1(Du_1)^2 \in L(X, \infty). \tag{2.4.23}$$

By (2.3.20) and (2.3.24), we have

$$\begin{aligned} q_3 &= (q/p)^{1/2}\{(1 + u_1^2)(1 - u_2^2)\}^{1/2} \\ &= (q/p)^{1/2}\{1 + u_1^2 - \tfrac{1}{4}(Du_1)^2/(1 + u_1^2)^2\}^{1/2} \\ &= (q/p)^{1/2}\{1 + u_1^2 - \tfrac{1}{4}(Du_1)^2\}^{1/2} + f, \end{aligned}$$

where f is $L(X, \infty)$ by (2.4.22) and (2.4.23). Hence the right-hand side of (2.4.8) is effectively

$$(pq)^{-1/4} \exp\left(\pm\int_X^x (q/p)^{1/2}\{1 + u_1^2 - \tfrac{1}{4}(Du_1)^2\}^{1/2}\,\mathrm{d}s\right), \tag{2.4.24}$$

where u_1 is defined by (2.3.25).

(ii) $M = 3$. In this case (2.4.2) is

$$u_1 = o(1), \quad Du_1 = o(1), \quad D^2u_1 = o(1) \qquad (x \to \infty) \tag{2.4.25}$$

and, leaving aside conditions which are redundant in the presence of (2.4.25), (2.4.3) requires that the expressions

$$D^3u_1, \quad Du_1(D^2u_1)^2, \quad u_1Du_1D^2u_1, \quad (Du_1)^3 \tag{2.4.26}$$

are all $L(X, \infty)$ when multiplied by $(q/p)^{1/2}$. Also, by (2.3.20) and (2.3.24), we have

$$\begin{aligned} q_4 &= (q/p)^{1/2}\{(1 + u_1^2)(1 - u_2^2)(1 + u_3^2)\}^{1/2} \\ &= (q/p)^{1/2}\{1 + u_1^2 - \tfrac{1}{4}(Du_1)^2/(1 + u_1^2)^2 + \tfrac{1}{4}(Du_2)^2/(1 - u_2^2)^2\}^{1/2} \\ &= (q/p)^{1/2}\{1 + u_1^2 - \tfrac{1}{4}(Du_1)^2/(1 + u_1^2)^2 + \tfrac{1}{16}(D^2u_1)^2\}^{1/2} + f, \end{aligned}$$

where f is $L(X, \infty)$ by (2.4.16), (2.4.25) and (2.4.26). Then (2.4.24) is modified accordingly.

When $M \geqslant 4$, it is lengthy to write down all the expressions (2.4.3) given by (2.4.4) and (2.4.5). We have only to note that $d_4 = 12$ and $d_5 = 29$. Of course, some of the expressions are redundant in the presence of (2.4.2).

The dichotomy condition (2.4.1) is satisfied when p and q are real-valued, as may be seen from (2.3.20) since, in this case, (2.3.24) shows that u_m^2 is real-valued for all m. Again, by (2.3.26), we can write (2.3.20) as $q_{M+1} = (q/p)^{1/2}\{1 + o(1)\}$ and it then follows that (2.4.1) is also satisfied when (2.2.13) holds. □

Example 2.4.1 Let $\psi(x)$ be real-valued and nowhere zero in $[a, \infty)$ with locally absolutely continuous derivatives up to order M. Let

$$\psi(x) \to \infty \qquad (2.4.27)$$

as $x \to \infty$ and let β (>0) be a constant. We consider the example of (2.1.1) in which $p(x) = 1$ and

$$q(x) = \psi^{\beta}(x) g\{\psi(x)\}, \qquad (2.4.28)$$

where $g(s)$ is periodic in s and nowhere zero. In addition to (2.4.27), let

$$\psi'(x) = o\{\psi^{\beta/2}(x)\} \qquad (2.4.29)$$

and

$$\psi^{(m)}(x) = O(\{\psi'(x)\}^m) \qquad (2 \leqslant m \leqslant M+1) \qquad (2.4.30)$$

as $x \to \infty$. It is easy to verify from (2.3.23), (2.3.25) and (2.4.28) that (2.4.30) gives

$$D^k u_1 = O(\psi^{-\beta/2}\psi')^{k+1} \qquad (0 \leqslant k \leqslant M).$$

It then follows from (2.4.29) that (2.4.2) holds and, further, that (2.4.3) holds if

$$\psi^{\beta/2}(\psi^{-\beta/2}\psi')^{M+1} \in L(a, \infty). \qquad (2.4.31)$$

In the case where

$$\psi(x) = x^{\gamma}, \qquad (2.4.32)$$

the conditions (2.4.27), (2.4.29), (2.4.30) and (2.4.31) give

$$0 < \gamma < \tfrac{1}{2}\beta\gamma + 1$$

and

$$M(\tfrac{1}{2}\beta\gamma + 1 - \gamma) > \gamma. \qquad (2.4.33)$$

This last inequality states that M must be suitably large and it therefore indicates the number of repeated diagonalizations that must be applied in § 2.3 to arrive at the Levinson form (2.3.17). We dealt with a similar situation in (1.7.33)–(1.7.35) and again in § 2.2 where (2.2.17) (with $\alpha = 0$ and a slight change of notation) is the case $M = 1$ of (2.4.33).

When $\gamma < 1$ in (2.4.32), the $q(x)$ in (2.4.28) can be described as slowly varying and, when $\gamma > 1$, as rapidly varying. Other possible choices of $\psi(x)$ besides (2.4.32) are the relatively uninteresting

$$\psi(x) = (\log x)^{\gamma} \qquad (\gamma > 1,\ M = 1),$$

which leads to a very slowly varying $q(x)$, and the more interesting

$$\psi(x) = \exp(x^{\gamma})$$

with $\gamma > 0$, $\beta > 2$, and $M(\frac{1}{2}\beta - 1) > 1$, which leads to a very rapidly varying $q(x)$.

2.5 THE LIOUVILLE–GREEN TRANSFORMATION

In this section, we pause in the development of the main theme of the chapter to discuss the transformation first introduced by Liouville and less explicitly by Green, both in 1837, under which (2.1.1) becomes an equation with leading coefficient unity and with a non-zero factor removed from $q(x)$. In this context, we write (2.1.1) as

$$\{p(x)y'(x)\}' - \rho(x)q_1(x)y(x) = 0, \tag{2.5.1}$$

where $\rho(x)$ is nowhere zero in $[a, \infty)$, and we consider the transformation of both the independent and dependent variables x and y defined by

$$t = \int_a^x \{\rho(s)/p(s)\}^{1/2}\,\mathrm{d}s, \qquad z(t) = \{p(x)\rho(x)\}^{1/4}y(x). \tag{2.5.2}$$

The transformation takes (2.5.1) into

$$\mathrm{d}^2z/\mathrm{d}t^2 - \{Q_1(t) + r(t)\}z(t) = 0, \tag{2.5.3}$$

where $Q_1(t) = q_1(x)$ and

$$r(t) = -p^{1/4}(x)\rho^{-3/4}(x)\frac{\mathrm{d}}{\mathrm{d}x}p(x)\frac{\mathrm{d}}{\mathrm{d}x}\{p(x)\rho(x)\}^{-1/4}. \tag{2.5.4}$$

On comparing (2.5.1) and (2.5.3), we see that the factor ρ has been removed from q_1 and replaced by an additive term $r(t)$, while the leading coefficient p has been replaced by unity.

Liouville originally introduced the transformation (2.5.2) in his development of the eigenvalue theory of the equation

$$\{p(x)y'(x)\}' + \{\lambda w(x) - s(x)\}y(x) = 0$$

on a finite interval $[a, b]$ and, with $\rho(x) = w(x)$, the transformed equation (2.5.3) has the coefficient of λ as unity. In the asymptotic theory of (2.1.1), however, the most obvious choice for $\rho(x)$ is

$$\rho(x) = |q(x)|, \tag{2.5.5}$$

and then $Q_1(t) = \exp\{\arg q(x)\}$ in (2.5.3). If $\arg q(x)$ is constant, the results in Example 1.9.1 can be applied to (2.5.3) and, from the cases $c^2 = \pm 1$ of Example 1.9.1(i), we obtain the following long-standing result.

Theorem 2.5.1 *Let p and q be real-valued and nowhere zero in $[a, \infty)$, with locally absolutely continuous first derivatives there. Let*

$$(q/p)^{1/2} \notin L(a, \infty) \tag{2.5.6}$$

and let

$$(pq)^{-1/4} \frac{\mathrm{d}}{\mathrm{d}x}\left(p \frac{\mathrm{d}}{\mathrm{d}x}(pq)^{-1/4}\right) \in L(a, \infty). \tag{2.5.7}$$

Then (2.1.1) *has solutions y_1 and y_2 such that* (2.2.10) *and* (2.2.11) *hold for y_1, with similar formulae for y_2 containing $-(q/p)^{1/2}$ in the exponential term.*

Before proving the theorem, we obtain a consequence of (2.5.6) and (2.5.7) which deals with the only point of difficulty in the proof. This result is given in a more general setting in the following lemma.

Lemma 2.5.1 *Let the functions f and g be real-valued and nowhere zero in $[a, \infty)$, with f and g' locally absolutely continuous in $[a, \infty)$. Let*

$$\int_a^\infty g(fg')' \,\mathrm{d}x \quad \textit{converge} \tag{2.5.8}$$

and let

$$(fg^2)^{-1} \notin L(a, \infty). \tag{2.5.9}$$

Then

$$fg'^2 \in L(a, \infty) \tag{2.5.10}$$

and, as $x \to \infty$,

$$fgg' = o(1). \tag{2.5.11}$$

Proof We can take it that $f > 0$ and $g > 0$ in $[a, \infty)$. Arguing by contradiction, we suppose that $fg'^2 \notin L(a, \infty)$. Since

$$(fgg')' = g(fg')' + fg'^2, \tag{2.5.12}$$

it then follows from (2.5.8) that

$$fgg' \to \infty \tag{2.5.13}$$

as $x \to \infty$. Hence $g' > 0$ in some interval $[X, \infty)$, and therefore g increases to a finite or infinite limit as $x \to \infty$. Then again (2.5.8) implies that $(fg')'$ has a convergent infinite integral, and hence

$$fg' \to l, \tag{2.5.14}$$

a finite limit, as $x \to \infty$. To show that $l \neq 0$, we write

$$fg' = l - \int_x^\infty g(fg')'g^{-1}\,ds = l + o(g^{-1}),$$

where we use (2.5.8) again and the monotonicity of g. Then $l \neq 0$ to tally with (2.5.13). Using (2.5.14), we now have

$$(fg^2)^{-1} \sim -l^{-1}(g^{-1})'$$

and then, since g^{-1} converges monotonically to a finite limit as $x \to \infty$, we obtain $(fg^2)^{-1} \in L(a, \infty)$. This contradicts (2.5.9), and therefore (2.5.10) is established.

To prove (2.5.11), we note that (2.5.12), (2.5.8) and (2.5.10) imply that

$$fgg' \to l_1,$$

a finite limit, as $x \to \infty$. Then

$$fg'^2 = \{l_1^2 + o(1)\}(fg^2)^{-1},$$

and it follows from (2.5.9) and (2.5.10) that $l_1 = 0$. This completes the proof of the lemma. □

On making the choice $f = p$, $g = (p\,|q|)^{-1/4}$, we find that (2.5.6) and (2.5.7) imply that

$$(pq)'/pq = o\{(q/p)^{1/2}\}. \tag{2.5.15}$$

Proof of Theorem 2.5.1 We can take it that $p > 0$ in $[a, \infty)$, and then (2.5.2) and (2.5.5) define a real variable t such that $t \to \infty$ as $x \to \infty$, by (2.5.6). Thus (2.5.3) is defined on the t-interval $[0, \infty)$ and, since q is real-valued and nowhere zero, we have $Q_1 = q/|q| = \pm 1$. By (2.5.2) and (2.5.4), the condition (2.5.7) implies that $r(t) \in L(0, \infty)$. On applying Example 1.9.1(i) to (2.5.3) and transforming back to y and x, we obtain solutions y_1 and y_2 of (2.1.1) such that

$$(pq)^{1/4}y_1 \sim \exp\left(\int_a^x (q/p)^{1/2}\,ds\right), \tag{2.5.16}$$

$$\{(pq)^{1/4}y_1\}' \sim (q/p)^{1/2}\exp\left(\int_a^x (q/p)^{1/2}\,ds\right), \tag{2.5.17}$$

with similar formulae for y_2 containing $-(q/p)^{1/2}$ in the exponential. Now (2.5.16) is already (2.2.10) but, to deduce (2.2.11) from (2.5.17), we write the left-hand side of (2.5.17) as

$$(pq)^{1/4}y_1' + \{(pq)^{1/4}\}'y_1$$

and, in the second term here, we use (2.5.16) and (2.5.15) to obtain the required result (2.2.11). □

The conditions (2.5.6) and (2.5.7) are broadly similar to the conditions (2.2.1), (2.2.2) and (2.2.8) which were imposed in Corollary 2.2.1. Indeed, in the case where $p = 1$ and q is real-valued, it is known that (2.5.6) and (2.5.7) imply not only (2.5.15) (which is (2.2.1)) but also (2.2.2) and (2.2.8) (Coppel 1965, p. 121). Thus, when $p = 1$, the conditions in Theorem 2.5.1 are a little more restrictive than the three previous conditions. The main restriction in Theorem 2.5.1, however, is that p and q are real-valued, and this difficulty is caused by the need to make t a real variable in (2.5.2). The choice (2.5.5) is already a compromise, and our purpose now is to show that it is possible in effect to choose $\rho = q$ and obtain Theorem 2.5.1 for complex-valued p and q. In this new theorem, however, we must be content with (2.5.17) since, as we shall see, Lemma 2.5.1 is not available for complex-valued functions.

We discuss this development of Theorem 2.5.1 in the wider context of a more general transformation than (2.5.2), which again takes (2.1.1) into an equation of the type (2.5.3). In (2.1.1), we write

$$y = uz, \tag{2.5.18}$$

where the function u is nowhere zero but otherwise at our disposal. After multiplying (2.1.1) by pu^3, we obtain

$$pu^2(pu^2z')' - \{pqu^4 - pu^3(pu')'\}z = 0.$$

The change of variable

$$t = \int_a^x (pu^2)^{-1}(s)\,\mathrm{d}s \tag{2.5.19}$$

gives

$$\mathrm{d}^2z/\mathrm{d}t^2 - \{pqu^4 - pu^3(pu')'\}z = 0. \tag{2.5.20}$$

The function u is chosen so that (2.5.20) assumes a simple form which is covered by Example 1.9.1. The most obvious choice is

$$u = (pq)^{-1/4}, \tag{2.5.21}$$

but more complicated choices allow some interplay between the two terms in the coefficient $\{\cdots\}$ of z. The choice in (2.5.2) is $u = (p\rho)^{-1/4}$. The general difficulty remains that t may appear as a complex variable in (2.5.19), and we now reformulate (2.5.18) and (2.5.19) in such a way that the change of variable from x to t is not explicitly made.

From (2.5.18), we have

$$py' = pu'z + puz' = pu'z + u^{-1}(pu^2z'), \tag{2.5.22}$$

and we note that the last term on the right would be $u^{-1}\,\mathrm{d}z/\mathrm{d}t$ if t were introduced as in (2.5.19). We define the vectors

$$Y = \begin{bmatrix} y \\ py' \end{bmatrix}, \qquad Z = \begin{bmatrix} z \\ pu^2z' \end{bmatrix},$$

the former being as in (2.1.3). Then (2.5.18) and (2.5.22) can be written

$$Y=\begin{bmatrix} u & 0 \\ pu' & u^{-1} \end{bmatrix} Z. \tag{2.5.23}$$

The transformation (2.5.23) takes (2.1.2) into

$$Z'=\begin{bmatrix} 0 & p^{-1}u^{-2} \\ qu^2-u(pu')' & 0 \end{bmatrix} Z, \tag{2.5.24}$$

and it is this system which corresponds to (2.5.20) but without the change of independent variable (2.5.19).

We can now give the extension of Theorem 2.5.1 to complex-valued p and q.

Theorem 2.5.2 *Let p and q be nowhere zero and have locally absolutely continuous first derivatives in $[a, \infty)$. Let the dichotomy condition hold in the form that either*

$$\left.\begin{aligned} &\operatorname{re}\int_t^x (q/p)^{1/2}(s)\,\mathrm{d}s \leqslant K_1 \\ \text{or}\qquad &\\ &\operatorname{re}\int_t^x (q/p)^{1/2}(s)\,\mathrm{d}s \geqslant K_2 \end{aligned}\right\} \tag{2.5.25}$$

for $a \leqslant t \leqslant x < \infty$, where K_1 and K_2 are constants. Also, let (2.5.7) hold. Then (2.1.1) has solutions y_1 and y_2 such that (2.5.16) and (2.5.17) hold for y_1, with similar formulae for y_2 containing $-(q/p)^{1/2}$ in the exponential.

Proof In (2.5.24), we take $u=(pq)^{-1/4}$ as in (2.5.21). Then $qu^2 = p^{-1}u^{-2}=(q/p)^{1/2}$ and, by (2.5.7), the remaining term $u(pu')'$ in (2.5.24) is $L(a, \infty)$. The transformation

$$Z=\begin{bmatrix} 1 & 1 \\ 1 & -1 \end{bmatrix} W \tag{2.5.26}$$

then takes (2.5.24) into

$$W'=(\Lambda+R)W, \tag{2.5.27}$$

where R is $L(a, \infty)$ and

$$\Lambda=(q/p)^{1/2}\,\mathrm{dg}\,(1, -1).$$

By (2.5.25), Λ satisfies the dichotomy condition L in § 1.3, and the Levinson theorem can then be applied to the W-system (2.5.27). The present theorem follows when we transform back to Y via (2.5.26) and (2.5.23). The only point to note is that (2.5.17) arises from the second component pu^2z' of Z because

$$(u^{-1}y)' = z' = (pu^2)^{-1}(pu^2z').$$

A sensitive example which is covered by Theorem 2.5.2 is

$$p(x) = \mathrm{e}^{\mathrm{i}x}, \qquad q(x) = \mathrm{e}^{3\mathrm{i}x}. \tag{2.5.28}$$

In this case $R = 0$ in (2.5.27), and (2.1.1) has the explicit solutions

$$\exp(-\mathrm{i}x \pm \mathrm{i}\mathrm{e}^{\mathrm{i}x}).$$

These solutions illustrate (2.5.16) and (2.5.17), but they do not satisfy the more refined formula (2.2.11) which was obtained in Theorem 2.5.1 when the coefficients are real-valued. We note that Lemma 2.5.1, which provided the link between (2.5.17) and (2.2.11), does not hold for complex-valued f and g, as the example $f = \mathrm{e}^{\mathrm{i}x}$, $g = \mathrm{e}^{-\mathrm{i}x}$ shows.

Further asymptotic results can be obtained from (2.5.24) by applying the other general theorems in Chapter 1 with alternative choices for u and, indeed, by iterating the transformation (2.5.23) which leads from (2.1.2) to (2.5.24). To give an indication of such results, we write (2.5.24) as

$$Z' = \left\{p^{-1}u^{-2}\begin{bmatrix} 0 & 1 \\ 1 & 0 \end{bmatrix} + \rho \begin{bmatrix} 0 & 0 \\ 1 & 0 \end{bmatrix}\right\} Z, \tag{2.5.29}$$

where

$$\rho = qu^2 - p^{-1}u^{-2} - u(pu')'. \tag{2.5.30}$$

The transformation (2.5.26) again gives a W-system (2.5.27) but now with

$$\Lambda = p^{-1}u^{-2}\,\mathrm{dg}\,(1, -1), \qquad R = \tfrac{1}{2}\rho \begin{bmatrix} 1 & 1 \\ -1 & -1 \end{bmatrix}.$$

Theorem 1.6.1, for example, is applicable to (2.5.27) if

$$pu^2\rho = o(1), \qquad (pu^2\rho)' \in L(a, \infty), \tag{2.5.31}$$

and then, in the notation of § 1.6,

$$\mu_1 = -\mu_2 = p^{-1}u^{-2}(1 + pu^2\rho)^{1/2}.$$

Asymptotic results in which q and $1/p$ are in some sense small as $x \to \infty$ can be obtained by taking

$$u = c\left(\int_a^x 1/p(s)\,\mathrm{d}s\right)^{\alpha}, \tag{2.5.32}$$

where c and α are constants with $c \neq 0$. In the case where $\alpha = \frac{1}{2}$ and $c^4 = 2$, (2.5.30) gives $\rho = qu^2$ and (2.5.31) becomes

$$pq\left(\int_a^x 1/p(s)\,\mathrm{d}s\right)^2 = o(1)$$

and

$$\frac{\mathrm{d}}{\mathrm{d}x}\left\{pq\left(\int_a^x 1/p(s)\,\mathrm{d}s\right)^2\right\} \in L(a, \infty).$$

An alternative proof of the asymptotic result based on these conditions will appear in Theorem 2.7.3, where the methods developed in §§ 1.10 and 2.7 will be used.

Finally, to outline the idea of iterating (2.5.23), we consider the choice (2.5.21) again and we re-write (2.5.24) as

$$Z_1' = \begin{bmatrix} 0 & q_1 \\ q_1(1+r_1) & 0 \end{bmatrix} Z_1, \tag{2.5.33}$$

where

$$q_1 = (q/p)^{1/2}, \quad u_1 = (pq)^{-1/4}, \quad r_1 = -q_1^{-1}u_1(pu_1')'.$$

Since the coefficient matrix in (2.5.33) has the same form as in (2.1.2), we can iterate (2.5.23) to obtain

$$Z_{m+1}' = \begin{bmatrix} 0 & q_{m+1} \\ q_{m+1}(1+r_{m+1}) & 0 \end{bmatrix} Z_{m+1}$$

after m iterations, where

$$\begin{aligned} q_{m+1} &= q_m(1+r_m)^{1/2}, \qquad u_{m+1} = (1+r_m)^{-1/4}, \\ r_{m+1} &= -q_{m+1}^{-1}u_{m+1}(q_m^{-1}u_{m+1}')' \\ &= \tfrac{1}{4}q_m^{-1}(1+r_m)^{-3/4}\{q_m^{-1}r_m'(1+r_m)^{-5/4}\}'. \end{aligned}$$

If M iterations produce a Z_{M+1}-system which has the Levinson form, we can then transform back to Y and obtain an asymptotic theorem for the solutions of (2.1.1) which is broadly similar to Theorem 2.4.1.

2.6 EQUATIONS OF EULER TYPE

So far in this chapter, we have considered the situation in § 2.1(i) with Λ taken to be large compared to R as $x \to \infty$. Now we examine what may be described as a borderline case where Λ and R have comparable size as $x \to \infty$. This case is specified by the condition (2.6.2) in the following theorem, and the simplest example is provided by the Euler equation in which

$$p(x) = (\text{const.})x^{\alpha}, \qquad q(x) = (\text{const.})x^{\alpha-2}. \tag{2.6.1}$$

The new feature which arises is that the asymptotic form of solutions involves the coefficients p and q in an algebraic manner rather than exponentially as in Theorem 2.2.1.

Theorem 2.6.1 *Let p and q be nowhere zero and have locally absolutely continuous first derivatives in* $[a, \infty)$. *Let*

$$(pq)'/pq = \kappa(q/p)^{1/2}(1+\phi), \tag{2.6.2}$$

where κ *is a non-zero constant with* $\kappa^2 \neq -16$,

$$\phi = o(1) \qquad (x \to \infty), \tag{2.6.3}$$

and

$$\phi' \in L(a, \infty). \tag{2.6.4}$$

Let

$$\mathrm{re}\,(q/p + r^2)^{1/2} \text{ have one sign in } [a, \infty), \tag{2.6.5}$$

where r is as in (2.2.4). *Then* (2.1.1) *has solutions* y_1 *and* y_2 *such that*

$$\left.\begin{aligned} y_1 &\sim (pq)^c \exp I(x), \\ py_1' &\sim c\kappa(pq)^{1/2+c} \exp I(x), \end{aligned}\right\} \tag{2.6.6}$$

$$\left.\begin{aligned} y_2 &\sim (pq)^{-1/2-c} \exp\{-I(x)\}, \\ py_2' &\sim -(c\kappa)^{-1}(pq)^{-c} \exp\{-I(x)\}, \end{aligned}\right\} \tag{2.6.7}$$

where

$$c = -\tfrac{1}{4} + \tfrac{1}{4}(1 + 16\kappa^{-2})^{1/2}, \tag{2.6.8}$$

and

$$I(x) = \frac{1}{4}\int_a^x \frac{(pq)'}{pq}\,\Phi\,\mathrm{d}t \tag{2.6.9}$$

with

$$\Phi = -16\kappa^{-2}(1 + 16\kappa^{-2})^{-1/2}\phi + O(\phi^2). \tag{2.6.10}$$

Proof As in the proof of Theorem 2.2.1, we diagonalize the matrix $\Lambda + R$ in (2.1.4). The eigenvalues μ_1 and μ_2 are again given by (2.2.7), and therefore (2.6.5) is the usual dichotomy condition. Here we use (2.6.2) to simplify the diagonalization and we start by writing

$$\Lambda + R = (q/p)^{1/2}(C + R_1)$$

where

$$C = \begin{bmatrix} 1 - \tfrac{1}{4}\kappa & \tfrac{1}{4}\kappa \\ \tfrac{1}{4}\kappa & -1 - \tfrac{1}{4}\kappa \end{bmatrix}, \qquad R_1 = -\tfrac{1}{4}\kappa\phi \begin{bmatrix} 1 & -1 \\ -1 & 1 \end{bmatrix}.$$

Since C is not a diagonal matrix, we have to express C in its diagonal form before applying Theorem 1.6.1 in the way that we did in § 2.2. We therefore write

$$T_1^{-1} C T_1 = \Lambda_1,$$

where Λ_1 is diagonal and

$$T_1 = \begin{bmatrix} \tfrac{1}{4}\kappa & -\tfrac{1}{4}\kappa \\ -1 + (1 + \kappa^2/16)^{1/2} & 1 + (1 + \kappa^2/16)^{1/2} \end{bmatrix},$$

T_1 being non-singular when $\kappa \neq 0$ and $\kappa^2 \neq -16$. The transformation

$$Z = T_1 W \tag{2.6.11}$$

gives

$$W' = (q/p)^{1/2}(\Lambda_1 + T_1^{-1} R_1 T_1)W. \tag{2.6.12}$$

By (2.6.3) and (2.6.4), the conditions (1.6.2) and (1.6.3) are satisfied by (2.6.12). The coefficient of W in (2.6.12) has eigenvalues μ_1 and μ_2, and therefore Theorem 1.6.1 shows that (2.6.12) has solutions

$$W_k(x) = \{e_k + o(1)\} \exp\left(\int_a^x \mu_k(t)\, \mathrm{d}t\right) \tag{2.6.13}$$

($k = 1, 2$). We note that, by (2.6.2), the expression (2.2.7) for μ_1 and μ_2 can also be written

$$\frac{(pq)'}{pq}(-\tfrac{1}{4} \pm \tfrac{1}{4}\{1 + 16\kappa^{-2}(1+\phi)^{-2}\}^{1/2}). \tag{2.6.14}$$

Then (2.6.6) and (2.6.7) follow when we transform back to y and py' via (2.6.11), (2.1.8), (2.1.5) and (2.1.3). To obtain (2.6.10), we note that the exact expression for Φ given by (2.6.13) and (2.6.14) is

$$\Phi = \{1 + 16\kappa^{-2}(1+\phi)^{-2}\}^{1/2} - \{1 + 16\kappa^{-2}\}^{1/2}$$

and we then expand in powers of ϕ up to ϕ^2. □

The dichotomy condition concerning (2.6.5) is clearly satisfied when p and q are real-valued, and it is also clearly satisfied (after a possible adjustment of the point $x = a$) when

$$-\pi + \omega \leqslant \arg\{(1 + \kappa^2/16)q/p\} \leqslant \pi - \omega$$

in $[a, \infty)$, where $\omega(>0)$ is fixed.

In addition to (2.6.1), a further class of coefficients for which (2.6.2) holds is given by

$$p(x) = Ax^{\alpha} \exp x^{\eta}, \qquad q(x) = Bx^{\beta} \exp x^{\eta}, \tag{2.6.15}$$

where α, β, η, A and B are constants with α, β and η real, $\eta > 0$, and

$$\beta - \alpha = 2(\eta - 1).$$

In full, (2.6.2) is

$$(\alpha + \beta)x^{-1} + 2\eta x^{\eta - 1} = \kappa(B/A)^{1/2} x^{(\beta - \alpha)/2}(1 + \phi),$$

giving

$$\kappa = 2\eta(A/B)^{1/2}$$

and

$$\phi(x) = \tfrac{1}{2}(\alpha + \beta)x^{-\eta}/\eta.$$

Then (2.6.9) and (2.6.10) give

$$I(x) = -4\kappa^{-2}(1+16\kappa^{-2})^{-1/2}(\alpha+\beta)\log x + (\text{const.}) + o(1)$$

and, after an adjustment of constant multiples, (2.6.6), (2.6.7) and (2.6.8) give

$$y_1(x) \sim x^{-(\alpha+\beta)(1+16\kappa^{-2})^{1/2}c} \exp(2cx^{\eta}),$$
$$y_2(x) \sim x^{-(\alpha+\beta)(1+16\kappa^{-2})^{1/2}(c+\frac{1}{2})} \exp\{(-2c-1)x^{\eta}\}.$$

2.7 SUBDOMINANT COEFFICIENT q

The remaining case of § 2.1(i) still to be considered is where Λ is small compared to R as $x \to \infty$, that is, q is in some sense subdominant. We start by re-writing (2.1.4) in a way that emphasizes R in preference to Λ. In (2.1.8), we have

$$T = D_0\Omega, \tag{2.7.1}$$

where

$$D_0 = \text{dg}\,(1, (pq)^{1/2}) \tag{2.7.2}$$

and Ω is the constant matrix

$$\Omega = \begin{bmatrix} 1 & 1 \\ 1 & -1 \end{bmatrix}.$$

It then follows from (2.1.9) that (2.1.4) is

$$Z' = (\Lambda - \Omega^{-1}\Lambda_1\Omega)Z, \tag{2.7.3}$$

where

$$\Lambda_1 = \text{dg}\,(0, \tfrac{1}{2}(pq)'/pq).$$

We can emphasize R, or equivalently Λ_1, rather than Λ in (2.7.3) by making the transformation

$$W = \Omega Z$$

to obtain the system

$$W' = (-\Lambda_1 + \Omega\Lambda\Omega^{-1})W.$$

By (2.1.5) and (2.7.1), the transformation from Y to W is simply

$$Y = D_0 W, \tag{2.7.4}$$

where D_0 is the diagonal matrix (2.7.2).

In (2.7.4), we have therefore arrived at an example of the procedure indicated in (2.1.10) with $D = D_0$ and W in place of Z. Accordingly, we now consider (2.1.1) in the context of § 2.1(ii), and we obtain the following theorem when (2.7.4) is used with a more general matrix D in place of D_0.

Theorem 2.7.1 *Let the function ρ be locally absolutely continuous and nowhere zero in $[a, \infty)$, and let ρ satisfy the dichotomy condition that either*

$$\text{(I)} \qquad |\rho(x)/\rho(t)| \leqslant K \tag{2.7.5}$$

or

$$\text{(II)} \qquad |\rho(x)/\rho(t)| \geqslant K \tag{2.7.6}$$

for $a \leqslant t \leqslant x < \infty$, where K (>0) is a constant. Also, let

$$\rho p^{-1} \qquad \text{and} \qquad \rho^{-1}q \tag{2.7.7}$$

both be $L(a, \infty)$. Then (2.1.1) *has solutions y_1 and y_2 such that*

$$y_1 \to 1, \qquad py_1' = o(\rho), \tag{2.7.8}$$

$$y_2 = o(\rho^{-1}), \qquad py_2' \to 1. \tag{2.7.9}$$

Proof As in (2.1.10), we make the transformation

$$Y = DZ \tag{2.7.10}$$

with

$$D = \text{dg}\,(1, \rho). \tag{2.7.11}$$

We obtain (2.1.11) with

$$D^{-1}D' = \text{dg}\,(0, \rho'/\rho)$$

and, by (2.7.7), we have

$$D^{-1}AD \in L(a, \infty).$$

Also, (2.7.5)–(2.7.6) are just the dichotomy condition for (2.1.11) in the form (1.3.11)–(1.3.12). It follows from Theorem 1.3.1 that there are solutions Z_1 and Z_2 of (2.1.11) such that

$$Z_1 = e_1 + o(1), \qquad Z_2 = \{e_2 + o(1)\}\rho^{-1},$$

and (2.7.8)–(2.7.9) follow when we transform back to Y in (2.7.10). □

We note that (2.7.5) and (2.7.6) are not mutually exclusive. However, any ρ which satisfies (2.7.5) is either bounded away from zero or tends to zero as $x \to \infty$. Similarly, any ρ which satisfies (2.7.6) is either bounded above or tends to infinity in modulus as $x \to \infty$. A simple choice for ρ in the theorem is $\rho(x) = 1/x$. Another choice, which is the one occurring in (2.7.2) and (2.7.4), is $\rho = (pq)^{1/2}$. In this case, (2.7.7) reduces to the single condition

$$(q/p)^{1/2} \in L(a, \infty).$$

It is also possible to obtain further asymptotic results for (2.1.1) by applying Theorems 1.5.1 and 1.6.1 to the system (2.1.11), but we shall not pursue these details here. We do observe, however, that it is a drawback in all these results that, as in (2.7.9), a true asymptotic formula for y_2 is not obtained immediately. This type of difficulty will also arise on future occasions. In the case of (2.7.9), we can go further if, for example, $p>0$ and $1/p \notin L(a, \infty)$. The formula for py_2' can then be integrated to give

$$y_2(x) \sim \int_a^x 1/p(s)\,\mathrm{d}s.$$

We move on to consider the results that are obtained by the method of § 1.10, with (2.1.2) regarded as a perturbation of the Jordan case. We write (2.1.2) in the form (1.10.2) and (1.10.3) by defining

$$J=\begin{bmatrix}0 & 1/p\\ 0 & 0\end{bmatrix}, \qquad R=\begin{bmatrix}0 & 0\\ q & 0\end{bmatrix}. \tag{2.7.12}$$

Then

$$\sigma(x)=\int_a^x 1/p(s)\,\mathrm{d}s \tag{2.7.13}$$

in (1.10.4), while the transformation (1.10.17) and the transformed system (1.10.15) become

$$Y=\begin{bmatrix}1 & 1\\ 0 & 1/\sigma\end{bmatrix}W \tag{2.7.14}$$

and

$$W'=(\Lambda+R_0)W, \tag{2.7.15}$$

where

$$\Lambda=\begin{bmatrix}0 & 0\\ 0 & \sigma'/\sigma\end{bmatrix}, \qquad R_0=q\sigma\begin{pmatrix}-1 & -1\\ 1 & 1\end{pmatrix}. \tag{2.7.16}$$

Theorem 1.10.1 is in this case simply the result of applying the Levinson theorem to (2.7.15), and the theorem becomes as follows.

Theorem 2.7.2 *Let σ be nowhere zero in some interval $[X, \infty)$, and let*

$$|\sigma(x)/\sigma(t)| \geqslant K \qquad (X \leqslant t \leqslant x < \infty), \tag{2.7.17}$$

where K (>0) is a constant. Let

$$q(x)\int_a^x 1/p(s)\,\mathrm{d}s \in L(a, \infty). \tag{2.7.18}$$

Then (2.1.1) *has solutions y_1 and y_2 such that*

$$y_1 \to 1, \qquad py_1' = o\left\{\left(\int_a^x 1/p(s)\,\mathrm{d}s\right)^{-1}\right\}, \tag{2.7.19}$$

$$y_2 \sim \int_a^x 1/p(s)\,\mathrm{d}s, \qquad py_2' \to 1. \tag{2.7.20}$$

We note that this theorem, in contrast to Theorem 2.7.1, does provide true asymptotic formulae for both y_1 and y_2. We therefore consider the system (2.7.15) further and give the result that is obtained when Theorem 1.6.1 is also applied to this system. The application of the theorems in § 1.5 will be dealt with later in § 2.8.

Theorem 2.7.3 *Let p and q be locally absolutely continuous in $[a, \infty)$. Let σ be defined by* (2.7.13) *and suppose that σ is nowhere zero in some interval $[X, \infty)$. Also, let*

(i)
$$pq\left(\int_a^x 1/p(s)\,\mathrm{d}s\right)^2 \to 0 \qquad (x \to \infty); \tag{2.7.21}$$

(ii)
$$\frac{\mathrm{d}}{\mathrm{d}x}\left\{pq\left(\int_a^x 1/p(s)\,\mathrm{d}s\right)^2\right\} \in L(a, \infty). \tag{2.7.22}$$

Let

$$\int_t^x \mathrm{re}\,\{(1+4pq\sigma^2)^{1/2}/p\sigma\}\,\mathrm{d}s \geq K \qquad (X \leq t \leq x < \infty), \tag{2.7.23}$$

where K is a constant. Then (2.1.1) *has solutions y_1 and y_2 such that*

$$y_1 \sim \mathrm{e}^{-\tau}, \qquad py_1' = o\left\{\left(\int_a^x 1/p(s)\,\mathrm{d}s\right)^{-1}\mathrm{e}^{-\tau}\right\}, \tag{2.7.24}$$

$$y_2 \sim \left(\int_a^x 1/p(s)\,\mathrm{d}s\right)e^{\tau}, \qquad py_2' \sim \mathrm{e}^{\tau}, \tag{2.7.25}$$

where

$$\tau(x) = \tfrac{1}{2}\int_a^x \{(1+4pq\sigma^2)^{1/2} - 1\}/p\sigma\,\mathrm{d}s. \tag{2.7.26}$$

Proof We apply Theorem 1.6.1 to (2.7.15) and (2.7.16). The conditions (2.7.21) and (2.7.22) are just (1.6.2) and (1.6.3) in the present situation. The eigenvalues of $\Lambda + R_0$ are

$$\mu_1 = -\frac{1}{2}\frac{\sigma'}{\sigma}\{(1+4pq\sigma^2)^{1/2} - 1\}$$

and

$$\mu_2 = \frac{\sigma'}{\sigma} + \frac{1}{2}\frac{\sigma'}{\sigma}\{(1+4pq\sigma^2)^{1/2} - 1\}.$$

Hence

$$\mathrm{re}\,(\mu_2 - \mu_1) = \mathrm{re}\,\{(1+4pq\sigma^2)^{1/2}/p\sigma\},$$

by (2.7.13), and therefore (2.7.23) is the dichotomy condition in the form (1.3.2). The conditions of Theorem 1.6.1 are now satisfied, and it follows that (2.7.15) has solutions

$$W_1 = \{e_1 + o(1)\}e^{-\tau}, \qquad W_2 = \{e_2 + o(1)\}\sigma e^{\tau}.$$

Then (2.7.24) and (2.7.25) follow when we transform back to Y, using (2.7.13) and (2.7.14). □

We note that, by (2.7.21), the form of $\tau(x)$ is

$$\tau(x) = \int_a^x \{q\sigma + O(pq^2\sigma^3)\}\,\mathrm{d}s.$$

Hence, if

$$pq^2\sigma^3 \in L(a, \infty) \tag{2.7.27}$$

in addition to (2.7.21) and (2.7.22), we can replace τ by

$$\int_a^x q(s)\sigma(s)\,\mathrm{d}s$$

in (2.7.24) and (2.7.25). Here (2.7.27) is the simplifying condition (1.6.35) in the present situation. We note also that (2.7.23) is certainly satisfied when p and q are real-valued.

As an example in which (2.7.21) and (2.7.22) hold, we take

$$p(x) = (\text{const.})x^{\alpha}, \qquad q(x) = (\text{const.})x^{-\beta}(\log x)^{-\gamma}, \tag{2.7.28}$$

with $\alpha < 1$ and $\gamma > 0$. Then (2.7.21) and (2.7.22) both hold when

$$\alpha + \beta \geqslant 2.$$

For comparison with Theorem 2.7.2, we observe that (2.7.18) also holds when $\alpha + \beta > 2$ but, when $\alpha + \beta = 2$, it holds only if $\gamma > 1$.

At the end of § 1.4, we noted that the $o(1)$ term in the asymptotic formula (1.3.5) can be estimated more explicitly, and can indeed be improved if the $L(a, \infty)$ condition on R is strengthened. Naturally, this observation also carries through to the various situations in which Theorem 1.3.1 is applied and, as an example, we give here the improvement which can be made to Theorem 2.7.2.

Theorem 2.7.4 *Let the conditions of Theorem* 2.7.2 *hold except that* (2.7.18) *is strengthened to*

$$q(x)\left(\int_a^x 1/p(s)\,\mathrm{d}s\right)^2 \in L(a, \infty). \tag{2.7.29}$$

Then (2.7.19) *and* (2.7.20) *are improved to*

$$\left.\begin{aligned} y_1 &= 1 + o(\sigma^{-1}), \qquad & py_1' &= o(\sigma^{-2}), \\ y_2 &= \sigma + O(1), \qquad & py_2' &= 1 + o(\sigma^{-1}). \end{aligned}\right\} \tag{2.7.30}$$

Proof When σ is bounded, Theorems 2.7.2 and 2.7.4 are almost identical. Therefore, leaving aside this trivial case, we assume that $|\sigma(x)| \to \infty$ as $x \to \infty$. Because of the form of Λ in (2.7.16), § 1.4(ii) is applicable with $\rho = |\sigma|$ and $c = 1$. In (1.4.29), we have $R = O(q\sigma)$ by (2.7.16), and hence

$$u_i(x) = O\{\sigma^{-1}(x)\}$$

by (2.7.29). This estimate is valid when $(i, k) = (1, 2)$ in (1.4.20). Also, (1.4.30) gives

$$u_i(x) = O\left(\int_x^\infty |q\sigma| \, \mathrm{d}t\right) = O\left(\sigma^{-1}\int_x^\infty |q\sigma^2| \, \mathrm{d}t\right)$$

by (2.7.17) (with x and t interchanged). Hence, for these $u_i(x)$, we have

$$u_i(x) = o\{\sigma^{-1}(x)\},$$

and this estimate is valid when $(i, k) = (1, 1)$, $(2, 1)$, $(2, 2)$ in (1.4.20). Hence, applying (1.4.23) to (2.7.15), we have solutions

$$W_1 = e_1 + o(\sigma^{-1}), \qquad W_2 = \left\{e_2 + \begin{pmatrix} u_1 \\ u_2 \end{pmatrix}\right\}\sigma, \tag{2.7.31}$$

where $u_1 = O(\sigma^{-1})$ and $u_2 = o(\sigma^{-1})$. On transforming back to Y by means of (2.7.14), we obtain (2.7.30). □

In the next and last of this group of theorems involving small q, we use the idea (1.11.7), where non-absolute convergence of the infinite integral is allowed. In § 1.11, we referred to the difficulty of obtaining an asymptotic theorem for (1.11.8) in general. However, in the case $n = 2$ that we have here, no such difficulties arise and the theorem is as follows.

Theorem 2.7.5 *Let σ be nowhere zero in some interval $[X, \infty)$ and let* (2.7.17) *hold. Let $\phi(x)$ be defined in $[a, \infty)$ by*

$$\phi(x) = -\int_x^\infty q(t) \, \mathrm{d}t \tag{2.7.32}$$

with the assumption that the integral converges. Let

$$\phi/p \in L(a, \infty), \qquad \phi^2\sigma/p \in L(a, \infty). \tag{2.7.33}$$

Then (2.1.1) *has solutions y_1 and y_2 such that*

$$y_1 \to 1, \qquad py_1' = \phi + o(\phi) + o(\sigma^{-1}), \tag{2.7.34}$$

$$y_2 \sim \sigma, \qquad py_2' = 1 + \phi\sigma + o(1) + o(\phi\sigma). \tag{2.7.35}$$

Proof Starting with (2.1.1), we make the transformation

$$Y = \begin{bmatrix} 1 & 0 \\ \phi & 1 \end{bmatrix} Z. \tag{2.7.36}$$

The term in q is eliminated since $\phi' = q$, by (2.7.32), and we obtain

$$Z' = p^{-1}\begin{bmatrix} \phi & 1 \\ -\phi^2 & -\phi \end{bmatrix} Z. \tag{2.7.37}$$

Then, regarding ϕ and ϕ^2 as small perturbations, we proceed as in (2.7.14) and write

$$Z = \begin{bmatrix} 1 & 1 \\ 0 & 1/\sigma \end{bmatrix} W. \tag{2.7.38}$$

This gives

$$W' = (\Lambda + R_0)W \tag{2.7.39}$$

with

$$\Lambda = \mathrm{dg}\,(0, \sigma'/\sigma)$$

and

$$R_0 = (\phi/p)\begin{bmatrix} 1 & 2 \\ 0 & -1 \end{bmatrix} + (\phi^2\sigma/p)\begin{bmatrix} 1 & 1 \\ -1 & -1 \end{bmatrix}. \tag{2.7.40}$$

By (2.7.33), R_0 is $L(a, \infty)$, and the theorem follows when we apply the Levinson theorem to (2.7.39) and transform back to Y via (2.7.38) and (2.7.36). □

To compare this theorem with Theorem 2.7.2, we note that when (2.7.18) holds, we have $\phi\sigma = o(1)$ and then the asymptotic formulae for y_1 and y_2 in the two theorems coincide. There is also a parallel situation to the one in Theorem 2.7.4 when (2.7.33) is strengthened to

$$\phi\sigma p \in L(a, \infty), \qquad \phi^2\sigma^2/p \in L(a, \infty). \tag{2.7.41}$$

Then, by (2.7.40), $\sigma R_0 \in L(a, \infty)$ and the improved formulae (2.7.31) hold for the solutions of (2.7.39). The transformation back to Y now gives

$$\left.\begin{aligned} y_1 &= 1 + o(\sigma^{-1}), \qquad & py_1' &= \phi + o(\phi\sigma^{-1}) + o(\sigma^{-2}), \\ y_2 &= \sigma + O(1), \qquad & py_2' &= 1 + \phi\sigma + O(\phi) + o(\sigma^{-1}). \end{aligned}\right\} \tag{2.7.42}$$

As a final comment on the various conditions, we note that (2.7.33) can be relaxed somewhat in the case where $\sigma(x)$ is real-valued and $\sigma(x) \to \infty$ as $x \to \infty$. Let us assume only that

$$\phi^2\sigma/p \in L(a, \infty). \tag{2.7.43}$$

We make a change of variable and define

$$u = \log \sigma(x).$$

Then (2.7.39) becomes

$$\mathrm{d}W/\mathrm{d}u = (\Lambda_0 + rC_1 + r^2C_2)W, \tag{2.7.44}$$

where $\Lambda_0 = \mathrm{dg}\,(0, 1)$, $r = \phi\sigma$, and C_1, C_2 are the constant matrices in (2.7.40). By (2.7.43), we have $r(u) \in L^2(u(a), \infty)$, and therefore we can apply the Hartman–Wintner theorem (Theorem 1.5.1) (with $p = 2$) to (2.7.44). Since $\mathrm{dg}\, C_1 = \mathrm{dg}\,(1, -1)$, we obtain (2.7.34) and (2.7.35) again but with an extra factor

$$\exp\left(\pm\int_a^x (\phi/p)\,\mathrm{d}t\right)$$

on the right-hand sides of the formulae for y_1 and y_2 respectively.

As a simple but important example on Theorem 2.7.5, we mention $p(x) = 1$ and

$$q(x) = x^{-\alpha}P(x), \tag{2.7.45}$$

where $\alpha > 0$ and $P(x)$ has period 2π with mean-value zero. Then an integration by parts in (2.7.32) shows that $\phi(x) = O(x^{-\alpha})$, and it follows that (2.7.33) holds when $\alpha > 1$. The asymptotic formulae (2.7.34) and (2.7.35) reduce to

$$y_1 \to 1, \qquad y_1' = o(x^{-1}), \qquad y_2 \sim x, \qquad y_2' \to 1.$$

The situation when $\alpha \leqslant 1$ is more complicated and will be taken up in § 4.13.

2.8 APPLICATION OF THE HARTMAN–WINTNER THEOREM

In § 2.2 we gave the result of applying Theorem 1.6.1 to the system (2.1.4), where Λ and R are defined by (2.1.7) and (2.1.9). It is also of interest to apply the Hartman–Wintner theorem (Theorem 1.5.1), since results with different features are obtained.

Theorem 2.8.1 *Let p and q be locally absolutely continuous and nowhere zero in* $[a, \infty)$, *and let there be a constant* δ (>0) *such that (with either choice of the square root)*

$$\mathrm{re}\,(q/p)^{1/2} \geqslant \delta \tag{2.8.1}$$

in $[a, \infty)$. *Also, let*

$$(pq)'/pq \in L^{\nu}(a, \infty) \tag{2.8.2}$$

for some ν *in* $(1, 2]$. *Then* (2.1.1) *has solutions* y_1 *and* y_2 *such that*

$$y_1 \sim (pq)^{-1/4} \exp\left(\int_a^x (q/p)^{1/2}\,\mathrm{d}t\right), \tag{2.8.3}$$

$$py_1' \sim (pq)^{1/4} \exp\left(\int_a^x (q/p)^{1/2}\,\mathrm{d}t\right), \tag{2.8.4}$$

with similar formulae for y_2 *containing* $-(q/p)^{1/2}$ *in the exponential term.*

Proof By (2.1.7) and (2.1.9), the conditions (2.8.1) and (2.8.2) are just (1.5.3) and (1.5.4) in the present situation: the exponent ν is introduced in (2.8.2) to avoid the conflict of two otherwise standard uses of p in (1.5.4) and (2.8.2). Also, the two diagonal elements of R in (2.1.9) are

$$r_{11} = r_{22} = -\tfrac{1}{4}(pq)'/pq,$$

and the theorem follows immediately when (1.5.5) is applied to (2.1.4). □

Although (2.8.3)–(2.8.4) are the same as (2.2.10)–(2.2.11), the conditions imposed here are different from those imposed in Corollary 2.2.1. In the first place, when p and q are real-valued, (2.8.1) does not allow $q/p<0$, whereas (2.2.9) does. On the other hand, no more smoothness than the absolute continuity of p and q is required in Theorem 2.8.1, whereas Corollary 2.2.1 requires the absolute continuity of p' and q'.

If (2.8.2) holds with $\nu>2$, we can use the extension of Theorem 1.5.1 which is stated as Theorem 1.5.2. Let us suppose now that ν lies in (2, 4], this being the case $M=2$ of (1.5.26). With Λ and R as in (2.1.7) and (2.1.9), the definition of Q in (1.5.16) and (1.5.19) gives

$$q_{12}(x) = -\mathrm{e}^{\mu(x)} \int_x^\infty \mathrm{e}^{-\mu(t)} r(t)\,\mathrm{d}t \tag{2.8.5}$$

and

$$q_{21}(x) = -\mathrm{e}^{-\mu(x)} \int_a^x \mathrm{e}^{\mu(t)} r(t)\,\mathrm{d}t, \tag{2.8.6}$$

where

$$\mu(x) = 2\int_a^x \{q(s)/p(s)\}^{1/2}\,\mathrm{d}s \tag{2.8.7}$$

and

$$r(x) = \tfrac{1}{4}(pq)'(x)/(pq)(x).$$

Also, by (1.5.12) and (1.5.22) (with exponent ν now), $|Q|^3\,|R|$ is $L(a,\infty)$. Hence (1.5.11) gives

$$R_1 = RQ - Q\,\mathrm{dg}\,R - QRQ + Q^2\,\mathrm{dg}\,R + S,$$

where S is $L(X,\infty)$. It is easy to check that

$$\mathrm{dg}\,R_1 = \mathrm{dg}\,RQ + \mathrm{dg}\,S.$$

Hence, ignoring $\mathrm{dg}\,S$, (1.5.27) gives

$$\Lambda_2 = \Lambda + \mathrm{dg}\,(-r + rq_{21},\ -r + rq_{12}). \tag{2.8.8}$$

Before substituting into Theorem 1.5.2, however, we note a connection between q_{12} and q_{21} as follows. From (2.8.5) and (2.8.6) we have

$$\begin{aligned} rq_{12} + rq_{21} &= -r\mathrm{e}^{\mu} \int_x^\infty \mathrm{e}^{-\mu(t)} r(t)\,\mathrm{d}t + r\mathrm{e}^{-\mu} \int_a^x \mathrm{e}^{\mu(t)} r(t)\,\mathrm{d}t \\ &= \phi', \end{aligned} \tag{2.8.9}$$

where

$$\phi(x) = -\int_a^x e^{\mu(t)} r(t)\, dt \int_x^\infty e^{-\mu(t)} r(t)\, dt.$$

By (2.8.2) and (2.8.7), we have

$$|\phi(x)| \leqslant \int_a^x e^{2\delta t}\, |r(t)|\, dt \int_x^\infty e^{-2\delta t}\, |r(t)|\, dt$$
$$= O\left(\int_x^\infty |r(t)|^\nu\, dt\right)^{1/\nu}$$

after an application of Hölder's inequality. Hence $\phi(x) = o(1)$ as $x \to \infty$, and therefore (2.8.9) gives

$$\int_a^x r(t)q_{12}(t)\, dt + \int_a^x r(t)q_{21}(t)\, dt = o(1).$$

It now follows from (2.8.8) and Theorem 1.5.2 that (2.1.1) has two solutions

$$y \sim (pq)^{-1/4} \exp\left(\pm\int_a^x \{(q/p)^{1/2} + r(t)q_{21}(t)\}\, dt\right), \tag{2.8.10}$$

where q_{21} is defined by (2.8.6). The corresponding formulae for py', with $(pq)^{1/4}$ in place of $(pq)^{-1/4}$, also hold as in (2.8.4).

We state this result formally as a theorem.

Theorem 2.8.2 *Let p and q be locally absolutely continuous and nowhere zero in* $[a, \infty)$, *and let* (2.8.1) *hold. Let* (2.8.2) *hold for some* ν *in* $(2, 4]$. *Then* (2.1.1) *has two solutions satisfying* (2.8.10), *where* q_{21} is defined by (2.8.6) and (2.8.7).

Values of ν throughout the range $(1, 4]$ are now covered by Theorems 2.8.1 and 2.8.2. The calculations could be continued further to cover greater values of ν in the manner indicated in Theorem 1.5.2.

As an example where (2.8.2) holds for some $\nu > 1$, we mention (2.2.16) again:

$$p(x) = x^\alpha f(x^\gamma), \qquad q(x) = x^\beta g(x^\eta),$$

except that now $f(t)$ and $g(t)$ need only be once differentiable. Then (2.8.1) implies that $\alpha < \beta$, and (2.8.2) holds if γ and η both lie in $[0, 1 - 1/\nu)$. Thus (2.8.2) holds for some $\nu > 1$ if γ and η lie in $[0, 1)$. The calculations required by Theorem 1.5.2 become more lengthy the nearer γ and η are to unity, that is, as p and q move from the class of slowly oscillating functions to that of regularly oscillating functions.

The theorems in § 1.5 can also be applied to the equation

$$y''(x) - \{k^2 + r(x)\}y(x) = 0, \tag{2.8.11}$$

where k is a constant with $\operatorname{re} k \neq 0$ and $r(x)$ is any function such that

$$r(x) \in L^{\nu}(a, \infty) \tag{2.8.12}$$

for some $\nu > 1$. The case where $1 < \nu \leqslant 2$ is already covered in Example 1.9.1(ii), and we now give the result for $2 < \nu \leqslant 4$.

Theorem 2.8.3 *Let k be a constant with* $\operatorname{re} k \neq 0$ *and let* (2.8.12) *hold for some ν in* (2, 4]. *Then* (2.8.11) *has two solutions y such that*

$$y \sim \exp\left\{\pm\left(kx + (2k)^{-1}\int_a^x r(t)\,\mathrm{d}t + (2k)^{-2}\int_a^x r_1(t)\,\mathrm{d}t\right)\right\}, \tag{2.8.13}$$

where

$$r_1(t) = -r(t)\mathrm{e}^{-2kt}\int_a^t r(s)\mathrm{e}^{2ks}\,\mathrm{d}s$$
$$+ k^{-1}r(t)\int_a^t r(s)\mathrm{e}^{2ks}\,\mathrm{d}s\int_t^\infty r(s)\mathrm{e}^{-2ks}\,\mathrm{d}s.$$

Proof We begin by writing (2.8.11) in the form (1.5.1) with

$$\Lambda = \mathrm{dg}\,(k, -k),$$

$$R = (2k)^{-1}r\begin{bmatrix} 1 & 1 \\ -1 & -1 \end{bmatrix}, \quad \begin{pmatrix} y \\ y' \end{pmatrix} = \begin{bmatrix} 1 & 1 \\ k & -k \end{bmatrix} Y. \tag{2.8.14}$$

Then, proceeding as we did in the working leading up to Theorem 2.8.2, in place of (2.8.5)–(2.8.7) we now have

$$q_{12}(x) = -(2k)^{-1}\int_x^\infty \mathrm{e}^{2k(x-t)}r(t)\,\mathrm{d}t,$$

$$q_{21}(x) = -(2k)^{-1}\int_a^x \mathrm{e}^{2k(t-x)}r(t)\,\mathrm{d}t,$$

and

$$\mu(x) = 2k(x-a).$$

In place of (2.8.8), we have

$$\Lambda_2 = \Lambda + (2k)^{-1}r\,\mathrm{dg}\,(1 + q_{21} + 2q_{12}q_{21}, -1 - q_{12} - 2q_{12}q_{21}). \tag{2.8.15}$$

Again, corresponding to (2.8.9), we have

$$rq_{12} - rq_{21} = \phi',$$

and

$$\phi(x) = (2k)^{-1}\int_a^x \mathrm{e}^{2kt}r(t)\,\mathrm{d}t\int_x^\infty \mathrm{e}^{-2kt}r(t)\,\mathrm{d}t$$

and $\phi(x) = o(1)$ as $x \to \infty$. Hence (2.8.15) gives

$$\Lambda_2 = \Lambda + \{(2k)^{-1}r + (2k)^{-2}r_1\} \operatorname{dg}(1, -1) - (2k)^{-1}\phi' \operatorname{dg}(0, 1).$$

Then (2.8.13) follows immediately from (2.8.14) and Theorem 1.5.2. □

Lastly, as indicated in § 2.7, we give the results that are obtained when the theorems in § 1.5 are applied to (2.7.15), with q regarded as subdominant. Since the Λ which appears in (2.7.16) may not satisfy (1.5.3), as for example when $p(x) = 1$, we make a change of independent variable and define

$$u = \log \sigma(x). \tag{2.8.16}$$

Here $\sigma(x)$ is given by (2.7.13) and therefore we require $p(x)$ to be real-valued and positive with $\sigma(x) \to \infty$ as $x \to \infty$. In terms of the new variable u, (2.7.15) is

$$\mathrm{d}W/\mathrm{d}u = (\Lambda_0 + pq\sigma^2\Gamma)W, \tag{2.8.17}$$

with

$$\Lambda_0 = \begin{bmatrix} 0 & 0 \\ 0 & 1 \end{bmatrix}, \qquad \Gamma = \begin{bmatrix} -1 & -1 \\ 1 & 1 \end{bmatrix}. \tag{2.8.18}$$

Theorem 2.8.4 *Let p be real-valued in $[a, \infty)$, with $p > 0$, and let*

$$\sigma(x) = \int_a^x 1/p(s)\,\mathrm{d}s \to \infty \tag{2.8.19}$$

as $x \to \infty$. Also, let

$$qp^{1-1/\nu}\left(\int_a^x 1/p(s)\,\mathrm{d}s\right)^{2-1/\nu} \in L^\nu(a, \infty) \tag{2.8.20}$$

for some ν in $(1, 2]$. *Then* (2.1.1) *has solutions y_1 and y_2 such that* (2.7.24) *and* (2.7.25) *hold, where now* (2.7.26) *is replaced by*

$$\tau(x) = \int_a^x q(s)\sigma(s)\,\mathrm{d}s. \tag{2.8.21}$$

Proof By (2.8.16) and (2.8.19), the condition (2.8.20) implies that

$$\int_b^\infty |pq\sigma^2|^\nu\,\mathrm{d}u < \infty$$

with some fixed lower limit b. We can therefore apply Theorem 1.5.1 to (2.8.17). Since $\operatorname{dg}\Gamma = \operatorname{dg}(-1, 1)$ in (2.8.18), the system (2.8.17) has solutions

$$W_1 = \{e_1 + o(1)\} \exp\left(-\int_b^u (pq\sigma^2)(v)\,\mathrm{d}v\right),$$

$$W_2 = \{e_2 + o(1)\} \exp\left(u + \int_b^u (pq\sigma^2)(v)\,\mathrm{d}v\right),$$

and the theorem follows when we transform back to Y and x, using (2.7.14) and (2.8.16). □

Theorem 2.8.5 *Let the conditions in Theorem* 2.8.4 *hold except that* ν *lies in* (2, 4] *in* (2.8.20). *Then* (2.1.1) *has solutions* y_1 *and* y_2 *such that* (2.7.24) *and* (2.7.25) *hold, where now* (2.7.26) *is replaced by*

$$\tau(x) = \int_a^x q(s)\sigma(s)\left(1 - \sigma^{-1}(s)\int_a^s (q\sigma^2)(t)\,\mathrm{d}t + 2\int_a^s (q\sigma^2)(t)\,\mathrm{d}t\int_s^\infty q(t)\,\mathrm{d}t\right)\mathrm{d}s. \quad (2.8.22)$$

Proof We define q_{ij} by (1.5.16) and (1.5.19), first in terms of (2.8.17) and the variable u, and then we obtain

$$q_{12}(x) = -\sigma^{-1}(x)\int_a^x (q\sigma^2)(t)\,\mathrm{d}t$$

and

$$q_{21}(x) = -\sigma(x)\int_x^\infty q(t)\,\mathrm{d}t,$$

on changing back to x by means of (2.8.16). Next, we have a similar expression to (2.8.15) which arises when (1.5.27) is applied to (2.8.17). Thus

$$\Lambda_2 = \Lambda_0 + pq\sigma^2\,\mathrm{dg}\,(-1 - q_{21} - 2q_{12}q_{21},\ 1 + q_{12} + 2q_{12}q_{21}). \quad (2.8.23)$$

Once again, as in (2.8.9), we have

$$q\sigma q_{12} - q\sigma q_{21} = \phi',$$

where

$$\phi(x) = \int_a^x (q\sigma^2)(t)\,\mathrm{d}t\int_x^\infty q(t)\,\mathrm{d}t.$$

It follows from (2.8.20) and a Hölder inequality that $\phi(x)\to 0$ as $x\to\infty$. Hence (2.8.23) gives

$$\Lambda_2 = \Lambda_0 + p\sigma\tau'\,\mathrm{dg}\,(-1, 1) + p\sigma\phi'\,\mathrm{dg}\,(1, 0)$$

with τ as in (2.8.22). The theorem follows when we apply Theorem 1.5.2 to (2.8.17) and then change back to the variable x. □

As an example on Theorems 2.8.4 and 2.8.5, we take

$$p(x) = x^\alpha, \qquad q(x) = O\{x^{-\beta}(\log x)^{-\gamma}\}.$$

This example is similar to (2.7.28) except that no differentiability of q is required. Then (2.8.19) holds if $\alpha < 1$ and (2.8.20) certainly holds if $\alpha + \beta > 2$. More to the point, however, is that (2.8.20) also holds when

$\alpha+\beta=2$ and $\gamma\nu>1$, and therefore the two theorems apply when $\gamma>\frac{1}{2}$ and $\gamma>\frac{1}{4}$ respectively. Smaller values of γ require further stages in the process developed in Theorem 1.5.2.

2.9 HIGHER-ORDER EQUATIONS

The ideas in §§ 2.1–8 can in principle be extended to equations of higher-order n (>2), based as they are on the general theorems of Chapter 1. The main question is whether the algebraic properties arising from an $n \times n$ matrix A in (2.1.2) can be worked out in sufficient detail. Here we consider the two-term differential equation

$$(p_{n-1} \cdots p_2(p_1 y')' \cdots)' - qy = 0 \tag{2.9.1}$$

which, although of a somewhat complicated appearance itself, can be written as a first-order system

$$Y' = AY \tag{2.9.2}$$

with A as the simple matrix

$$A = \begin{bmatrix} 0 & p_1^{-1} & \cdots & 0 \\ & & & \\ & & & p_{n-1}^{-1} \\ q & & & 0 \end{bmatrix} \tag{2.9.3}$$

and unmarked entries are zero. The components of Y are

$$y, p_1 y', \ldots .$$

As in the case of (1.9.12) and (1.9.14), we can regard (2.9.2) as a more general formulation of (2.9.1) which avoids differentiability conditions on the p_j. All that is needed in (2.9.2) and (2.9.3) at this stage is that q and the p_j^{-1} should be locally Lebesgue integrable. A special case of (2.9.1) is the equation

$$(py^{(m)})^{(l)} - qy = 0, \tag{2.9.4}$$

which is obtained by taking all except one of the p_j equal to unity. Here $l \geqslant 1$ and $m \geqslant 1$, and the Sturm–Liouville equation (2.1.1) is in turn the case $l = m = 1$.

We begin the asymptotic analysis of (2.9.2) by obtaining first the transformation theory corresponding to § 2.1(i) and then the generalization of the Liouville–Green formulae in Corollary 2.2.1. The characteristic equation of A in (2.9.3) is easily seen to be

$$\lambda^n - q/(p_1 \cdots p_{n-1}) = 0. \tag{2.9.5}$$

Hence, defining

$$P = \{q/(p_1 \cdots p_{n-1})\}^{1/n} \tag{2.9.6}$$

with any fixed determination of the nth root, we obtain the eigenvalues λ_k of A as

$$\lambda_k = \omega_k P \qquad (1 \leq k \leq n), \tag{2.9.7}$$

where ω_k is the kth root of unity,

$$\omega_k = \exp\{2(k-1)\pi i/n\}.$$

An eigenvector v_k corresponding to λ_k has components

$$1,\ \lambda_k p_1,\ \lambda_k^2 p_1 p_2,\ \ldots,\ \lambda_k^{n-1} p_1 \cdots p_{n-1}. \tag{2.9.8}$$

We can then express A in its diagonal form

$$T^{-1}AT = \Lambda, \tag{2.9.9}$$

where

$$D_0 = \mathrm{dg}\,(1, P_1, P_1P_2, \ldots, P_1 \cdots P_{n-1}), \qquad (2.9.12)$$

and

$$T = (v_1 \cdots v_n).$$

We also note that, by (2.9.7) and (2.9.8),

$$T = D_0 \Omega, \tag{2.9.11}$$

where

$$D_0 = \mathrm{dg}\,(1, P_1, P_1P_2, \ldots, P_1 \cdots P_{n-1}), \tag{2.9.12}$$

$$P_j = Pp_j, \tag{2.9.13}$$

and Ω is the constant matrix whose entry in the (j, k) position is indicated by

$$\Omega = (\omega_k^{j-1}) \qquad (1 \leq j, k \leq n). \tag{2.9.14}$$

By (2.9.9), the transformation

$$Y = TZ \tag{2.9.15}$$

takes (2.9.2) into

$$Z' = (\Lambda + R)Z, \tag{2.9.16}$$

where

$$R = -T^{-1}T' = -\Omega^{-1}\Lambda_1\Omega \tag{2.9.17}$$

and

$$\begin{aligned}\Lambda_1 &= D_0^{-1}D_0' \\ &= \mathrm{dg}\,(0, P_1'/P_1, (P_1P_2)'/P_1P_2, \ldots, (P_1 \cdots P_{n-1})'/(P_1 \cdots P_{n-1})).\end{aligned} \tag{2.9.18}$$

The calculations leading to (2.9.16) are valid provided that q and the p_j are all nowhere zero and locally absolutely continuous in $[a, \infty)$. As in § 2.2, the aim now is to apply Theorem 1.6.1 to deal with the situation where Λ is large compared to R. However, because of the difficulty of calculating the eigenvalues μ_k of $\Lambda + R$ in the $n \times n$ case, we apply only the simplified form of Theorem 1.6.1 given by (1.6.35) and (1.6.37), to obtain the asymptotic result for (2.9.1) which corresponds to Corollary 2.2.1.

Theorem 2.9.1 *Let $q, p_1, \ldots, p_{n-1}$ be nowhere zero and have locally absolutely continuous derivatives in $[a, \infty)$. For $1 \leqslant j \leqslant n-1$, let*

$$\text{(i)} \qquad (Pp_j)'/Pp_j = o(P) \qquad (x \to \infty) \tag{2.9.19}$$

$$\text{(ii)} \qquad \{(Pp_j)'\}^2/P^3p_j^2 \in L(a, \infty) \tag{2.9.20}$$

$$\text{(iii)} \qquad \{(Pp_j)'/P^2p_j\}' \in L(a, \infty), \tag{2.9.21}$$

where P is defined by (2.9.6). *Also, let*

$$\operatorname{re}\{(\omega_j - \omega_k)P\} \text{ have one sign in } [a, \infty) \tag{2.9.22}$$

for each j and k. Then (2.9.1) *has solutions y_k $(1 \leqslant k \leqslant n)$ such that*

$$y_k \sim P^{-(n-1)/2}(p_1^{n-1}p_2^{n-2} \cdots p_{n-2}^2 p_{n-1})^{-1/n} \exp\left(\omega_k \int_a^x P(t)\,\mathrm{d}t\right). \tag{2.9.23}$$

Proof We apply Theorem 1.6.1 to (2.9.16). By (2.9.10), (2.9.13) and (2.9.18), the conditions (2.9.19) and (2.9.21) are just (1.6.2) and (1.6.3). Also, (2.9.20) is the simplifying condition (1.6.35). Subject to the dichotomy condition on $\lambda_k + r_{kk}$, we can therefore use (1.6.37) with r_{kk} as the kth diagonal entry of the matrix R defined by (2.9.17). To calculate r_{kk}, we note that, by (2.9.14),

$$\Omega^{-1} = n^{-1}(\omega_j^{-(k-1)}), \tag{2.9.24}$$

and therefore

$$\begin{aligned} r_{kk} &= -\frac{1}{n}\left(\frac{P_1'}{P_1} + \frac{(P_1P_2)'}{P_1P_2} + \cdots + \frac{(P_1 \cdots P_{n-1})'}{P_1 \cdots P_{n-1}}\right) \\ &= -\frac{1}{n}(P_1^{n-1}P_2^{n-2} \cdots P_{n-1})'/(P_1^{n-1}P_2^{n-2} \cdots P_{n-1}). \end{aligned} \tag{2.9.25}$$

By (2.9.7) and the fact that r_{kk} is independent of k, (2.9.22) is the necessary dichotomy condition on $\lambda_k + r_{kk}$ in the form (1.3.13).

It now follows from (1.6.37) and (2.9.25) that there are solutions Z_k $(1 \leqslant k \leqslant n)$ of (2.9.16) such that

$$Z_k = (P_1^{n-1}P_2^{n-2} \cdots P_{n-1})^{-1/n}\{e_k + o(1)\} \exp\left(\int_a^x \omega_k P(t)\,\mathrm{d}t\right).$$

We then substitute for the P_j from (2.9.13) and transform back to Y via (2.9.15) and (2.9.11). The required formula (2.9.23) follows when the first component y of Y is considered. There are also corresponding asymptotic formulae involving derivatives of y_k, and these are obtained when the other components p_1y', $p_2(p_1y')'$, ... are considered. □

We make the following remarks concerning Theorem 2.9.1.

(a) The dichotomy condition (2.9.22) is certainly satisfied when q and the p_j are all real-valued in $[a, \infty)$. More generally, (2.9.22) delineates the closed sectors

$$r\pi/n \leqslant \arg P \leqslant (r+1)\pi/n \quad (0 \leqslant r \leqslant 2n-1)$$

of the complex plane, within one of which the values of P must be confined for all x in $[a, \infty)$.

(b) As in (2.2.14), the simplest example covered by the theorem is

$$p_j(x) = (\text{const.})x^{\alpha_j}, \qquad q(x) = (\text{const.})x^{\beta}. \tag{2.9.26}$$

Here (2.9.6) gives $P(x) = (\text{const.})x^{\gamma}$, where

$$\gamma = \{\beta - (\alpha_1 + \cdots + \alpha_{n-1})\}/n,$$

and the conditions (2.9.19)–(2.9.21) are all satisfied if

$$\alpha_1 + \cdots + \alpha_{n-1} - \beta < n. \tag{2.9.27}$$

There are also examples corresponding to (2.2.16) and (2.2.18).

(c) We note the form which the theorem takes in the case of the equation (2.9.4). Here $p_m = p$ and the other p_j are unity, and $n = l + m$. Then (2.9.6) gives $P = (q/p)^{1/n}$, and the conditions (2.9.19)–(2.9.21) become

(i)′ p'/p and q'/q are $o(P)$
(ii)′ $P^{-1}(p'/p)^2$ and $P^{-1}(q'/q)^2$ are $L(a, \infty)$
(iii)′ $(P^{-1}p'/p)'$ and $(P^{-1}q'/q)'$ are $L(a, \infty)$.

The asymptotic formula (2.9.23) becomes

$$y_k \sim (p/q)^{(n-1)/2n} p^{-1/n} \exp\left(\omega_k \int_a^x (q/p)^{1/n}\, dt\right). \tag{2.9.28}$$

2.10 HIGHER-ORDER EQUATIONS OF EULER TYPE

The previous section dealt with the case where, in (2.9.16), Λ is large compared to R as $x \to \infty$. A borderline situation corresponding to the one considered in § 2.6 arises when Λ and R have similar size for large x. Generalizing (2.6.2), we suppose now that there are non-zero constants κ_j $(1 \leqslant j \leqslant n-1)$ such that

$$(Pp_j)'/Pp_j = \kappa_j P(1 + \phi_j), \tag{2.10.1}$$

where

$$\phi_j = o(1) \qquad (x \to \infty), \tag{2.10.2}$$

$$\phi_j' \in L(a, \infty), \tag{2.10.3}$$

and P is defined by (2.9.6). By (2.9.10), (2.9.13), (2.9.17) and (2.9.18), we then have

$$\Lambda + R = P\{\Lambda_0 - \Omega^{-1}(M_0 + R_0)\Omega\}, \tag{2.10.4}$$

where

$$\Lambda_0 = \mathrm{dg}\,(\omega_1, \ldots, \omega_n),$$

$$M_0 = \mathrm{dg}\,(0, \kappa_1, \kappa_1 + \kappa_2, \ldots, \kappa_1 + \cdots + \kappa_{n-1}), \tag{2.10.5}$$

and

$$R_0 = \mathrm{dg}\,(0, \kappa_1\phi_1, \ldots, \kappa_1\phi_1 + \cdots + \kappa_{n-1}\phi_{n-1}). \tag{2.10.6}$$

By (2.9.14) and (2.9.24), it is easy to check that

$$\Omega\Lambda_0\Omega^{-1} = A_0,$$

where A_0 has entries unity in the $(n, 1)$ and $(i, i+1)$ positions $(1 \leqslant i \leqslant n-1)$ and zeros elsewhere. Hence (2.10.4) gives

$$\Lambda + R = P\Omega^{-1}(C_0 - R_0)\Omega, \tag{2.10.7}$$

where again

$$C_0 = A_0 - M_0. \tag{2.10.8}$$

It follows from (2.10.7), (2.9.11), (2.9.15) and (2.9.16) that the transformation

$$Y = D_0 W \tag{2.10.9}$$

takes (2.9.2) into

$$W' = P(C_0 - R_0)W. \tag{2.10.10}$$

In order to apply Theorem 1.6.1, we have to express the constant matrix C_0 in its diagonal form. The characteristic equation of C_0 is easily found to be

$$\det\,(\nu I - C_0) = \nu \prod_{r=1}^{n-1} \{\nu + (\kappa_1 + \cdots + \kappa_r)\} - 1 = 0, \tag{2.10.11}$$

by (2.10.8) and (2.10.5). Let the eigenvalues be ν_k $(1 \leqslant k \leqslant n)$, with corresponding eigenvectors v_k, and we suppose that the κ_r make the ν_k all distinct. Then, with

$$T_1 = (v_1 \cdots v_n), \tag{2.10.12}$$

the transformation

$$W = T_1 V$$

takes (2.10.10) into

$$V' = P(\tilde{\Lambda} - T_1^{-1}R_0T_1)V, \tag{2.10.13}$$

where

$$\tilde{\Lambda} = \mathrm{dg}\,(\nu_1, \ldots, \nu_n).$$

By (2.10.2) and (2.10.3), the conditions (1.6.2) and (1.6.3) are satisfied by (2.10.13) and hence, subject to a dichotomy condition, Theorem 1.6.1 shows that there are solutions

$$V_k(x) = \{e_k + o(1)\} \exp\left(\int_a^x \mu_k(t)\,\mathrm{d}t\right) \qquad (1 \leqslant k \leqslant n), \tag{2.10.14}$$

where the μ_k are the eigenvalues of the coefficient matrix $P(\tilde{\Lambda} - T_1^{-1}R_0T_1)$. We note that the μ_k are also the eigenvalues of $P(C_0 - R_0)$ because of a similarity transformation.

If in addition the simplifying condition (1.6.35) is satisfied by (2.10.13), we can replace μ_k by

$$P(\nu_k + O(\phi)),$$

where $\phi = \max_j |\phi_j|$ and we have used (1.6.37) and (2.10.6). Also by (2.10.6), this condition is $P\phi_j^2 \in L(a, \infty)$ $(1 \leqslant j \leqslant n-1)$. Then, by (2.10.1) and after an adjustment of a constant multiple, (2.10.14) gives

$$V_k = (Pp_k)^{\nu_k/\kappa_k}\{e_k + o(1)\} \exp I_k(x),$$

where $I_k = O(P\phi)$. By (2.10.9) and (2.10.12), the transformation back to Y gives solutions of (2.9.2) such that

$$Y_k = (Pp_k)^{\nu_k/\kappa_k} D_0\{v_k + o(1)\} \exp I_k,$$

where D_0 is defined by (2.9.12). Thus, as in § 2.6, the solutions of (2.9.2) involve the p_j in an algebraic manner. In general, the ν_k can only be found explicitly from (2.10.11) when $n \leqslant 4$. Accordingly we leave the details in this section as they stand, noting only that, in the example (2.9.26), the borderline case (2.10.1) occurs when $\alpha_1 + \cdots + \alpha_{n-1} - \beta = n$.

2.11 SUBDOMINANT COEFFICIENT q

To complete our discussion of (2.9.1), we develop the result which corresponds to the one proved for second-order equations with a subdominant q in Theorem 2.7.2. However, whereas Theorem 2.7.2 was simply a re-statement of Theorem 1.10.1 for a particular case, here we need some extension of the ideas in § 1.10. We write (2.9.2) as

$$Y' = (J + R)Y, \tag{2.11.1}$$

where R has the entry q in the $(n, 1)$ position and zeros elsewhere, while J has the entries p_j^{-1}. In passing, we note that J becomes an elementary matrix of Jordan type (1.10.1) when the p_j all coincide. Proceeding as in (1.2.12) and (1.10.13), we remove the matrix J from (2.11.1) by making the transformation

$$Y = \Phi Z, \tag{2.11.2}$$

where Φ is a fundamental matrix for the system

$$U' = JU.$$

Then (2.11.1) becomes

$$Z' = \Phi^{-1} R \Phi Z. \tag{2.11.3}$$

The most suitable choice of Φ for our purposes is the upper triangular matrix $\Phi = (\phi_{jk})$ defined by

$$\phi_{jk}(x) = 0 \qquad (j > k),$$
$$\phi_{jj}(x) = 1 \qquad (1 \leqslant j \leqslant n),$$

and

$$\phi_{jk}(x) = \int_a^x p_j^{-1}(t)\phi_{j+1,k}(t)\,\mathrm{d}t \qquad (1 \leqslant j \leqslant k-1,\ 2 \leqslant k \leqslant n). \tag{2.11.4}$$

Thus ϕ_{jk} in (2.11.4) is the repeated integral involving $p_j, \ldots, p_{k-1}$ obtained by integrating p_{k-1}^{-1} first and finishing with p_j^{-1}. We note that Φ reduces to the form (1.10.9) when the p_j all coincide.

To work out the inverse matrix Φ^{-1}, we define the functions ψ_{jk} by

$$\psi_{jk}(x) = 0 \qquad (j > k),$$
$$\psi_{jj}(x) = 1 \qquad (1 \leqslant j \leqslant n),$$

and

$$\psi_{jk}(x) = \int_a^x p_{k-1}^{-1}(t)\psi_{j,k-1}(t)\,\mathrm{d}t \qquad (j+1 \leqslant k \leqslant n,\ 1 \leqslant j \leqslant n-1). \tag{2.11.5}$$

Again, ψ_{jk} in (2.11.5) is the repeated integral involving $p_j, \ldots, p_{k-1}$, this time starting the integration with p_j^{-1} and finishing with p_{k-1}^{-1}. From (2.11.4), (2.11.5), and an integration by parts, we have

$$\begin{aligned}
\phi_{jk} &= \int_a^x \psi'_{j,j+1}\phi_{j+1,k}\,\mathrm{d}t \\
&= \psi_{j,j+1}\phi_{j+1,k} - \int_a^x \psi_{j,j+1}p_{j+1}^{-1}\phi_{j+2,k}\,\mathrm{d}t \\
&= \psi_{j,j+1}\phi_{j+1,k} - \int_a^x \psi'_{j,j+2}\phi_{j+2,k}\,\mathrm{d}t
\end{aligned}$$

for $j \leqslant k-1$. Continuing the integration by parts, we obtain

$$\phi_{jk} - \psi_{j,j+1}\phi_{j+1,k} + \psi_{j,j+2}\phi_{j+2,k} - \cdots + (-1)^{j+k}\psi_{jk}\phi_{kk} = 0.$$

Since also $\psi_{jj} = 1$, it follows that Φ^{-1} is the upper triangular matrix defined by

$$\Phi^{-1} = ((-1)^{j+k}\psi_{jk})$$

for all j and k in $[1, n]$.

Bearing in mind the definition of R, we now have

$$\Phi^{-1}R\Phi = qD_2ED_1$$

in (2.11.3), where

$$\begin{aligned} D_1 &= \mathrm{dg}\,(\phi_{1k}), \\ D_2 &= \mathrm{dg}\,((-1)^{k+n}\psi_{kn}), \end{aligned} \tag{2.11.6}$$

with $1 \leqslant k \leqslant n$, and E has all entries unity. Next we make the further transformation

$$Z = D_1^{-1}W \tag{2.11.7}$$

to obtain

$$W' = (D_1'D_1^{-1} + qD_1D_2E)W. \tag{2.11.8}$$

We shall of course arrange that D_1 is non-singular, and then (2.11.8) has a form similar to (1.10.15), now with $\Lambda = D_1'D_1^{-1}$. We also note at this point that D_1D_2 has zero trace, that is,

$$\sum_{k=1}^{n} (-1)^{k+n}\phi_{1k}\psi_{kn} = 0. \tag{2.11.9}$$

This follows because the left-hand side is the $(1, n)$ entry of $\Phi\Phi^{-1}$, which is the identity matrix.

By (2.11.6), it is only the entries ϕ_{1k} and ψ_{kn} of Φ and Φ^{-1} that occur in (2.11.8), and therefore we specify these entries in full now. By (2.11.4), ϕ_{1k} is the repeated integral σ_k given by

$$\sigma_k(x) = \int_a^x p_1^{-1}(t_1)\,\mathrm{d}t_1 \int_a^{t_1} \cdots \int_a^{t_{k-2}} p_{k-1}^{-1}(t_{k-1})\,\mathrm{d}t_{k-1} \tag{2.11.10}$$

for $k \geqslant 2$ and, of course, we define $\sigma_1(x) = 1$. Also, by (2.11.5), ψ_{kn} is the repeated integral τ_k given by

$$\tau_k(x) = \int_a^x p_{n-1}^{-1}(t_{n-1})\,\mathrm{d}t_{n-1} \int_a^{t_{n-1}} \cdots \int_a^{t_{k+1}} p_k^{-1}(t_k)\,\mathrm{d}t_k \tag{2.11.11$_1$}$$

for $k \leqslant n-1$, and $\tau_n(x) = 1$. On inverting the order of integration in (2.11.11$_1$), we can also write

$$\tau_k(x) = \int_a^x p_k^{-1}(t_k)\,\mathrm{d}t_k \int_{t_k}^x \cdots \int_{t_{n-2}}^x p_{n-1}^{-1}(t_{n-1})\,\mathrm{d}t_{n-1}, \tag{2.11.11$_2$}$$

and similarly for σ_k.

Theorem 2.11.1 *Let σ_k $(2 \leqslant k \leqslant n)$ be nowhere zero in some interval $[X, \infty)$. Also, for $X \leqslant t \leqslant x < \infty$,* let

(A)
$$|\sigma_k(x)/\sigma_k(t)| \geqslant K \tag{2.11.12}$$

and

(B) *for all unequal j and k in $[2, n]$, either*

$$|\sigma_k(x)\sigma_j(t)/\sigma_k(t)\sigma_j(x)| \leqslant K_1, \tag{2.11.13}$$

or

$$|\sigma_k(x)\sigma_j(t)/\sigma_k(t)\sigma_j(x)| \geqslant K, \tag{2.11.14}$$

where K and K_1 are positive constants. Finally, let

$$q\sigma_k\tau_k \in L(a, \infty) \tag{2.11.15}$$

for $1 \leqslant k \leqslant n$. Then (2.9.1) has solutions y_k $(1 \leqslant k \leqslant n)$ such that, as $x \to \infty$,

$$y_k(x) \sim \sigma_k(x). \tag{2.11.16}$$

Proof Since the σ_k are nowhere zero in $[X, \infty)$, D_1 is non-singular there, and therefore (2.11.8) is valid in this interval. We now apply the Levinson theorem, Theorem 1.3.1, to (2.11.8). By (2.11.6) and (2.11.15), the Levinson condition (1.3.3) holds. Also, here we have

$$\Lambda = D_1' D_1^{-1} = \text{dg}\,(\sigma_k'/\sigma_k),$$

and therefore (2.11.13) and (2.11.14) are just the dichotomy condition for the $\lambda_k = \sigma_k'/\sigma_k$ $(k \neq 1)$ in the form (1.3.11) and (1.3.12). Further, since $\sigma_1(x) = 1$, (2.11.12) is the dichotomy condition (1.3.12) for $\lambda_k - \lambda_1$. We note that the condition (1.3.11) would only be of relevance to $\lambda_k - \lambda_1$ if $\sigma_k(x) \to 0$ as $x \to \infty$ and, by choice of a in (2.11.10), it can be arranged that this does not occur.

It follows from Theorem 1.3.1 that (2.11.8) has solutions W_k of the form (1.3.5) with $\lambda_k = \sigma_k'/\sigma_k$. On carrying out the integration in the exponential and transforming back to Y via (2.11.7) and (2.11.2), we find that (2.11.1) has solutions

$$Y_k = \sigma_k \Phi D_1^{-1}\{e_k + o(1)\}. \tag{2.11.17}$$

By (2.11.6), the diagonal terms in D_1 form the first row in Φ, and hence the first component in (2.11.17) is

$$y_k = \sigma_k(1 \cdots 1)\{e_k + o(1)\} = \sigma_k\{1 + o(1)\}.$$

This completes the proof of (2.11.16). □

We note that the conditions (2.11.15) are not quite independent. By (2.11.9) we have

$$\sum_{k=1}^{n} (-1)^{k+n}\sigma_k\tau_k = 0,$$

and therefore one of the conditions is dependent on the others. We also note that the conditions are all implied by the single condition

$$q \prod_{j=1}^{n-1} \int_a^x |p_j^{-1}(s)| \, ds \in L(a, \infty). \tag{2.11.18}$$

This follows from the definition of σ_k and τ_k in (2.11.10) and (2.11.11). A further point is that the fixed limit of integration a in (2.11.4), (2.11.5), (2.11.10), and (2.11.11) can be replaced throughout by ∞ provided that all the integrals converge.

Example 2.11.1 The simplest example on the theorem is where the $p_j(x)$ all coincide: $p_j(x) = p(x)$ say. Then (2.11.10) and (2.11.11) give

$$\sigma_k(x) = \frac{1}{(k-1)!} \left(\int_a^x p^{-1}(s) \, ds \right)^{k-1},$$

$$\tau_k(x) = \frac{1}{(n-k)!} \left(\int_a^x p^{-1}(s) \, ds \right)^{n-k}.$$

The conditions (2.11.15) all coincide and become

$$q \left(\int_a^x p^{-1}(s) \, ds \right)^{n-1} \in L(a, \infty).$$

The dichotomy condition A and B is certainly satisfied if p is real-valued and positive. More generally, we can have $p^{-1} = p_1 + ip_2$, where p_1 (>0) and p_2 are real-valued and

$$\int_a^x p_2(s) \, ds = o\left(\int_a^x p_1(s) \, ds \right) \qquad (x \to \infty).$$

Again, the fixed limit of integration a can be replaced by ∞ in this example if p^{-1} has a convergent infinite integral.

Example 2.11.2 In the case of (2.9.4), where $p_m = p$ and the other p_j are unity, the forms of σ_k and τ_k are a little more complicated. The conditions (2.11.15) do not now coincide but, by (2.11.18), they are all implied by

$$x^{n-2} q(x) \int_a^x p^{-1}(s) \, ds \in L(a, \infty).$$

Theorem 2.11.1 covers the two cases where all the p_j are relatively small as $x \to \infty$ (finite a in (2.11.4) and (2.11.5)) and all the p_j are relatively large as $x \to \infty$ ($a = \infty$). The less regular situation where a mixture of these types of p_j occurs is not covered. In this situation, however, we can use a simpler result which corresponds to Theorem 2.7.1 but obtaining only less precise asymptotic formulae of the kind (2.7.8) and (2.7.9). We indicate the details briefly.

Corresponding to (2.7.10), we make a transformation

$$Y = DZ \tag{2.11.19}$$

where

$$D = \mathrm{dg}\,(1, \rho_1, \ldots, \rho_{n-1}).$$

Then (2.9.2) becomes

$$Z' = (-D^{-1}D' + D^{-1}AD)Z. \tag{2.11.20}$$

By (2.9.3), the Levinson condition (1.3.3) is satisfied by $D^{-1}AD$ if the ρ_j can be chosen so that

$$\rho_{n-1}^{-1}q, \quad \rho_{j-1}^{-1}p_j^{-1}\rho_j \quad (1 \leqslant j \leqslant n-1) \tag{2.11.21}$$

are all $L(a, \infty)$, where $\rho_0 = 1$. Then we obtain solutions Y_k of (2.9.2) such that

$$Y_k = \rho_{k-1}^{-1}D\{e_k + o(1)\}.$$

In terms of the components of Y_k, we therefore obtain a true asymptotic formula for the kth component, but only o-formulae for the other components. On taking the first component y_k of Y_k, we find that (2.9.1) has solutions such that

$$y_1 \to 1, \quad y_k = o(\rho_{k-1}^{-1}) \quad (2 \leqslant k \leqslant n). \tag{2.11.22}$$

A simple choice for ρ_j is $\rho_j(x) = x^{-j}$, and then (2.11.21) becomes

$$x^{n-1}q, \qquad x^{-1}p_j^{-1} \in L(a, \infty).$$

We also mention the example

$$\left(\mathrm{e}^x \frac{\mathrm{d}^2}{\mathrm{d}x^2}\right)^m y + qy = 0, \tag{2.11.23}$$

which occurs in a physical application. Here we have

$$p_1 = p_3 = \cdots = p_{2m-1} = 1, \qquad p_2 = \cdots = p_{2m-2} = \mathrm{e}^x.$$

Then (2.11.21) is satisfied if, for some $\alpha > 2m-1$,

$$\mathrm{e}^{-(m-1)x}x^{\alpha}q \in L(a, \infty)$$

and the choice for the ρ_j is

$$\rho_{2k+1}(x) = \mathrm{e}^{kx}x^{2(m-k-1)(1+\delta)-\alpha} \qquad (0 \leqslant k \leqslant m-1),$$
$$\rho_{2k}(x) = \mathrm{e}^{kx}x^{(2m-2k-1)(1+\delta)-\alpha} \qquad (1 \leqslant k \leqslant m-1),$$

where $0 < \delta < \{\alpha - (2m-1)\}/(2m-2)$.

NOTES AND REFERENCES

2.1 The transformation based on (2.1.8) goes back to Hartman and Wintner (1955, pp. 80–1) in the case $p = 1$. The second-order equation can also be investigated in the form $y'' + py' + qy = 0$ by means of the system (1.9.3), and the nature of the solutions depends on the relative sizes of p^2 and q (Walker, 1971*c*).

2.2 The formulae (2.2.10)–(2.2.11), which are the earliest of those in this section, are due to Wintner (1947*b*) in the case $p = 1$, and they were proved by means of the Liouville–Green transformation of § 2.5; see also Coppel (1965, pp. 118–20) for general p. The formulae are also proved by Pinto (1985) under alternative conditions which involve non-absolutely convergent integrals arising from a transformation based on (1.11.7).

The formulae (2.2.5)–(2.2.6) are due to Hartman and Wintner (1955, pp. 82–6), again with $p = 1$. Different proofs were given for the 'elliptic' and 'hyperbolic' cases, but both proofs allowed q to be complex-valued. Atkinson (1957/8) gave a simpler proof based on the same ideas as those in the text, except that a change of independent variable was also involved, with a consequent restriction to real-valued coefficients. The diagonalization which is implicit in the proof of Theorem 2.2.1 is worked out by Eastham (1983*a*, pp. 110–22) and Gingold (1983).

In the related WKBJ theory, there are formulae similar to (2.2.10)–(2.2.11) which give approximate solutions in the situation where x is fixed but q contains a large parameter. This theory has a longer history, of which an account is given by McHugh (1971) and, of the earlier papers, we mention Jeffreys (1924).

2.3 The method of repeated diagonalization is also successful in other aspects of the theory of (2.1.1). The method itself, together with applications in spectral and oscillation theory, has been developed by Harris (1978, 1982, 1983, 1984).

2.4 Theorem 2.4.1 is due to Eastham (1987). Earlier asymptotic results of this kind were obtained by Ben-Artzi (1980, 1983), who used the Liouville–Green transformation, and by Cassell (1985, 1986, 1988), who used the Prüfer transformation and other integral equation techniques. Both authors developed the idea of a repeated process and, although there are differences of detail, their conditions on the higher-order derivatives of p and q are broadly similar to those in Theorem 2.4.1. Ben-Artzi (1980, 1983) obtained the asymptotic formulae in the context of the spectral theory of the Schrödinger operator and therefore restricted the analysis to a real-valued q.

With the choice (2.4.32), Example 2.4.1 was investigated by Atkinson *et al.* (1977, §§ 6–7) using a repeated Liouville–Green transformation.

2.5 The papers in which the Liouville–Green transformation originated are those of Liouville (1837, especially pp. 22–3) and Green (1837); see also the historical accounts by Lutzen (1984, p. 343) and Schlissel (1977, pp. 311–4). Recent applications of the transformation to spectral theory are discussed by Everitt (1974, pp. 338–52, 1982).

Theorem 2.5.1 is due to Wintner (1947b) in the case $p = 1$; see also Barbuti (1954). We refer to Coppel (1965, pp. 118–20) for general p and the original proof of Lemma 2.5.1.

Theorem 2.5.2 is due to Cassell (1988, Theorem 1). The formulation of (2.1.1) as an integral equation by Cassell (1988) and Titchmarsh (1962, § 5.8) also avoids the change of variable (2.5.19). The asymptotic analysis of this integral equation is in effect a proof of the Levinson theorem for (2.5.27).

Explicit bounds for the error terms in the Liouville–Green asymptotic formulae may be obtained from the general estimates (1.4.24) and (1.4.25). However, in the case of (2.1.1) with $p = 1$, more refined bounds are given by Olver (1961, 1974, Chapter 6), Taylor (1978, 1982) and, with an alternative approach based on the Riccati transformation, Smith (1986). The asymptotic formulae can also be refined for more specific types of q such as polynomials (Hsieh and Sibuya, 1966). An interesting extension of the Liouville–Green formula in a quite different direction is given by Sultanaev (1984).

The example (2.5.28) is due to Cassell (1988, § 2).

The system (2.5.29) is in effect the basis for the paper by Read (1980) and, of the results in this paper, Corollary 2.3 can be obtained from (2.5.29) by making a transformation of the type (1.11.7). Read (1980) extends his range of results by replacing u by ue^v in (2.5.30) and making suitable choices for v as well as u.

Cassell (1988) gives the condition (2.5.31) together with a discussion of the results arising from the choice (2.5.32) and details of the iterated transformation which starts at (2.5.33).

2.6 Theorem 2.6.1 was in effect proved by Hartman and Wintner (1955, pp. 82–6) and Atkinson (1957/8) in the case $p = 1$. Here, however, we draw attention to the true nature of the theorem as representing a borderline case in which asymptotic formulae of algebraic type appear.

2.7 Results of the kind in this section go back to Bôcher (1900) who proved Theorem 2.7.2 in the case $p = 1$. Estimates for the error terms in the asymptotic formulae of Theorems 2.7.1 and 2.7.2 are given by Ràb (1972). Theorem 2.7.3 is due to Cassell (1988, § 3(e)) and Eastham (1988, § 4). In the case $p = 1$, a more precise version of Theorem 2.7.4 is due to Hille (1948) who established the explicit inequalities

$$|y_1 - 1| < \exp\left(\int_x^\infty t\,|q(t)|\,dt\right) - 1$$

and

$$|y_2 - x| < \exp\left(\int_x^\infty t^2 |q(t)| \,\mathrm{d}t\right) - 1,$$

provided that the infinite integrals converge.

In the case $p = 1$, the results based on (2.7.33) and (2.7.41) are similar to those of Hallam (1970, § 2). Also in the case $p = 1$, the results based on (2.7.43) are due to Hartman and Wintner (1953). An extension of Theorem 2.7.5 to non-linear second-order equations is given by Naito (1984).

For other specialized results with subdominant q, we refer to Wintner (1948, 1949*a*), Hartman (1982, pp. 375–81), Trench (1986), Šimša (1987), and Cassell (1988, § 3).

2.8 Theorem 2.8.1 is due to Hartman and Wintner (1955, p. 80); see also Coppel (1965, p. 128) and Eastham (1983*a*, p. 120). A similar result, with real-valued coefficients, is given by Pinto (1985, Theorem 6). The extension in Theorem 2.8.2 appears to be new here.

Theorem 2.8.3 was first proved by Bellman (1950, 1953, pp. 130–3) for the case $\nu = 3$. The proof given in the text is due to Harris and Lutz (1977, Example 4) again in the case $\nu = 3$. Further analysis is given by Ascoli (1953*b*) who obtains an iterative formula to determine the exponential factor corresponding to the one in (2.8.13) but valid for all $\nu > 1$.

Theorem 2.8.4 is due to Hartman and Wintner (1953) in the case $p = 1$. The extension given in Theorem 2.8.5 appears to be new here.

2.9 The equation (2.9.4) was first introduced by Hinton (1968). The asymptotic formula (2.9.28) is due to Fedorjuk (1966*a*) (in the case $p = 1$) and Hinton (1968); see also Gingold (1985*a*, § 3). Theorem 2.9.1 is due to Eastham (1983*b*). The earlier papers of Švec (1958, 1963) gave partial asymptotic information in the cases $n = 3$, 4. We also mention the paper of Braaksma (1971) which deals with (2.9.4) when $p(x) = 1$ and $q(x) = x^m$ in the wider context where x is complex and which provides earlier references for this case.

2.10 In the cases $n = 3$ and $n = 4$ (with all $p_j = 1$) the necessary eigenvalues and the resulting asymptotic formulae are worked out in detail by Hinton (1972).

2.11 Theorem 2.11.1 is due to Eastham (1988). It is also possible to apply the other general theorems of Chapter 1, besides the Levinson theorem, to (2.11.8) and (2.11.20). Some results of this kind, arising from Theorem 1.11.1, are given by Hinton (1984). The results of Eastham (1983*b*, § 4) are in essence covered by (2.11.19) and (2.11.22).

If we write $y = x^{\beta} z$, $p_j(x) = x^{\beta}$ (all j), $q(x) = (\text{const.})x^{-2\beta}$ in (2.9.1), the equation becomes one considered by Paris and Wood (1984, § 2, 1986, pp. 206–7) who obtained the asymptotic form of solutions using properties of generalized hypergeometric functions and Meijer G-functions. Here, Theorems 2.9.1 and 2.11.1 (with $a = \infty$ in the latter, if appropriate) cover the two cases where $\beta(n+1) - n$ is negative and positive respectively.

The case $m = 3$ of (2.11.23) arises in the theory of the gas centrifuge (Wood and Morton 1980), and attention was drawn to this example in the present context by Hinton (1984).

A matrix analogue of (2.9.1) can be formulated by replacing p_j^{-1} and q by $m \times m$ matrices P_j and Q in (2.9.3), and then Y becomes an mn-component vector. The ideas based on (2.11.2) continue to apply and they lead to a result similar to Theorem 2.11.1. The case $n = 2$ and $P_1 = I$ gives an asymptotic result of Shreve (1971, Lemma 1.2). Other more detailed results for higher-order matrix equations are given by Hallam (1970, 1971), Rasmussen (1979), and Tomastik (1984).

3
Equations of self-adjoint type

3.1 INTRODUCTION

In this chapter we develop further the application of the theorems in Chapter 1 to higher-order equations, and we consider first the even-order equation of self-adjoint type

$$(p_0 y^{(m)})^{(m)} + (p_1 y^{(m-1)})^{(m-1)} + \cdots + p_m y = 0. \tag{3.1.1}$$

This equation was introduced in (1.9.12), and we recall the system formulation

$$Y' = AY \tag{3.1.2}$$

in which, as stated in (1.9.13), the $2m \times 2m$ matrix A is defined by

$$\left.\begin{aligned} a_{i,i+1} &= 1 \qquad && (1 \leqslant i \leqslant 2m-1,\ i \neq m), \\ a_{m,m+1} &= 1/p_0, && \\ a_{2m-j+1,j} &= -p_{m-j+1} && (1 \leqslant j \leqslant m). \end{aligned}\right\} \tag{3.1.3}$$

The main complication which arises, as compared to the theory in § 2.9, is that the eigenvalues of A can no longer be determined explicitly in terms of the coefficients in the differential equation. Under suitable conditions on the coefficients, however, the size of the eigenvalues can be estimated with sufficient accuracy to allow the theory to proceed. A complete analysis would involve the consideration of a large number of cases, specified by the relative sizes of the coefficients, and we shall therefore confine the account to a few cases of particular interest. The basic conditions which guarantee the existence of solutions for (3.1.2) are that $1/p_0$ and p_r $(1 \leqslant r \leqslant m)$ should be locally Lebesgue integrable in $[a, \infty)$, and these conditons are therefore assumed throughout.

As on previous occasions in §§ 2.2 and 2.9, we develop the idea of expressing A in its diagonal form

$$T^{-1}AT = \Lambda \tag{3.1.4}$$

and making the transformation

$$Y = TZ \tag{3.1.5}$$

to obtain the system

$$Z' = (\Lambda + R)Z, \tag{3.1.6}$$

where

$$R = -T^{-1}T'. \tag{3.1.7}$$

Preparatory to the main analysis in this chapter, we note that, by (3.1.3), the characteristic equation $\det(\lambda I - A) = 0$ takes the convenient form

$$p_0\lambda^{2m} + p_1\lambda^{2m-2} + \cdots + p_m = 0. \tag{3.1.8}$$

We denote the eigenvalues and eigenvectors of A by λ_k and v_k $(1 \leqslant k \leqslant 2m)$, where the first component of v_k is unity. In the situations that we consider, the λ_k will be distinct. It is easy to check from the equation $Av_k = \lambda_k v_k$ that v_k has components

$$\left.\begin{array}{ll} \lambda_k^{l-1} & (1 \leqslant l \leqslant m), \\ p_0\lambda_k^{l-1} + p_1\lambda_k^{l-3} + \cdots + p_{l-m-1}\lambda_k^{2m+1-l} & (m+1 \leqslant l \leqslant 2m), \end{array}\right\} \tag{3.1.9}$$

and then the matrix T in (3.1.4) is given by

$$T = (v_1 \cdots v_{2m}). \tag{3.1.10}$$

As already noted in § 1.9, the matrix A in (3.1.2) and (3.1.3) has the form (1.2.5), with the b_{ij} $(i + j = 2m + 2)$ all unity except for $b_{m+1,m+1} = 1/p_0$. The inverse matrix T^{-1} can therefore be written down from Lemmas 1.2.1 and 1.2.2. Thus T^{-1} has the rows

$$M_j^{-1}w_j, \tag{3.1.11}$$

where w_j has the entries

$$\left.\begin{array}{ll} p_0\lambda_j^{2m-l} + p_1\lambda_j^{2m-l-2} + \cdots + p_{m-l}\lambda_j^l & (1 \leqslant l \leqslant m) \\ \lambda_j^{2m-l} & (m+1 \leqslant l \leqslant 2m) \end{array}\right\} \tag{3.1.12}$$

and, by Lemma 1.2.2 and (3.1.8),

$$M_j = 2mp_0\lambda_j^{2m-1} + (2m-2)p_1\lambda_j^{2m-3} + \cdots + 2p_{m-1}\lambda_j. \tag{3.1.13}$$

We can now go on to work out the matrix $T^{-1}T'$ for use in (3.1.6) and, in the following lemma, we require the function of two variables $M(x, \lambda)$ defined by

$$M(x, \lambda) = \sum_{r=0}^{m-1} (2m - 2r)p_r(x)\lambda^{2m-2r-1}. \tag{3.1.14}$$

We note that (3.1.13) gives $M_j(x) = M(x, \lambda_j(x))$.

Lemma 3.1.1 *The* (j, k) *entries* τ_{jk} *in* $T^{-1}T'$ *are given by*

(i) $$\tau_{jj} = \tfrac{1}{2}M_j'/M_j; \tag{3.1.15}$$

(ii) $$\tau_{jk} = (\lambda_j - \lambda_k)^{-1}M_j^{-1}\sum_{r=0}^{m} p_r'(\lambda_j\lambda_k)^{m-r} \qquad (j \neq k). \tag{3.1.16}$$

Proof (i) By (3.1.9), (3.1.11), and (3.1.12), we have

$$M_j\tau_{jj} = \lambda_j'\left\{\sum_{l=2}^{m}\left(\sum_{r=0}^{m-l} p_r\lambda_j^{2m-l-2r}\right)(l-1)\lambda_j^{l-2}\right.$$
$$\left. + \sum_{l=m+1}^{2m}\left(\sum_{r=0}^{l-m-1}(l-2r-1)p_r\lambda_j^{l-2r-2}\right)\lambda_j^{2m-l}\right\}$$
$$+ \sum_{l=m+1}^{2m}\left(\sum_{r=0}^{l-m-1} p_r'\lambda_j^{l-2r-1}\right)\lambda_j^{2m-1}.$$

On changing the order of the summations, we obtain

$$M_j\tau_{jj} = \lambda_j'\left\{\sum_{r=0}^{m-1}\left(\sum_{l=2}^{m-r}(l-1)\right)p_r\lambda_j^{2m-2r-2}\right.$$
$$\left. + \sum_{r=0}^{m-1}\left(\sum_{l=r+1}^{m}(l+m-2r-1)\right)p_r\lambda_j^{2m-2r-2}\right\}$$
$$+ \sum_{r=0}^{m-1}(m-r)p_r'\lambda_j^{2m-2r-1} \tag{3.1.17}$$
$$= \lambda_j'\sum_{r=0}^{m-1}(2m-2r-1)(m-r)p_r\lambda_j^{2m-2r-2} + \sum_{r=0}^{m-1}(m-r)p_r'\lambda_j^{2m-2r-1}$$
$$= \frac{1}{2}\left(\frac{\partial M}{\partial\lambda}\lambda' + \frac{\partial M}{\partial x}\right)_{\lambda=\lambda_j} = \tfrac{1}{2}M_j',$$

and (3.1.15) follows.

(ii) When $j \neq k$, (3.1.17) is replaced by

$$M_j\tau_{jk} = \lambda_k'\left\{\sum_{r=0}^{m-1}\left(\sum_{l=2}^{m-r}(l-1)\lambda_j^{2m-l-2r}\lambda_k^{l-2}\right)p_r\right.$$
$$\left. + \sum_{r=0}^{m-1}\left(\sum_{l=r+1}^{m}(l+m-2r-1)\lambda_j^{m-l}\lambda_k^{l+m-2r-2}\right)p_r\right\}$$
$$+ \sum_{r=0}^{m-1}\left(\sum_{l=r+1}^{m}\lambda_j^{m-l}\lambda_k^{l+m-2r-1}\right)p_r'. \tag{3.1.18}$$

In the coefficient of λ_k', the two l-summations combine to give a standard summation

$$\sum_{l=2}^{2m-2r}(l-1)\lambda_j^{2m-l-2r}\lambda_k^{l-2} = (\lambda_j-\lambda_k)^{-2}(\lambda_j^{2m-2r}-\lambda_k^{2m-2r})$$
$$-(\lambda_j-\lambda_k)^{-1}(2m-2r)\lambda_k^{2m-2r-1}.$$

On multiplying this last equation by p_r and summing for $0 \leqslant r \leqslant m-1$, we find that the term involving $(\lambda_j-\lambda_k)^{-2}$ is zero since λ_j and λ_k are solutions of (3.1.8). Hence the term involving λ'_k in (3.1.18) is

$$-(\lambda_j-\lambda_k)^{-1}\lambda'_k\sum_{r=0}^{m-1}(2m-2r)p_r\lambda_k^{2m-2r-1}. \tag{3.1.19}$$

The final term in (3.1.18) again involves a standard l-summation, and this term is therefore

$$-(\lambda_j-\lambda_k)^{-1}\left\{\sum_{r=0}^{m-1}p'_r\lambda_k^{2m-2r}-\sum_{r=0}^{m-1}p'_r(\lambda_j\lambda_k)^{m-r}\right\}. \tag{3.1.20}$$

Now, apart from the factor $(\lambda_j-\lambda_k)^{-1}$, (3.1.19) combines with the first summation in (3.1.20) to make

$$\frac{\mathrm{d}}{\mathrm{d}x}\left(\sum_{r=0}^{m-1}p_r(x)\lambda_k^{2m-2r}(x)\right)=-p'_m(x),$$

by (3.1.8) with $\lambda=\lambda_k$. Hence (3.1.18) reduces to

$$M_j\tau_{jk}=(\lambda_j-\lambda_k)^{-1}\sum_{r=0}^{m}p'_r(\lambda_j\lambda_k)^{m-r},$$

and (3.1.16) follows. □

3.2 EIGENVALUES OF THE SAME MAGNITUDE

The form of R given by (3.1.7) and Lemma 3.1.1 is in general much more complicated than in the second-order case (2.1.9). There is however an important class of equations (3.1.1) for which the eigenvalues $\lambda_k(x)$ all have the same order of magnitude as $x\to\infty$, and then the analysis of (3.1.6) simplifies considerably to make possible an immediate extension of Corollary 2.2.1. This result is given in the following theorem, in which we require the function P defined by

$$P=(p_m/p_0)^{1/2m}. \tag{3.2.1}$$

Theorem 3.2.1 *Let p_r $(0\le r\le m)$ have locally absolutely continuous first derivatives in $[a,\infty)$, and let p_0 and p_m be nowhere zero in $[a,\infty)$. Also, let*

(i) $$p_r(x)/\{p_0(x)P^{2r}(x)\}\to c_r \tag{3.2.2}$$

as $x\to\infty$, where c_r is a finite limit;
(ii) *the polynomial*

$$g(\xi)=c_0\xi^{2m}+c_1\xi^{2m-2}+\cdots+c_m \tag{3.2.3}$$

have 2m distinct roots ξ_k $(1 \leq k \leq 2m)$;

(iii) $$p_r' p_0^{-1} P^{-2r-1} = o(1) \qquad (x \to \infty); \tag{3.2.4}$$

(iv) $$p_r'' p_0^{-1} P^{-2r-1} \in L(a, \infty); \tag{3.2.5}$$

(v) $$p_r'^2 p_0^{-2} P^{-4r-1} \in L(a, \infty). \tag{3.2.6}$$

Finally, let

$$\operatorname{re}\{\lambda_j(x) - \lambda_k(x)\} \tag{3.2.7}$$

have one sign in $[a, \infty)$ *for each unequal pair* j, k *in* $[1, 2m]$, *where the* λ_k *are the solutions of* (3.1.8). *Then* (3.1.1) *has solutions* $y_k(x)$ $(1 \leq k \leq 2m)$ *such that, as* $x \to \infty$,

$$y_k \sim (p_0 p_m^{2m-1})^{-1/4m} \exp\left(\int_a^x \lambda_k(t)\, dt\right). \tag{3.2.8}$$

Proof First we use (3.2.2) to determine the asymptotic form of the $\lambda_k(x)$. On substituting

$$\lambda = P\mu \tag{3.2.9}$$

in (3.1.8), we obtain

$$\sum_{r=0}^{m} p_r P^{2m-2r} \mu^{2m-2r} = 0$$

and hence, by (3.2.2),

$$\sum_{r=0}^{m} c_r \{1 + o(1)\} \mu^{2m-2r} = 0.$$

By (3.2.3), this equation has solutions μ_k such that $\mu_k \to \xi_k$ as $x \to \infty$ and then, by (3.2.9), we obtain

$$\lambda_k \sim P\xi_k. \tag{3.2.10}$$

The aim now is to apply Theorem 1.6.1 to the system (3.1.6) and, to do this, we require asymptotic estimates for M_j and M_j', which occur in Lemma 3.1.1, and for λ_j'. First, from (3.1.13) and (3.2.2), we obtain immediately

$$M_j \sim \lambda_j^{-1} p_0 P^{2m} \xi_j\, \partial g / \partial \xi_j$$

and then, by (3.2.2) again,

$$M_j \sim p_0 P^{2m-1}\, \partial g / \partial \xi_j. \tag{3.2.11}$$

Next, on taking $\lambda = \lambda_j(x)$ in (3.1.8) and differentiating, we have

$$\lambda_j' = -M_j^{-1} \sum_{r=0}^{m} p_r' \lambda_j^{2m-2r} \tag{3.2.12}$$

$$= O\left(|P/p_0| \sum_{r=0}^{m} |p_r'|\, |P|^{-2r}\right), \tag{3.2.13}$$

by (3.2.10) and (3.2.11). Again, on differentiating (3.1.13), we have

$$M_j' = \lambda_j' \sum_{r=0}^{m-1} (2m-2r)(2m-2r-1)p_r\lambda_j^{2m-2r-2} + \sum_{r=0}^{m-1} (2m-2r)p_r'\lambda_j^{2m-2r-1}$$

$$= O\left(|P|^{2m-1} \sum_{r=0}^{m} |p_r'|\,|P|^{-2r}\right), \qquad (3.2.14)$$

by (3.2.2), (3.2.10), and (3.2.13).

We also require an estimate for M_j''. We differentiate (3.2.12) and use (3.2.10), (3.2.11), (3.2.13), and (3.2.14) to obtain

$$\lambda_j'' = O\left(|P/p_0^2| \sum_{r=0}^{m} |p_r'|^2\,|P|^{-4r}\right) + O\left(|P/p_0| \sum_{r=0}^{m} |p_r''|\,|P|^{-2r}\right).$$

Then we differentiate (3.1.13) twice and use the above estimates for λ_j, λ_j', λ_j'' to obtain

$$M_j'' = O\left(|P^{2m-1}/p_0| \sum_{r=0}^{m} |p_r'|^2\,|P|^{-4r}\right)$$

$$+ O\left(|P|^{2m-1} \sum_{r=0}^{m} |p_r''|\,|P|^{-2r}\right) \qquad (3.2.15)$$

after a straighforward calculation.

We can now verify the conditions (1.6.1) and (1.6.2) in Theorem 1.6.1. In the situation under consideration here, these conditions (with a slight change of notation) are

$$(\lambda_i - \lambda_l)^{-1}\tau_{jk} \to 0 \qquad (3.2.16)$$

and

$$\{(\lambda_i - \lambda_l)^{-1}\tau_{jk}\}' \in L(a, \infty) \qquad (3.2.17)$$

for all unequal i and l, where the τ_{jk} are given by (3.1.15) and (3.1.16). When $j \neq k$, we have

$$(\lambda_i - \lambda_l)^{-1}\tau_{jk} = O\left(P^{-2}M_j^{-1} \sum_{r=0}^{m} |p_r'|\,|P|^{2m-2r}\right)$$

$$= o(P^{2m-1}p_0M_j^{-1}) = o(1),$$

by (3.2.10), (3.2.4), and (3.2.11). The same result holds for $j = k$, by (3.1.15) and (3.2.14). Thus (3.2.16) is satisfied. In the case of (3.2.17), we carry out the differentiation on the left-hand side and substitute for the various derivatives of M_j and the λs from (3.2.13), (3.2.14), and (3.2.15). After a simple calculation, we obtain

$$\{(\lambda_i - \lambda_l)^{-1}\tau_{jk}\}' = O\left(P^{-1}p_0^{-1} \sum_{r=0}^{m} |p_r''|\,|P|^{-2r}\right)$$

$$+ O\left(P^{-1}p_0^{-2} \sum_{r=0}^{m} |p_r'|^2\,|P|^{-4r}\right),$$

and the right-hand side is $L(a, \infty)$ by (3.2.5) and (3.2.6).

We note further that the simplifying condition (1.6.35) is also satisfied. In the situation here, this condition is

$$(\lambda_i - \lambda_l)^{-1}\tau_{jk}^2 \in L(a, \infty)$$

and, by (3.1.15), (3.1.16), (3.2.10), (3.2.11), and (3.2.14), the condition reduces to

$$P^{-1}p_0^{-2}\sum_{r=0}^{m} |p_r'|^2\,|P|^{-4r} \in L(a, \infty),$$

which is guaranteed by (3.2.6). Hence, subject to some form of the dichotomy condition, we can apply Theorem 1.6.1 in the form (1.6.37) to obtain solutions Z_k of (3.1.6) such that

$$Z_k(x) = \{e_k + o(1)\}\exp\left(\int_a^x \{\lambda_k(t) - \tfrac{1}{2}M_k'(t)/M_k(t)\}\,dt\right),$$

where we have used (3.1.15) for the diagonal entries in R. On carrying out the integration involving M_k, transforming back to Y via (3.1.5) and using (3.2.11), we obtain solutions Y_k of (3.1.2) such that

$$Y_k = (p_0P^{2m-1})^{-1/2}T\{e_k + o(1)\}\exp\left(\int_a^x \lambda_k(t)\,dt\right) \qquad (3.2.18)$$

after adjustment of a constant multiple. Then (3.2.8) follows when we take the first component, bearing in mind that the entries in the first row of T in (3.1.10) are all unity, and recalling the definition of P in (3.2.1).

It remains to show that the functions

$$\lambda_k - \tfrac{1}{2}M_k'/M_k \qquad (= \tilde{\lambda}_k, \text{ say})$$

satisfy the dichotomy condition L in § 1.3. Suppose for example that, in (3.2.7), j and k are such that $\text{re}\,\{\lambda_j(x) - \lambda_k(x)\} \geqslant 0$ in $[a, \infty)$. Then, referring to (1.3.2), we have

$$\begin{aligned}\int_t^x \text{re}\,\{\tilde{\lambda}_j(s) - \tilde{\lambda}_k(s)\}\,ds &\geqslant \tfrac{1}{2}\,\text{re}\int_t^x \left(\frac{M_k'(s)}{M_k(s)} - \frac{M_j'(s)}{M_j(s)}\right)ds \\ &= \tfrac{1}{2}\log|M_k(x)M_j(t)/M_k(t)M_j(x)| \\ &\geqslant (\text{const.})\end{aligned}$$

since, by (3.2.11), M_k/M_j converges to a non-zero constant at ∞. This completes the proof of Theorem 3.2.1. □

We note that (3.2.7) is certainly satisfied (perhaps with a fresh choice of the point a) if

$$\text{re}\,\{(\xi_j - \xi_k)P\} \qquad (3.2.19)$$

is nowhere zero in $[a, \infty)$. This follows from (3.2.10). Turning to the conditions on the p_r in the theorem, the conditions (3.2.4)–(3.2.6) represent some restriction on the oscillatory nature of the p_r in terms of the size of P, and this limitation is similar to the one discussed in § 2.2(c). We also note that (3.2.2) is significant only for $1 \leqslant r \leqslant m-1$, and it then restricts the size of these p_r in terms of p_0 and p_m. As an example, we take

$$p_r(x) = (\text{const.})x^{\alpha_r} \qquad (0 \leqslant r \leqslant m). \tag{3.2.20}$$

Then the conditions on the p_r in the theorem are all satisfied when

$$\alpha_m - \alpha_0 > -2m \tag{3.2.21}$$

and

$$\alpha_r \leqslant \{r\alpha_m + (m-r)\alpha_0\}/m \qquad (1 \leqslant r \leqslant m-1). \tag{3.2.22}$$

The theorem covers the case where some of the c_r are zero in (3.2.2), that is, some p_r are small compared to p_0 and p_m in a pointwise sense. With a slight extension of the argument we can also deal with the situation where the relative smallness of some of the p_r is taken in an $L(a, \infty)$ sense. No conditions are required on the derivatives of such p_r in the following theorem.

Theorem 3.2.2 *Let σ be a set of integers in* $[1, m-1]$. *Let* (*i*)–(*v*) *of Theorem* 3.2.1 *hold for $r \notin \sigma$ except that, in place of* (3.2.3),

$$g(\xi) = \sum_{r \notin \sigma} c_r \xi^{2m-2r}. \tag{3.2.23}$$

Let the conditions involving (3.2.7) *hold, where the λ_k are now the solutions of*

$$\sum_{r \notin \sigma} p_r \lambda^{2m-2r} = 0. \tag{3.2.24}$$

Finally, for $r \in \sigma$, let

$$p_r p_0^{-1} P^{-2r+1} \in L(a, \infty). \tag{3.2.25}$$

Then (3.2.8) *continues to hold with the new λ_k.*

Proof We write (3.1.2) as

$$Y' = (A_1 + A_2)Y,$$

where the p_r ($r \in \sigma$) have been separated out to form the $(2m - r + 1, r)$ entries of the matrix A_2, of which the other entries are zero. Then, as in (3.1.4) and (3.1.5), we express A_1 in its diagonal form $T_1^{-1}A_1T_1 = \Lambda_1$ and make the transformation $Y = T_1Z$ to obtain the system

$$Z' = (\Lambda_1 - T_1^{-1}T_1' - T_1^{-1}A_2T_1)Z. \tag{3.2.26}$$

Here T_1 is defined by (3.1.9) and (3.1.10) except that the p_r ($r \in \sigma$) do not appear. The formula (3.2.10) continues to hold with the new λ_k and ξ_k, and it is then simple to check that the entries involving a particular p_r in $T_1^{-1}A_2T_1$ are $O(p_r p_0^{-1} P^{-2r+1})$. Hence (3.2.25) implies that

$$T_1^{-1}A_2T_1 \in L(a, \infty).$$

Thus (3.2.26) is a system of the form (1.6.42), and Theorem 1.6.1 can be applied as before. □

In the example (3.2.20), the condition (3.2.25) is

$$\alpha_r < \{r\alpha_m + (m-r)\alpha_0 - \tfrac{1}{2}(\alpha_m - \alpha_0 + 2m)\}/m \qquad (3.2.27)$$

for $r \in \sigma$ and, because of (3.2.21), this is a stronger condition than (3.2.22) for these r.

In the case where σ consists of all the integers in $[1, m-1]$, (3.2.23) and (3.2.24) are simply

$$g(\xi) = \xi^{2m} + 1 \qquad \text{and} \qquad p_0\lambda^{2m} + p_m = 0.$$

Thus

$$\lambda_k = \omega_k(-p_m/p_0)^{1/2m}, \qquad (3.2.28)$$

where the ω_k are the $2m$th roots of unity.

As a final remark in this section, we note that (3.2.8) was obtained by considering the first component in (3.2.18). Other components can also be selected and then, by (1.9.15), asymptotic formulae for the quasi-derivatives of y_k are obtained. Now the quasi-derivatives (1.9.15) coincide with the ordinary derivatives for $1 \leqslant \nu \leqslant m+1$, except for a factor p_0 when $\nu = m+1$. Hence, from (3.2.18) and (3.1.9), we obtain

$$y_k^{(\nu-1)} \sim \lambda_k^{\nu-1}(p_0 p_m^{2m-1})^{-1/4m} \exp\left(\int_a^x \lambda_k(t)\, \mathrm{d}t\right) \qquad (3.2.29)$$

for $1 \leqslant \nu \leqslant m+1$.

3.3 EIGENVALUES OF DIFFERING MAGNITUDES

In this section, we impose a fresh set of conditions on the coefficients p_r under which (3.1.8) can again be solved asymptotically for the λ_k. In contrast to (3.2.10) however, the λ_k now have differing orders of magnitude as $x \to \infty$, and the analysis of (3.1.6) becomes more complicated. One difference is that the final asymptotic formulae for (3.1.1) necessarily arise in terms of various derivatives of the solutions, and therefore the selection of the appropriate component in the equation corresponding to (3.2.18) assumes a greater significance than previously. In the following lemma, we introduce the new conditions on the p_r and determine the λ_k asymptotically.

Lemma 3.3.1 *Let all the p_r be nowhere zero in $[a, \infty)$, and let*

$$p_{r+1}/p_r = o(p_r/p_{r-1}) \qquad (x \to \infty, \quad 1 \leqslant r \leqslant m-1). \tag{3.3.1}$$

Then, with the λ_k written in non-increasing order of magnitude as $x \to \infty$,

$$\lambda_k^2 \sim -p_K/p_{K-1} \qquad (k = 2K-1, 2K). \tag{3.3.2}$$

Proof From (3.3.1) we have

$$p_r/p_{r-1} = o(p_\nu/p_{\nu-1})$$

for $r > \nu$. On taking the product of these equations with ν fixed and r varying, we obtain

$$p_r/p_\nu = o\{(p_\nu/p_{\nu-1})^{r-\nu}\},$$

which we rewrite as

$$p_r(p_\nu/p_{\nu-1})^{\nu-r-1} = o(p_{\nu-1}). \tag{3.3.3}$$

Again, for $r < \nu - 1$, (3.3.1) gives

$$p_\nu/p_{\nu-1} = o(p_{r+1}/p_r)$$

and, taking the product as before, we find that (3.3.3) holds for $r < \nu - 1$ also. Thus we have

$$p_r(p_\nu/p_{\nu-1})^{\nu-r-1} = o(p_{\nu-1}) \qquad (r \neq \nu - 1, \nu). \tag{3.3.4}$$

We can now prove (3.3.2) for a specific K. On substituting

$$\lambda^2 = -(p_K/p_{K-1})(1 + \eta_K) \tag{3.3.5}$$

in (3.1.8) and dividing by $(-p_K/p_{K-1})^{m-K+1}(1+\eta_K)^{m-K}$, we obtain

$$\left(\sum_{r=0}^{K-2} + \sum_{r=K+1}^{m}\right) p_r(-p_K/p_{K-1})^{K-r-1}(1+\eta_K)^{K-r} + p_{K-1}\eta_K = 0$$

as an equation to be solved for η_K. By (3.3.4) with $\nu = K$, this equation for η_K takes the form

$$\eta_K = \left(\sum_{0}^{K-2} + \sum_{K+1}^{m}\right) o(1)(1+\eta_K)^{K-r}.$$

There is certainly a solution $\eta_K = o(1)$, and then (3.3.2) follows from (3.3.5). □

We note that (3.3.1) and (3.3.2) imply that

$$\lambda_j = o(\lambda_k) \tag{3.3.6}$$

when $j = 2J - 1$, $2J$ and $J > K$. We also require a second lemma which involves M_j and various powers of the λ_j.

Lemma 3.3.2 *Let K be a given integer in* $[1, m]$. *Then*

$$\lambda_j^{m-K} M_j^{-1/2} = o(\lambda_l^{m-K} M_l^{-1/2}), \tag{3.3.7}$$

where $j = 2J-1,\ 2J$ *and* $l = 2L-1,\ 2L$ with either $J < L \leq K$ or $J > L \geq K$.

Proof It is sufficient to consider the two cases

$$J+1 = L \leq K \qquad \text{and} \qquad J-1 = L \geq K.$$

By (3.1.13), (3.3.2) and (3.3.4) (with $\nu = J$), we have

$$M_j \sim -2p_J \lambda_j^{2m-2J-1}. \tag{3.3.8}$$

Hence, in the first case where $J+1 = L$, (3.3.7) becomes

$$p_J^{-1}(p_J/p_{J-1})^{J-K+\frac{1}{2}} = o\{p_{J+1}^{-1}(p_{J+1}/p_J)^{J-K+\frac{3}{2}}\},$$

where we have used (3.3.2) again. Then a re-arrangement gives

$$(p_{J+1}/p_J)^{K-J-\frac{1}{2}} = o\{(p_J/p_{J-1})^{K-J-\frac{1}{2}}\}$$

and, when $K > J$, this is an immediate consequence of (3.3.1). The proof for the second case is almost identical. □

We are now in a position to establish asymptotic formulae for the solutions of (3.1.1) under the conditions (3.3.1), but first we indicate why the matrix T, as defined in (3.1.10), is no longer entirely suitable in the present situation. The choice $K = m$ in (3.3.7) gives

$$M_j^{-1/2} = o(M_l^{-1/2}) \tag{3.3.9}$$

for $J < L$. Consequently, comparing τ_{jk} and τ_{kj} in (3.1.16), we see that one is large compared to the other and, although the smaller may be suitable for the application of Theorem 1.6.1 to (3.1.6), the larger will generally not be. In order to balance the sizes of τ_{jk} and τ_{kj}, we shall replace T by another matrix T_1 whose columns are suitable scalar multiples of the v_j in (3.1.9). In the theorem which follows, we require the function δ defined by

$$\delta = \max_r |p_{r+1}p_{r-1}/p_r^2|, \tag{3.3.10}$$

and then $\delta = o(1)$ by (3.3.1).

Theorem 3.3.1 *Let* p_r $(0 \leq r \leq m)$ *be nowhere zero with locally absolutely continuous first derivatives in* $[a, \infty)$. *Let* (3.3.1) *hold, and let*

(i) $$\delta^{1/2} p_r'/p_r \in L(a, \infty) \tag{3.3.11}$$

(ii) $$(p_{r-1}p_r)'/(p_{r-1}p_r) = o\{(p_r/p_{r-1})^{1/2}\} \tag{3.3.12}$$

(iii) $$\{p_{r-1}^{-1/2} p_r^{-3/2}(p_{r-1}p_r)'\}' \in L(a, \infty) \tag{3.3.13}$$

(iv) $$p_{r-1}^{-3/2} p_r^{-5/2}(p_{r-1}p_r)'^2 \in L(a, \infty). \tag{3.3.14}$$

Finally, let

$$\mathrm{re}\,\{\lambda_j(x) - \lambda_k(x)\} \tag{3.3.15}$$

have one sign in $[a, \infty)$ *for each unequal pair* j, k *in* $[1, 2m]$, *where the* λ_k *are the solutions of* (3.1.8), *ordered according to* (3.3.6). *Then* (3.1.1) *has solutions* $y_k(x)$ $(k = 2K-1,\ 2K,\ 1 \leqslant K \leqslant m)$ *such that, as* $x \to \infty$,

$$y_k^{(m-K)} \sim (p_{K-1}p_K)^{-1/4} \exp\left(\int_a^x \lambda_k(t)\,\mathrm{d}t\right). \tag{3.3.16}$$

Proof Let M be the diagonal matrix

$$M = \mathrm{dg}\,(M_1^{-1/2}, \ldots, M_{2m}^{-1/2}) \tag{3.3.17}$$

and let T be as in (3.1.10). We make the transformation

$$Y = T_1 Z \tag{3.3.18}$$

in (3.1.2), where

$$T_1 = TM. \tag{3.3.19}$$

The columns of T_1 are still eigenvectors of A, and therefore the transformed system is

$$Z' = (\Lambda + R_1)Z, \tag{3.3.20}$$

where

$$\begin{aligned} R_1 &= -T_1^{-1}T_1' \\ &= -M^{-1}T^{-1}T'M - M^{-1}M', \end{aligned} \tag{3.3.21}$$

by (3.3.19). It follows from (3.3.17) and Lemma 3.1.1 that $T_1^{-1}T_1'$ is a skew-symmetric matrix whose entries σ_{jk} are given by

$$\sigma_{jj} = 0 \tag{3.3.22}$$

and, for $j \neq k$,

$$\sigma_{jk} = (\lambda_j - \lambda_k)^{-1} M_j^{-1/2} M_k^{-1/2} \sum_{r=0}^{m} p_r'(\lambda_j\lambda_k)^{m-r}. \tag{3.3.23}$$

We have now to estimate the size of the σ_{jk} and we begin by noting that, with δ defined by (3.3.10), the o-estimates in the proof of Lemma 3.3.1 are, more precisely, $O(\delta)$ estimates. Thus (3.3.4) becomes

$$p_r(p_\nu/p_{\nu-1})^{\nu-r-1} = O(\delta p_{\nu-1}) \qquad (r \neq \nu - 1,\ \nu) \tag{3.3.24}$$

and, consequently, (3.3.2) and (3.3.8) become

$$\lambda_K^2 = -(p_K/p_{K-1})\{1 + O(\delta)\} \tag{3.3.25}$$

and

$$M_j = -2p_J\lambda_j^{2m-2J-1}\{1 + O(\delta)\}. \tag{3.3.26}$$

There are two cases to consider concerning σ_{jk}. The first is where j and

k are as in (3.3.6) and the second is where $j = k + 1 = 2J$, making $\lambda_j = -\lambda_k$. In the first case, by (3.3.23) and (3.3.8), we have

$$\sigma_{jk} = O\left(p_J^{-1/2} p_K^{-1/2} \lambda_j^{-m+J+\frac{1}{2}} \lambda_k^{-m+K-\frac{1}{2}} \sum_{r=0}^{m} |p_r'|\, |\lambda_j \lambda_k|^{m-r}\right).$$

Now

$$\begin{aligned} p_r'(\lambda_j \lambda_k)^{m-r} &= (p_r'/p_r)(p_r \lambda_j^{2m-2r} p_r \lambda_k^{2m-2r})^{1/2} \\ &= (p_r'/p_r) O\{(\delta p_J \lambda_j^{2m-2J} p_K \lambda_k^{2m-2K})^{1/2}\} \end{aligned}$$

by (3.3.24), and δ (rather than δ^2) arises because of the possibilities $r = J = K + 1$ and $r = K = J - 1$. Hence we obtain

$$\sigma_{jk} = O(\delta^{1/2} \lambda_j^{1/2} \lambda_k^{-1/2} \max_r |p_r'/p_r|), \tag{3.3.27}$$

and therefore

$$\sigma_{jk} \in L(a, \infty) \tag{3.3.28}$$

for the j and k under consideration, by (3.3.6) and (3.3.11).

Next we consider σ_{jk} when $\lambda_j = -\lambda_k$. By (3.1.13), we then have $M_j = -M_k$ and, with the square roots in (3.2.23) chosen so that

$$M_{2J}^{1/2} = \mathrm{i} M_{2J-1}^{1/2}, \tag{3.3.29}$$

we obtain

$$\begin{aligned} \sigma_{2J,2J-1} &= -\sigma_{2J-1,2J} \\ &= -\tfrac{1}{4}\mathrm{i} p_J^{-1} \lambda_{2J}^{-2m+2J} \{1 + O(\delta)\} \sum_{r=0}^{m} p_r'(-1)^{m-r} \lambda_{2J}^{2m-2r}, \end{aligned}$$

on using (3.3.26). Except for $r = J - 1$ and $r = J$, the terms arising from the summation here are $L(a, \infty)$, as for (3.3.28). Then, apart from the $L(a, \infty)$ terms, we have

$$\begin{aligned} \sigma_{2J,2J-1} &= -\tfrac{1}{4}\mathrm{i} p_J^{-1}(-1)^{m-J}\{1 + O(\delta)\}\{p_J' - p_{J-1}' \lambda_{2J}^2\} \\ &= -\tfrac{1}{4}\mathrm{i}(-1)^{m-J}\{1 + O(\delta)\}[p_J'/p_J + \{1 + O(\delta)\} p_{J-1}'/p_{J-1}] \\ &= -\tfrac{1}{4}\mathrm{i}(-1)^{m-J}(p_{J-1}p_J)'/(p_{J-1}p_J), \end{aligned} \tag{3.3.30}$$

again neglecting additive terms which, by (3.3.11), are $L(a, \infty)$.

By (3.3.28) and (3.3.30), we have established that the system (3.3.20) has the form

$$Z' = (\Lambda + R_2 + S)Z, \tag{3.3.31}$$

where $S \in L(a, \infty)$ and, in terms of 2×2 blocks,

$$R_2 = \mathrm{dg}\,(\rho_1, \ldots, \rho_m)$$

with

$$\rho_J = \begin{bmatrix} 0 & -r_J \\ r_J & 0 \end{bmatrix} \tag{3.3.32}$$

and

$$r_J = \tfrac{1}{4}\mathrm{i}(-1)^{m-J}(p_{J-1}p_J)'/(p_{J-1}p_J). \tag{3.3.33}$$

We can now take advantage of the block structure of (3.3.31) to proceed to the Levinson form in the same way as we did in the proof of Corollary 2.2.1 for second-order equations. To indicate the details, we note that r_J has a similar form to r in (2.2.4) and that the conditions (3.3.12)–(3.3.14) correspond to (2.2.1), (2.2.2) and (2.2.8). We therefore express each block

$$\begin{bmatrix} \lambda_{2J-1} & -r_J \\ r_J & \lambda_{2J} \end{bmatrix} \tag{3.3.34}$$

(in which $\lambda_{2J} = -\lambda_{2J-1}$) in its diagonal form by means of a matrix $I + Q_J$, with $Q_J = o(1)$. Then, in (3.3.31), we make the transformation

$$Z = (I + Q)W \tag{3.3.35}$$

with

$$Q = \mathrm{dg}\,(Q_1, \dots, Q_m).$$

The conditions (3.3.12)–(3.3.14) guarantee that we obtain a system

$$W' = (\Lambda + S)W$$

of the Levinson form, and therefore solutions

$$W_k = \{e_k + o(1)\} \exp\left(\int_a^x \lambda_k(t)\,\mathrm{d}t\right). \tag{3.3.36}$$

We have now to transform back to Y and it is here that the difficulty arises which leads to the appearance of derivatives in (3.3.16). By (3.3.18), (3.3.19) and (3.3.35), the transformation back to Y gives solutions Y_k of (3.1.2) such that

$$Y_k = TM\{e_k + o(1)\} \exp\left(\int_a^x \lambda_k(t)\,\mathrm{d}t\right), \tag{3.3.37}$$

where M is as in (3.3.17). If, as on previous occasions, the first component in (3.3.37) is selected, we obtain solutions y_k of (3.1.1) such that

$$y_k = \left\{M_k^{-1/2} + \sum_{j \neq k} o(M_j^{-1/2})\right\} \exp\left(\int_a^x \lambda_k(t)\,\mathrm{d}t\right). \tag{3.3.38}$$

However, because of (3.3.9), the factor $\{\cdots\}$ here is a true asymptotic factor only when $k = 2m - 1$ and $k = 2m$; otherwise, it is no more precise than $o(M_{2m}^{-1/2})$. Instead, we select the $(m - K + 1)$th component of (3.3.37). Then, by definition of T in (3.1.9) and (3.1.10), (3.3.38) is replaced by

$$y_k^{(m-K)} = \left\{\lambda_k^{m-K} M_k^{-1/2} + \sum_{j \neq k} o(\lambda_j^{m-K} M_j^{-1/2})\right\} \exp\left(\int_a^x \lambda_k(t)\,dt\right) \quad (3.3.39)$$

and, by (3.3.7), this is a true asymptotic formula

$$y_k^{(m-K)} \sim \lambda_k^{m-K} M_k^{-1/2} \exp\left(\int_a^x \lambda_k(t)\,dt\right). \quad (3.3.40)$$

Finally, by (3.3.8) and (3.3.2), we have

$$\lambda_k^{m-K} M_k^{-1/2} \sim (\text{const.}) p_K^{-1/2} \lambda_k^{1/2} \sim (\text{const.})(p_K p_{K-1})^{-1/4}, \quad (3.3.41)$$

and (3.3.16) follows after adjustment of a constant multiple. □

The simplest situation in which the dichotomy condition is satisfied in terms of (3.3.15) is where

$$|\arg(-p_r/p_{r-1})| \leqslant \pi - \omega \quad (3.3.42)$$

for all r and a fixed $\omega > 0$. By (3.3.2) we then have

$$c_1 |p_K/p_{K-1}|^{1/2} \leqslant |\text{re}\,\lambda_k| \leqslant c_2 |p_K/p_{K-1}|^{1/2},$$

where c_1 (>0) and c_2 are constants, and it follows from (3.3.6) that re $\{\lambda_j(x) - \lambda_k(x)\}$ has one sign in some interval $[a, \infty)$. In particular, (3.3.42) covers the case where the p_r are real-valued with alternating sign. In addition, we can allow all real-valued p_r, irrespective of their signs, if we note that η_K in (3.3.5) is then also real-valued and hence so is each λ_k^2. Thus either re $\lambda_k = 0$ or re $\lambda_k = \lambda_k$ and, once again, (3.3.15) has one sign in $[a, \infty)$. Let us also note that, in the example (3.2.20), the other conditions imposed in the theorem, which are (3.3.1) and (3.3.11)–(3.3.14), are all satisfied if

$$\alpha_0 - \alpha_1 < \alpha_1 - \alpha_2 < \cdots < \alpha_{m-1} - \alpha_m < 2. \quad (3.3.43)$$

The theorem does of course leave open the question of whether asymptotic formulae for the y_k themselves can be obtained either by integration of (3.3.16) or by some other refinement of the analysis. In certain situations, progress can be made along these lines, and we shall return to this question in the context of one of the main applications of these formulae in §§ 3.8–3.11. At this stage, we note only that (3.3.16) can be supplemented by

$$y_k^{(m-K)} = o\left\{(p_{K-1} p_K)^{-1/4} \exp\left(\int_a^x \lambda_k(t)\,dt\right)\right\} \quad (3.3.44)$$

for values of k other than $2K-1$ and $2K$. This follows immediately from (3.3.39) and (3.3.7).

The analysis in this section can be extended to deal with the situation where some of the p_r in (3.1.1) are identically zero, and we indicate the details briefly. Suppose that the non-zero coefficients are specified by

$$r = r_s \qquad (0 = r_0 < r_1 < \cdots < r_S = m),$$

and write $q_s = p_r$ $(r = r_s)$. Then (3.1.8) is

$$q_0\lambda^{2m} + \cdots + q_s\lambda^{2m-2r} + \cdots + q_S = 0. \tag{3.3.45}$$

If we now replace (3.3.1) by

$$(q_{s+1}/q_s)^{1/\delta(s+1)} = o\{(q_s/q_{s-1})^{1/\delta(s)}\} \qquad (1 \leqslant s \leqslant S-1), \tag{3.3.46}$$

where

$$\delta(s) = r_s - r_{s-1},$$

we can again solve (3.3.45) asymptotically. In place of (3.3.2), we obtain

$$\lambda_k^{2\delta(K)} \sim -q_K/q_{K-1} \qquad (2r_{K-1} + 1 \leqslant k \leqslant 2r_K,\ 1 \leqslant K \leqslant S),$$

with the $2\delta(K)$th roots of (-1) appearing when we solve for the λ_k themselves.

In place of (3.3.34), we obtain $2\delta(J) \times 2\delta(J)$ blocks which resemble a special case of the matrix $\Lambda + R$ in (2.9.16), the case being

$$n = 2\delta(J), \qquad q = -q_J, \qquad p_{\delta(J)} = q_{J-1},$$

and the other $p_j = 1$, where the notation of § 2.9 is stated first in each of these equations. As in § 2.9, the diagonalization process of Theorem 1.6.1 can be applied to each block to determine the transformation (3.3.35) in the present situation. The final asymptotic result corresponding to (3.3.16) is

$$y_k^{(m-\nu)} \sim (q_{K-1}q_K)^{-1/4}(q_K/q_{K-1})^{(r_K + r_{K-1} + 1 - 2\nu)/4\delta(K)}$$
$$\times \exp\left(\int_a^x \lambda_k(t)\,dt\right), \tag{3.3.47}$$

where $2r_{K-1} + 1 \leqslant k \leqslant 2r_K$, $r_{K-1} + 1 \leqslant \nu \leqslant r_K$, and $1 \leqslant K \leqslant S$.

In the case where (3.1.1) is a three-term equation

$$(p_0 y^{(m)})^{(m)} + (p_r y^{(m-r)})^{(m-r)} + p_m y = 0 \tag{3.3.48}$$

with a fixed r in $[1, m-1]$, we have $S = 2$ in (3.3.45) and

$$q_0 = p_0, \quad q_1 = p_r, \quad q_2 = p_m.$$

Then (3.3.46) becomes

$$p_m^r p_0^{m-r} = o(p_r^m)$$

and, giving ν its maximum permitted values in (3.3.47), we obtain solutions y_k such that

$$y_k^{(m-r)} \sim p_r^{-1/2}(p_r/p_0)^{1/4r} \exp\left(\int_a^x \lambda_k(t)\,\mathrm{d}t\right) \qquad (1 \leqslant k \leqslant 2r), \tag{3.3.49}$$

$$y_k \sim p_m^{-1/2}(p_m/p_r)^{1/4(m-r)} \exp\left(\int_a^x \lambda_k(t)\,\mathrm{d}t\right) \qquad (2r+1 \leqslant k \leqslant 2m). \tag{3.3.50}$$

3.4 SMALL EIGENVALUES

In the proof of Theorem 3.3.1, the formulae (3.3.31)–(3.3.33) were obtained on the basis of the conditions (3.3.1) and (3.3.11) only. The remaining conditions (3.3.12)–(3.3.14) were then used in the diagonalization of (3.3.34) with λ_{2J} regarded as dominating r_J. Here we consider the complementary case where λ_{2J} is subdominant.

Theorem 3.4.1 *Let the conditions in Theorem* 3.3.1 *hold except that* (3.3.12)–(3.3.14) *hold only for* $1 \leqslant r \leqslant L$ *with a given* L *in* $[1, m-1]$. *For* $L+1 \leqslant r \leqslant m$, *let*

$$(p_r/p_{r-1})^{1/2} \in L(a, \infty). \tag{3.4.1}$$

Let the $2m$ *functions* $\tilde{\lambda}_j$ *satisfy the dichotomy condition in* § 1.3, *where*

$$\tilde{\lambda}_j = \begin{cases} \lambda_j & (1 \leqslant j \leqslant 2L) \\ \tfrac{1}{4}(-1)^j(p_{J-1}p_J)'/(p_{J-1}p_J) \end{cases} \tag{3.4.2}$$

$$(j = 2J-1, 2J;\ L+1 \leqslant J \leqslant m). \tag{3.4.3}$$

Then (3.1.1) *has solutions* $y_k(x)$ $(k = 2K-1, 2K)$ *such that* (3.3.16) *holds for* $1 \leqslant K \leqslant L$, *and*

$$y_{2K-1}^{(m-K)} = o\{(p_{K-1}p_K)^{-1/2}\}, \qquad y_{2K}^{(m-K)} = 1 + o(1) \tag{3.4.4}$$

for $L+1 \leqslant K \leqslant m$.

Proof We have (3.3.31)–(3.3.33) as before, and we diagonalize those blocks (3.3.34) in which $1 \leqslant J \leqslant L$, also as before. In the remaining blocks (3.3.34) with $L+1 \leqslant J \leqslant m$, λ_{2J-1} and λ_{2J} are $L(a, \infty)$ by (3.3.2) and (3.4.1), and these terms can be transferred to the matrix S in (3.3.31). We are left with the blocks

$$r_J\begin{bmatrix} 0 & -1 \\ 1 & 0 \end{bmatrix}$$

which can be expressed in the diagonal form

$$r_J\begin{bmatrix} 1 & -\mathrm{i} \\ -\mathrm{i} & 1 \end{bmatrix}^{-1}\begin{bmatrix} 0 & -1 \\ 1 & 0 \end{bmatrix}\begin{bmatrix} 1 & -\mathrm{i} \\ -\mathrm{i} & 1 \end{bmatrix} = r_J\begin{bmatrix} \mathrm{i} & 0 \\ 0 & -\mathrm{i} \end{bmatrix}. \tag{3.4.5}$$

We then make the transformation (3.3.35), where the Q_J $(1 \leqslant J \leqslant L)$ are as before, and

$$Q_{L+1} = \cdots = Q_m = \begin{bmatrix} 0 & -\mathrm{i} \\ -\mathrm{i} & 0 \end{bmatrix}. \tag{3.4.6}$$

Bearing in mind the dichotomy condition imposed on the $\tilde{\lambda}_j$ in (3.4.2) and (3.4.3), we obtain (3.3.36) with λ_k $(1 \leqslant k \leqslant 2L)$ as before while, by (3.4.5), λ_{2K-1} and λ_{2K} $(L+1 \leqslant K \leqslant m)$ are replaced by

$$\mathrm{i}r_K, \qquad -\mathrm{i}r_K. \tag{3.4.7}$$

Concentrating on these last values of K, it follows from (3.4.7) and (3.3.33) that the integration in (3.3.36) can be carried out. Then, by (3.4.6), (3.3.35), and (3.3.17)–(3.3.19), the transformation back to Y gives asymptotic formulae which correspond to (3.3.40) in the form

$$y_{2K-1}^{(m-K)} = [\lambda_{2K-1}^{m-K} M_{2K-1}^{-1/2} - \mathrm{i}\lambda_{2K}^{m-K} M_{2K}^{-1/2}\{1+o(1)\}](p_{K-1}p_K)^{-\frac{1}{4}\eta},$$
$$y_{2K}^{(m-K)} = [-\mathrm{i}\lambda_{2K-1}^{m-K} M_{2K-1}^{-1/2} + \lambda_{2K}^{m-K} M_{2K}^{-1/2}\{1+o(1)\}](p_{K-1}p_K)^{\frac{1}{4}\eta},$$

where $\eta = (-1)^{m-K}$. By (3.3.29) and the fact that $\lambda_{2K-1} = -\lambda_{2K}$, we obtain

$$\left.\begin{aligned} y_{2K-1}^{(m-K)} &= \lambda_{2K}^{m-K} M_{2K}^{-1/2}\, \mathrm{i}\{\eta - 1 + o(1)\}(p_{K-1}p_K)^{-\frac{1}{4}\eta}, \\ y_{2K}^{(m-K)} &= \lambda_{2K}^{m-K} M_{2K}^{-1/2}\{\eta + 1 + o(1)\}(p_{K-1}p_K)^{\frac{1}{4}\eta}. \end{aligned}\right\} \tag{3.4.8}$$

When $m-K$ is even, we have $\eta = 1$, and then (3.4.4) follows from (3.4.8) and (3.3.41) after adjustment of a constant multiple. When $m-K$ is odd, we have $\eta = -1$ and we interchange the two solutions in (3.4.8) to obtain (3.4.4) again. □

The dichotomy condition on the $\tilde{\lambda}_j$ is a little complicated because it involves a mixture of the two types (3.4.2) and (3.4.3). We can clarify the condition by pointing out that, when $\tilde{\lambda}_j$ and $\tilde{\lambda}_k$ are both of the type (3.4.2), it is sufficient that $\mathrm{re}\,(\lambda_j - \lambda_k)$ has one sign in $[a, \infty)$, as in (3.3.15). When $\tilde{\lambda}_j$ and $\tilde{\lambda}_k$ are both of the type (3.4.3), we have a situation like the one in (1.3.10), and it is sufficient that

$$|p_{J-1}p_J/p_{K-1}p_K| \tag{3.4.9}$$

is monotonic in $[a, \infty)$. When $\tilde{\lambda}_j$ and $\tilde{\lambda}_k$ are of different types (3.4.2) and (3.4.3) respectively, it is sufficient that $\mathrm{re}\,\lambda_j$ is nowhere zero in $[a, \infty)$ and, as $x \to \infty$,

$$(p_{K-1}p_K)'/(p_{K-1}p_K) = o(\mathrm{re}\,\lambda_j).$$

In the example (3.2.20), the conditions of Theorem 3.4.1 mean that we have a situation where

$$\alpha_0 - \alpha_1 < \cdots < \alpha_{L-1} - \alpha_L < 2 < \alpha_L - \alpha_{L+1} < \cdots < \alpha_{m-1} - \alpha_m \tag{3.4.10}$$

in which, as compared to (3.3.43), the value 2 is inserted within the sequence of inequalities.

In Theorem 3.4.1, the condition (3.4.1) is imposed for r in a certain restricted range. The case where (3.4.1) holds for all r in $[1, m]$ can also be considered, but in a somewhat wider setting which we now introduce. The product of all the conditions (3.4.1) for $1 \leqslant r \leqslant m$ implies that

$$(p_m/p_0)^{1/2m} \in L(a, \infty). \tag{3.4.11}$$

Thus, in the notation of (3.2.1), $P \in L(a, \infty)$ and we have a situation complementary to the one considered in Theorem 3.2.1, where P was regarded as dominant. Similar complementary cases have occurred previously in §§ 2.2 and 2.7 for the Sturm–Liouville equation, and in §§ 2.9 and 2.11 for the higher-order two-term equation. Leaving aside (3.4.1) now, we can deal with (3.4.11) directly as on the previous occasions by making a suitable transformation

$$Y = DZ \tag{3.4.12}$$

with D diagonal. The results are given in terms of the quasi-derivatives (1.9.15) in the following theorem and corollary.

Theorem 3.4.2 *Let the functions ρ_j $(1 \leqslant j \leqslant 2m)$ be locally absolutely continuous and nowhere zero in $[a, \infty)$. For each unequal pair j and k, suppose that either*

(I) $$|\rho_k(x)\rho_j(t)/\rho_k(t)\rho_j(x)| \leqslant K' \tag{3.4.13}$$

or

(II) $$|\rho_k(x)\rho_j(t)/\rho_k(t)\rho_j(x)| \geqslant K'', \tag{3.4.14}$$

for $a \leqslant t \leqslant x < \infty$, where K' and K'' (>0) are constants. Also, let

$$\rho_j^{-1}\rho_{j+1} \qquad (1 \leqslant j \leqslant 2m-1, j \neq m), \tag{3.4.15}$$

$$\rho_m^{-1}\rho_{m+1}p_0^{-1}, \tag{3.4.16}$$

and

$$\rho_{2m-j+1}^{-1}\rho_j p_{m-j+1} \qquad (1 \leqslant j \leqslant m) \tag{3.4.17}$$

all be $L(a, \infty)$. Then (3.1.1) *has solutions $y_k(x)$ $(1 \leqslant k \leqslant 2m)$ such that*

$$y_k^{[k-1]} = 1 + o(1), \quad y_k^{[\nu-1]} = o(\rho_\nu/\rho_k) \qquad (\nu \neq k). \tag{3.4.18}$$

Proof We make the transformation (3.4.12), with

$$D = \mathrm{dg}\,(\rho_1, \dots, \rho_{2m}). \tag{3.4.19}$$

Then (3.1.2) becomes

$$Z' = (-D^{-1}D' + D^{-1}AD)Z. \tag{3.4.20}$$

By (3.1.3) and (3.4.15)–(3.4.17), we have

$$D^{-1}AD \in L(a, \infty).$$

Also, the entries in the diagonal matrix $D^{-1}D'$ are ρ_j'/ρ_j, and therefore (3.4.13) and (3.4.14) are just the dichotomy condition in the form (1.3.11) and (1.3.12). It follows from Theorem 1.3.1 that (3.4.20) has solutions Z_k of the form (1.3.5) with $\lambda_k = -\rho_k'/\rho_k$. On carrying out the integration in the exponential and transforming back to Y via (3.4.19) and (3.4.12), we find that (3.1.2) has solutions

$$Y_k = \rho_k^{-1} D\{e_k + o(1)\},$$

and (3.4.18) follows since the components of Y_k are the quasi-derivatives of y_k. □

Corollary 3.4.1 *Let the function ϕ be locally absolutely continuous and nowhere zero in $[a, \infty)$. Let N be an integer in $[0, 2m-1]$ and suppose that, for some constant κ (>0) and $a \leqslant t \leqslant x < \infty$,*

(A)
$$|\phi(x)| \leqslant \kappa\, |\phi(t)|; \tag{3.4.21}$$

(B) *for integers l in $[1, N]$,*

$$|(p_0\phi^l)(x)| \geqslant \kappa |(p_0\phi^l)(t)|; \tag{3.4.22}$$

(C) *for integers l in $[N+1, 2m-1]$,*

$$(p_0\phi^l)(x) \to 0 \quad \text{as } x \to \infty \tag{3.4.23}$$

with

$$|(p_0\phi^l)(x)| \leqslant \kappa\, |(p_0\phi^l)(t)|. \tag{3.4.24}$$

Also, let

$$\phi \in L(a, \infty) \tag{3.4.25}$$

and

$$p_r p_0^{-1} \phi^{-2r+1} \in L(a, \infty) \qquad (1 \leqslant r \leqslant m). \tag{3.4.26}$$

Then (3.1.1) *has solutions y_k $(1 \leqslant k \leqslant 2m)$ such that*

$$y_k^{[k-1]} = 1 + o(1). \tag{3.4.27}$$

Also, for $\nu \neq k$,

$$y_k^{[\nu-1]} = \begin{cases} o(\phi^{\nu-k}) & (\nu \leqslant m) \\ o(p_0\phi^{\nu-k}) & (\nu \geqslant m+1) \end{cases} \tag{3.4.28}$$

when $k \leqslant m$ while, if $k \geqslant m+1$, it is $p_0 y_k^{[\nu-1]}$ which satisfies (3.4.28).

Proof In Theorem 3.4.2 we take

$$\rho_j = \begin{cases} \phi^{j-1} & (1 \leqslant j \leqslant m) \\ p_0\phi^{j-1} & (m+1 \leqslant j \leqslant 2m). \end{cases} \tag{3.4.29}$$

Then (3.4.25) implies that both (3.4.15) and (3.4.16) are $L(a, \infty)$, while

(3.4.26) implies that (3.4.17) is $L(a, \infty)$. It remains to check the dichotomy condition involving (3.4.13) and (3.4.14). First, (3.4.21) is just (3.4.13) for the two cases $j<k\leqslant m$ and $m+1\leqslant j<k$. Also, (3.4.22) is (3.4.14) for $j\leqslant m$, $k\geqslant m+1$ and $l=k-j\leqslant N$. Again, (3.4.24) is just (3.4.13) for $j\leqslant m$, $k\geqslant m+1$ and $l=k-j\geqslant N+1$. The corollary now follows from (3.4.18) and (3.4.29). □

The condition (3.4.21) is certainly satisfied when $|\phi|$ is decreasing, while (3.4.22), (3.4.23), and (3.4.24) are satisfied when $|p_0\phi^l|$ is increasing for $1\leqslant l\leqslant N$ and decreasing to zero for $N+1\leqslant l\leqslant 2m-1$. In particular, p_0 and ϕ could be powers of x.

Returning to (3.4.11), let us suppose that this condition holds and, in the notation (3.2.1), make the choice

$$\phi = P = (p_m/p_0)^{1/2m}. \tag{3.4.30}$$

Then (3.4.25) holds and the same condition is repeated in the case $r=m$ of (3.4.26). The significant range for r in (3.4.26) is therefore $1\leqslant r\leqslant m-1$, and then this condition is the case $\sigma=[1, m-1]$ of (3.2.25). Thus we have a situation complementary to the one in Theorem 3.2.2 in terms of the dominance or otherwise of P. In the example (3.2.20) with powers of x, the conditions of Corollary 3.4.1 are satisfied by (3.4.30) if

$$\alpha_m - \alpha_0 < -2m \tag{3.4.31}$$

and, as in (3.2.27),

$$\alpha_r < \{r\alpha_m + (m-r)\alpha_0 - \tfrac{1}{2}(\alpha_m - \alpha_0 + 2m)\}/m. \tag{3.4.32}$$

An alternative choice to (3.4.30) is

$$\phi = (p_J/p_0)^{1/2J} \tag{3.4.33}$$

with a given J in $[1, m]$. This condition (3.4.25) is again repeated in the case $r=J$ of (3.4.26). In the example (3.2.20), the conditions are now

$$\alpha_J - \alpha_0 < -2J \tag{3.4.34}$$

and

$$\alpha_r < \{r\alpha_J + (J-r)\alpha_0 - \tfrac{1}{2}(\alpha_J - \alpha_0 + 2J)\}/J \tag{3.4.35}$$

in place of (3.4.31) and (3.4.32).

Before leaving this section, we make a further comment on the blocks (3.3.34). In Theorems 3.3.1 and 3.4.1 we have dealt with the two cases where λ_{2J} $(=-\lambda_{2J-1})$ is dominant and subdominant as compared to r_J. A third possibility is that λ_{2J} and r_J have comparable size as $x\to\infty$, and then we have the Euler type of situation considered in §2.6. A final transformation of the kind discussed in § 2.6 would reduce the block to the Levinson form. We deal with this situation in more detail for the fourth-order equation in the next section.

3.5 THE FOURTH-ORDER EQUATION

The fourth-order case of (3.1.1) is

$$(p_0 y'')'' + (p_1 y')' + p_2 y = 0 \tag{3.5.1}$$

and, as there are only three coefficients involved here, it is possible to add further detail to the theory and discuss a number of special features. The characteristic equation (3.1.8) is now

$$p_0\lambda^4 + p_1\lambda^2 + p_2 = 0 \tag{3.5.2}$$

and it has the advantage that it can be solved explicitly to give

$$\left.\begin{aligned} \lambda_1 &= \{(-p_1 - \Delta)/2p_0\}^{1/2}, \qquad & \lambda_2 &= -\lambda_1, \\ \lambda_3 &= \{(-p_1 + \Delta)/2p_0\}^{1/2}, \qquad & \lambda_4 &= -\lambda_3 \end{aligned}\right\} \tag{3.5.3}$$

where

$$\Delta = (p_1^2 - 4p_0 p_2)^{1/2}. \tag{3.5.4}$$

The eigenvalues are distinct in some interval $[a, \infty)$ if p_0, p_2 and Δ are all nowhere zero there.

The first feature in our discussion of (3.5.1) concerns Theorem 3.3.1. Here we re-examine the proof of this theorem and show that it is possible to avoid the modified transformation (3.3.18). The original transformation (3.1.5), with T as in (3.1.10), is sufficient and it leads to asymptotic formulae for the solutions y_k themselves, without the need to introduce derivatives as in (3.3.16). We indicate the main reason for this special feature of (3.5.1) after the proof of the theorem which follows. In condition (c) of this theorem, we require a minor strengthening of the conditions in Theorem 3.3.1. As in (3.3.10), we define

$$\delta = |p_0 p_2 / p_1^2|. \tag{3.5.5}$$

Theorem 3.5.1 *Let p_r $(0 \leq r \leq 2)$ be nowhere zero with locally absolutely continuous first derivatives in $[a, \infty)$. Let*

$$p_2/p_1 = o(p_1/p_0) \tag{3.5.6}$$

as $x \to \infty$, and let

(a) $$\delta^{1/2} p_r'/p_r \in L(a, \infty); \tag{3.5.7}$$

(b) (3.3.12)–(3.3.14) *hold for $r = 2$, that is,*

$$(p_1 p_2)'/(p_1 p_2) = o\{(p_2/p_1)^{1/2}\}, \tag{3.5.8}$$

$$\{p_1^{-1/2} p_2^{-3/2} (p_1 p_2)'\}' \in L(a, \infty), \tag{3.5.9}$$

$$p_1^{-3/2} p_2^{-5/2} (p_1 p_2)'^2 \in L(a, \infty); \tag{3.5.10}$$

(c) *for $r=0$ and $r=1$,*

$$p_r'/p_r = o\{(p_1/p_0)^{1/2}\}, \tag{3.5.11}$$

$$(p_0/p_1)^{1/2} p_r^{-1} p_r'' \in L(a, \infty), \tag{3.5.12}$$

$$(p_0/p_1)^{1/2} p_r^{-2} p_r'^2 \in L(a, \infty); \tag{3.5.13}$$

(d) re λ_1 *and* re λ_3 *each have one sign in* $[a, \infty)$;

(e)
$$\int_t^x \operatorname{re}\{\pm\lambda_1(s) \pm \lambda_3(s)\}\,ds + \tfrac{1}{4}\log \delta(x)/\delta(t) \tag{3.5.14}$$

be either bounded above or bounded below for $a \leq t \leq x < \infty$ *and each choice of the* $\pm$ *sign.*

Then (3.5.1) *has solutions* y_k *such that*

$$y_k \sim (p_1^3/p_0)^{-1/4} \exp\left(\int_a^x \lambda_k(t)\,dt\right) \qquad (k=1, 2), \tag{3.5.15}$$

$$y_k \sim (p_1 p_2)^{-1/4} \exp\left(\int_a^x \lambda_k(t)\,dt\right) \qquad (k=3, 4). \tag{3.5.16}$$

Proof By (3.5.6), we can choose the square root in (3.5.4) so that $\Delta \sim p_1$ as $x \to \infty$, and then

$$\Delta = p_1\{1 + O(\delta)\} \tag{3.5.17}$$

with δ as in (3.5.5). From (3.5.3) we obtain

$$\left.\begin{aligned} \lambda_1 &= \mathrm{i}\,(p_1/p_0)^{1/2}\{1+O(\delta)\},\\ \lambda_3 &= \mathrm{i}\,(p_2/p_1)^{1/2}\{1+O(\delta)\},\end{aligned}\right\} \tag{3.5.18}$$

in accordance with (3.3.25). Also, by (3.1.13) and (3.5.3),

$$\left.\begin{aligned} M_1 &= -2\lambda_1\Delta = -M_2\\ \text{and}\qquad M_3 &= 2\lambda_3\Delta = -M_4,\end{aligned}\right\} \tag{3.5.19}$$

leading to

$$\left.\begin{aligned} M_1 &= -2\,\mathrm{i}\,(p_1^3/p_0)^{1/2}\{1+O(\delta)\}\\ \text{and}\qquad M_3 &= 2\,\mathrm{i}\,(p_1 p_2)^{1/2}\{1+O(\delta)\},\end{aligned}\right\} \tag{3.5.20}$$

again in accordance with (3.3.26).

We now consider the transformed system (3.1.6) with T given by (3.1.10), and we have therefore to estimate the entries τ_{jk} as specified in Lemma 3.1.1. We could of course use the estimates already obtained for σ_{jk} in § 3.3, and then (3.3.21) gives $\tau_{jk} = M_k^{1/2} M_j^{-1/2} \sigma_{jk}$ $(j \neq k)$. However, it is simpler to make direct use of (3.1.16) which, in the present situation, is

$$\tau_{jk} = (\lambda_j - \lambda_k)^{-1} M_j^{-1}(p_0'\lambda_j^2\lambda_k^2 + p_1'\lambda_j\lambda_k + p_2'). \tag{3.5.21}$$

We define

$$s = \delta \sum_{r=0}^{2} |p_r'/p_r|, \tag{3.5.22}$$

which is certainly $L(a, \infty)$ by (3.5.7). It follows immediately from (3.5.18), (3.5.20) and (3.5.21) that

$$\begin{aligned}
\tau_{12} &= \tau_{21} = \tfrac{1}{4}(p_0p_1)'/(p_0p_1) + O(s),\\
\tau_{34} &= \tau_{43} = -\tfrac{1}{4}(p_1p_2)'/(p_1p_2) + O(s),\\
\tau_{13} &= \tau_{24} = -\tfrac{1}{2}(p_1'/p_1)(p_0p_2/p_1^2)^{1/2} + O(s),\\
\tau_{23} &= \tau_{14} = \tfrac{1}{2}(p_1'/p_1)(p_0p_2/p_1^2)^{1/2} + O(s),\\
\tau_{31} &= \tau_{42} = -\tfrac{1}{2}p_1'/p_1 + O(\delta^{-1/2}s),\\
\tau_{32} &= \tau_{43} = -\tfrac{1}{2}p_1'/p_1 + O(\delta^{-1/2}s).
\end{aligned}$$

In addition, on differentiating (3.5.19) and using (3.5.3) and (3.5.4), we obtain from (3.1.15)

$$\tau_{11} = \tau_{22} = \frac{3}{4}\frac{p_1'}{p_1} - \frac{1}{4}\frac{p_0'}{p_0} + O(s)$$

and

$$\tau_{33} = \tau_{44} = \tfrac{1}{4}(p_1p_2)'/(p_1p_2) + O(s).$$

Now s and $\delta^{-1/2}s$ are $L(a, \infty)$, by (3.5.7) and (3.5.22), and therefore altogether the system (3.1.6) can be written

$$Z' = (\Lambda + R_1 + S)Z, \tag{3.5.23}$$

where S is $L(a, \infty)$ and

$$R_1 = \begin{bmatrix} \xi & \eta & 0 & 0 \\ \eta & \xi & 0 & 0 \\ \zeta & \zeta & \theta & -\theta \\ \zeta & \zeta & -\theta & \theta \end{bmatrix}, \tag{3.5.24}$$

with

$$\left.\begin{aligned}
\xi &= -\frac{3}{4}\frac{p_1'}{p_1} + \frac{1}{4}\frac{p_0'}{p_0}, & \eta &= -\frac{1}{4}\frac{(p_0p_1)'}{p_0p_1},\\
\zeta &= \frac{1}{2}\frac{p_1'}{p_1}, & \theta &= -\frac{1}{4}\frac{(p_1p_2)'}{p_1p_2}.
\end{aligned}\right\} \tag{3.5.25}$$

Now (3.5.23) has the form (1.6.42), and we could at this point apply Theorem 1.6.1 to (3.5.23) to establish (3.5.15) and (3.5.16). Here, however, we follow rather the method of proof in Theorem 1.6.1 and take advantage of the special form of R_1 in (3.5.24) to obtain the asymptotic results under slightly less restrictive conditions than Theorem 1.6.1 itself entails. We therefore start by expressing $\Lambda + R_1$ in its diagonal form

$$(I+Q)^{-1}(\Lambda + R_1)(I+Q) = \Lambda_1, \tag{3.5.26}$$

where $\Lambda_1 = \mathrm{dg}\,(\mu_k)$ and the μ_k are the eigenvalues of $\Lambda + R_1$, given by

$$\left.\begin{aligned}\mu_1 &= \lambda_1 + \xi + \eta^2/\{\lambda_1 + (\lambda_1^2+\eta^2)^{1/2}\},\\ \mu_2 &= \lambda_2 + \xi - \eta^2/\{\lambda_1 + (\lambda_1^2+\eta^2)^{1/2}\},\end{aligned}\right\} \tag{3.5.27}$$

$$\left.\begin{aligned}\mu_3 &= \theta + (\lambda_3^2+\theta^2)^{1/2},\\ \mu_4 &= \theta - (\lambda_3^2+\theta^2)^{1/2}.\end{aligned}\right\} \tag{3.5.28}$$

By (3.5.25), (3.5.18), (3.5.11), and (3.5.8), we have

$$\mu_k \sim \lambda_k \qquad (1 \leqslant k \leqslant 4)$$

and further, by (3.5.13) and (3.5.10),

$$\mu_k = \lambda_k + \xi + s_k \qquad (k=1,\,2), \tag{3.5.29}$$

$$\mu_k = \lambda_k + \theta + s_k \qquad (k=3,\,4), \tag{3.5.30}$$

where the expressions s_k are $L(a,\infty)$.

Considering the eigenvectors of $\Lambda + R_1$, we can take Q in (3.5.26) to have the form

$$Q = \begin{bmatrix} 0 & u_2 & 0 & 0\\ u_1 & 0 & 0 & 0\\ v_1 & v_2 & 0 & -u\\ w_1 & w_2 & u & 0\end{bmatrix} \tag{3.5.31}$$

and, as in (1.6.8) and (1.6.9), we have to show that

$$Q = o(1), \qquad Q' \in L(a,\infty). \tag{3.5.32}$$

First we have

$$u = -\theta/\{\lambda_3 + (\lambda_3^2+\theta^2)^{1/2}\} = o(1), \tag{3.5.33}$$

by (3.5.25), (3.5.18), and (3.5.8). Also, $u' = O\{(\theta/\lambda_3)'\}$ which, after a calculation using (3.5.3), (3.5.7), and (3.5.9), is $L(a,\infty)$. Next,

$$u_1 = \eta/\{\lambda_1 + (\lambda_1^2+\eta^2)^{1/2}\},$$

and similar reasoning using (3.5.11)–(3.5.13) gives $u_1 = o(1)$ and $u_1' \in L(a,\infty)$. Again u_2 can be treated similarly. Turning to v_1, we have

$$v_1 = \frac{\zeta(\mu_1 - 2\theta - \lambda_4)}{(\mu_1-\theta)^2 - \lambda_4^2 - \theta^2}(1+u_1). \tag{3.5.34}$$

On dividing numerator and denominator here by λ_1^2 and using (3.5.27), we obtain $v_1 = o(1)$ and

$$v_1' = O\{(\zeta/\lambda_1)'\} + O\{(\xi/\lambda_1)'\zeta/\lambda_1\} + O\{(\eta/\lambda_1)'\eta\zeta/\lambda_1^2\}$$
$$+ O\{(\theta/\lambda_1)'\zeta/\lambda_1\} + O\{(\lambda_4/\lambda_1)'\zeta/\lambda_1\} + O(u_1'\zeta/\lambda_1). \quad (3.5.35)$$

As in earlier stages of the proof, all the O-terms are $L(a, \infty)$ as a consequence of (3.5.7)–(3.5.13), and similar reasoning applies to w_1, v_2 and w_2.

We have now established (3.5.32), and it follows from (3.5.26), (3.5.29) and (3.5.30) that the transformation

$$Z = (I + Q)W$$

takes (3.1.6) into

$$W' = (\Lambda_2 + S)W, \quad (3.5.36)$$

where $S \in L(X, \infty)$ and

$$\Lambda_2 = \mathrm{dg}\,(\lambda_1 + \xi, \lambda_2 + \xi, \lambda_3 + \theta, \lambda_4 + \theta).$$

Now conditions (d) and (e) in the statement of the theorem imply that the dichotomy condition (1.3.1)–(1.3.2) holds in the case of Λ_2. Hence, applying Theorem 1.3.1 to the W-system and transforming back to Y, we find that there are solutions Y_k $(1 \leqslant k \leqslant 4)$ of (3.1.2) such that

$$Y_k = T\{e_k + o(1)\} \exp\left(\int_a^x (\lambda_k + \phi_k)\,\mathrm{d}t\right),$$

where $\phi_1 = \phi_2 = \xi$ and $\phi_3 = \phi_4 = \theta$. By (3.5.25), the integrations involving ξ and θ can be carried out, and then the first component of Y_k yields (3.5.15) and (3.5.16). This completes the proof of the theorem. □

We can now indicate why the above proof, based on the transformation (3.1.5) and (3.1.10), extends in only a limited way to equations (3.1.1) of order higher than the fourth. Let us take the sixth-order case $(m = 3)$ for instance, and examine again the entries τ_{jk} of $T^{-1}T'$. By (3.1.16), we now have

$$\tau_{jk} = (\lambda_j - \lambda_k)^{-1}M_j^{-1}\{p_0'(\lambda_j\lambda_k)^3 + p_1'(\lambda_j\lambda_k)^2 + p_2'\lambda_j\lambda_k + p_3'\}.$$

Since the λ_j occur in $\pm$ pairs, let us divide $T^{-1}T'$ into the nine 2×2 blocks in each of which the τ_{jk} have similar size. Subject to (3.3.1) and (3.3.11) (for $0 \leqslant r \leqslant 3$), it is easy to check that the $(1, 2)$, $(1, 3)$, and $(2, 3)$ blocks are $L(a, \infty)$, and these terms can therefore be moved out to form part of an S matrix as in (3.5.23) and (3.5.24). Again, apart from $L(a, \infty)$ terms, the diagonal $(1, 1)$, $(2, 2)$, and $(3, 3)$ blocks are similar to the corresponding ones in (3.5.24). The $(3, 2)$ block has terms asymptotic to (const.)p_2'/p_2, which are similar to the ζ entries in (3.5.24). A new feature appears in the (2.1) block where there are terms asymptotic to

$$(\text{const.})(\lambda_1/\lambda_3)p_1'/p_1. \quad (3.5.37)$$

Also, in the (3, 1) block, there are terms asymptotic to

$$(\text{const.})(\lambda_1\lambda_5/\lambda_3^2)p_1'/p_1. \tag{3.5.38}$$

Now the factor λ_1/λ_3 in (3.5.37) is large as $x\to\infty$, by (3.3.6), while the factor $\lambda_1\lambda_5/\lambda_3^2$ in (3.5.38) may or may not be large. It would still be possible to apply Theorem 1.6.1 to the system (3.5.23) in the present case, and so obtain asymptotic formulae for the y_k $(1\leqslant k\leqslant 6)$ themselves, but the conditions in Theorem 1.6.1 imply a restriction on the size of the ratios λ_1/λ_3 and $\lambda_1\lambda_5/\lambda_3^2$. Thus we would be in a situation not greatly different from the one covered in § 3.2 where the λ_j all have the same size asymptotically.

So far in this section, we have considered Λ to be large compared to R_1 in (3.5.23)–(3.5.25). Now we move on to a situation of Euler type which arises when the smaller pair of eigenvalues, λ_3 and λ_4, have comparable size to θ as $x\to\infty$. There is also a second such situation which arises when the larger pair, λ_1 and λ_2, have comparable size to ξ and η, but we confine the discussion here to the first case. The following theorem is an extension of Theorem 2.6.1 to the fourth-order equation (3.5.1).

Theorem 3.5.2 *Let* (3.5.6), (3.5.7), *and* (3.5.11)–(3.5.13) *hold and, in place of* (3.5.8)–(3.5.10), *let*

$$(p_1p_2)'/p_1p_2=\kappa(p_2/p_1)^{1/2}(1+\phi), \tag{3.5.39}$$

where κ is a non-zero constant with $\kappa^2\neq 16$,

$$\phi=o(1)\qquad (x\to\infty), \tag{3.5.40}$$

and

$$\phi'\in L(a,\infty). \tag{3.5.41}$$

Let

$$\operatorname{re}\lambda_1 \qquad \text{and} \qquad \operatorname{re}(\lambda_3^2+\theta^2)^{1/2} \tag{3.5.42}$$

each have one sign in $[a,\infty)$, and let

$$\int_t^x \operatorname{re}[\pm\lambda_1(s)\pm\{\lambda_3^2(s)+\theta^2(s)\}^{1/2}]\,ds+\tfrac14\log\delta(x)/\delta(t) \tag{3.5.43}$$

be either bounded above or bounded below for $a\leqslant t\leqslant x<\infty$ and each choice of the $\pm$ sign. Then (3.1.1) *has solutions y_k such that* (3.5.15) *holds for $k=1$, 2, and*

$$\left.\begin{aligned} y_3&\sim(p_1p_2)^c\exp I(x),\\ y_4&\sim(p_1p_2)^{-\frac12-c}\exp\{-I(x)\},\end{aligned}\right\} \tag{3.5.44}$$

where

$$c=-\tfrac14+\tfrac14(1-16\kappa^{-2})^{1/2}, \tag{3.5.45}$$

and

$$I(x)=\frac14\int_a^x\frac{(p_1p_2)'}{p_1p_2}\,\Phi\,dt \tag{3.5.46}$$

with

$$\Phi = 16\kappa^{-2}(1 - 16\kappa^{-2})^{-1/2}\phi + O(\phi^2). \tag{3.5.47}$$

Proof The formulae (3.5.23)–(3.5.29) and (3.5.31)–(3.5.35) continue to hold as before except that, because of (3.5.39) and (3.5.40), in (3.5.33) we now have

$$u \to \tfrac{1}{4}\kappa/\{\mathrm{i} + (\tfrac{1}{16}\kappa^2 - 1)^{1/2}\} = \gamma, \tag{3.5.48}$$

say, and consequently $Q \to Q_0$ in (3.5.32), where Q_0 has $(3, 4)$ entry $-\gamma$, $(4, 3)$ entry γ, and other entries zero. Now

$$\det(I + Q_0) = \gamma^2 + 1 \neq 0,$$

and hence $(I + Q)^{-1}$ exists in some interval $[X, \infty)$, as required in (3.5.26). In the W-system (3.5.36), we now have

$$\Lambda_2 = \mathrm{dg}\,(\lambda_1 + \xi, \lambda_2 + \xi, \mu_3, \mu_4),$$

where μ_3 and μ_4 are given by (3.5.28). Then (3.5.42) and (3.5.43) imply that Λ_2 once again satisfies the dichotomy condition, and therefore Theorem 1.3.1 can be applied to (3.5.36) again. The transformation back to Y is

$$Y = T\{I + Q_0 + o(1)\}W.$$

Bearing in mind the form of Q_0, we obtain the solutions y_1 and y_2 satisfying (3.5.15) as in the proof of Theorem 3.5.1. For y_3 and y_4, we obtain

$$y_3 \sim (1 + \gamma)\exp\left(\int_X^x \mu_3(t)\,\mathrm{d}t\right)$$

and (3.5.49)

$$y_4 \sim (1 - \gamma)\exp\left(\int_X^x \mu_4(t)\,\mathrm{d}t\right).$$

By (3.5.48), $\gamma \neq \pm 1$ and we can therefore adjust y_3 and y_4 to replace the factors $1 \pm \gamma$ by unity. Next, by (3.5.28), (3.5.18), (3.5.39) and (3.5.7), we have

$$\mu_3 = \theta + (-p_2/p_1 + \theta^2)^{1/2} + s_3,$$

where s_3 is $L(a, \infty)$, and similarly for μ_4. The exponential factors in (3.5.49) are therefore similar to those in (2.6.13), with p_1, p_2 and $-\kappa$ in place of the previous p, $-q$ and κ. The proof of (3.5.44)–(3.5.47) is now completed in the same way as for the corresponding formulae in (2.6.6)–(2.6.10). □

As in (3.3.41), the dichotomy condition (d)–(e) in Theorem 3.5.1 is certainly satisfied when

$$|\arg(-p_1/p_0)| \leq \pi - \omega$$

and

$$|\arg(-p_2/p_1)| \leq \pi - \omega \tag{3.5.50}$$

for a fixed $\omega > 0$, and it is also satisfied in the case of real-valued p_r with the following two provisos. First, when the p_r all have the same sign, the λ_k are all pure imaginary and then (3.5.14) is an extra condition on $\delta(x)$ which is satisfied if, for example, $\delta(x)$ decreases to zero as $x \to \infty$. Second, when p_1 has the same sign as p_0 but the opposite sign to p_2, we have re $\lambda_1 = 0$ and re $\lambda_3 = \lambda_3$. Then, in addition to (3.5.8), we require

$$p_r'/p_r = o\{(p_2/p_1)^{1/2}\} \qquad (0 \leq r \leq 2)$$

in order to satisfy (3.5.14).

In Theorem 3.5.2, we have to consider (3.5.42)–(3.5.43) and, corresponding to (3.5.50), these conditions are satisfied when

$$|\arg(-p_1/p_0)| \leq \pi - \omega$$

and

$$|\arg\{-p_2(1 - \tfrac{1}{16}\kappa^2)/p_1\}| \leq \pi - \omega.$$

Again, when the p_r are real-valued and re $\lambda_1 = 0$, (3.5.43) represents an extra condition, either on $\delta(x)$ itself when re $(\lambda_3^2 + \theta^2)^{1/2} = 0$, or on a modification of $\delta(x)$ otherwise.

Let us now draw together the various theorems in §§ 3.2–3.5 in terms of the fourth-order equation (3.1.1) with the coefficients

$$p_r(x) = a_r x^{\alpha_r} \qquad (0 \leq r \leq 2), \tag{3.5.51}$$

where the a_r and α_r are constants with real α_r. We can describe the applicability of the theorems in terms of certain sectors in the (α_1, α_2)-plane and a fixed value of α_0. Let A be the point $(\alpha_0 - 2, \alpha_0 - 4)$ in this plane, and let L_g denote the half-rays emanating from A with various gradients g as follows. The definitions proceed clockwise around A.

L_2 and L_1: the half-rays with $\alpha_1 > \alpha_0 - 2$

L_∞: the half-ray with $\alpha_2 < \alpha_0 - 4$

L_4, $L_{3/2}$ and L_0: the half-rays with $\alpha_1 < \alpha_0 - 2$.

Then, referring in the first instance to the open sectors bounded by these half-rays, the theorems apply as follows.

Theorem 3.2.1: sector (L_0, L_2)
Theorems 3.3.1 and 3.5.1: sector (L_2, L_1)
Theorem 3.4.1: sector (L_1, L_∞)
Corollary 3.4.1 (3.4.33): sector $(L_\infty, L_{3/2})$
Corollary 3.4.1 (3.4.30): sector (L_4, L_0).

In identifying these sectors, we use (3.2.21)–(3.2.22), (3.3.43), (3.4.10),

(3.4.31)–(3.4.32), and (3.4.34)–(3.4.35), all with $m = 2$. In addition, values of α_1 and α_2 on the half-ray L_2 are covered by Theorem 3.2.1 provided that $a_1^2 \neq 4a_0a_2$, this being the condition for the polynomial (3.2.3) to have distinct roots when $m = 2$. The half-rays not covered so far are L_1, L_∞, and L_0, and they all represent situations of Euler type. Theorem 3.5.2 deals with L_1 provided that $(\alpha_1 + \alpha_2)^2 a_1/a_2 \neq 16$, which again is a condition on a_1 and a_2 to disallow a multiple eigenvalue λ_k. The corresponding theorem in which the larger pair of eigenvalues, λ_1 and λ_2, have comparable size to ξ and η would deal with L_∞. Also, L_0 represents the Euler-type situation corresponding to Theorem 3.2.1 when $P \sim (\text{const.})\tau_{jk}$ in place of (3.2.16)–(3.2.17). Finally, at the point A itself, (3.5.1) is the standard Euler equation with explicit solutions, and no asymptotic analysis is required.

3.6 ODD-ORDER EQUATIONS

The equation (3.1.1) can be augmented by the addition to the left-hand side of odd-order differential expressions

$$\tfrac{1}{2}\{(q_{m-j}y^{(j)})^{(j+1)} + (q_{m-j}y^{(j+1)})^{(j)}\},$$

as discussed in (1.9.19) and (1.9.22). These expressions are self-adjoint if q_{m-j} is pure imaginary but, once again, we do not need to make this assumption at this stage. The methods already developed in §§ 3.1–4 continue to apply to the augmented equation, but some points of difference emerge when the expression may have order $2m + 1$ or higher. Here we consider the case where j takes the values $0, 1, \ldots, m$, so that (3.1.1) is replaced by the equation

$$\sum_{j=0}^{m} [\tfrac{1}{2}\{(q_{m-j}y^{(j)})^{(j+1)} + (q_{m-j}y^{(j+1)})^{(j)}\} + (p_{m-j}y^{(j)})^{(j)}] = 0 \qquad (3.6.1)$$

of order $2m + 1$. The corresponding system formulation

$$Y' = AY \qquad (3.6.2)$$

is given by (1.9.23) with, as usual, y appearing as the first component of Y.

In this section we concentrate on the ideas developed in §§ 3.1 and 3.3, and we apply them to (3.6.1), omitting, however, detailed calculations which are similar to those given earlier. It follows from (1.9.23) that the characteristic equation $\det(\lambda I - A) = 0$ is now

$$q_0\lambda^{2m+1} + p_0\lambda^{2m} + q_1\lambda^{2m-1} + \cdots + p_{m-1}\lambda^2 + q_m\lambda + p_m = 0. \quad (3.6.3)$$

Corresponding to (3.1.9), an eigenvector v_k associated with an eigenvalue

λ_k of A has components

$$\lambda_k^{l-1} \qquad (1 \leqslant l \leqslant m),$$

$$q_0^{1/2}\lambda_k^m,$$

$$q_0\lambda_k^{l-1} + p_0\lambda_k^{l-2} + q_1\lambda_k^{l-3} + \cdots + p_{l-m-2}\lambda_k^{2m-l+2} + \tfrac{1}{2}q_{l-m-1}\lambda_k^{2m-l+1} \qquad (m+2 \leqslant l \leqslant 2m+1),$$

and the matrix T in (3.1.4) is now

$$T = (v_1 \cdots v_{2m+1}). \tag{3.6.4}$$

The inverse matrix T^{-1} can again be obtained from Lemmas 1.2.1 and 1.2.2 and, in place of (3.1.13), we now have

$$M_j = (2m+1)q_0\lambda_j^{2m} + 2mp_0\lambda_j^{2m-1} + \cdots + 2p_{m-1}\lambda_j + q_m. \tag{3.6.5}$$

Then, corresponding to (3.1.15) and (3.1.16), the entries τ_{jk} in $T^{-1}T'$ are given by

$$\tau_{jj} = \tfrac{1}{2}M_j'/M_j \tag{3.6.6}$$

and, for $j \neq k$,

$$\tau_{jk} = (\lambda_j - \lambda_k)^{-1}M_j^{-1}\left(\sum_{r=0}^{m} \{\tfrac{1}{2}q_r'(\lambda_j\lambda_k)^{m-r}(\lambda_j + \lambda_k) + p_r'(\lambda_j\lambda_k)^{m-r}\}\right). \tag{3.6.7}$$

Moving on to the situation which is analagous to (3.3.1), we suppose that the p_r and q_r are all nowhere zero in $[a, \infty)$, with

and
$$\left.\begin{aligned} p_r/q_r &= o(q_r/p_{r-1}) \\ q_r/p_{r-1} &= o(p_{r-1}/q_{r-1}) \end{aligned}\right\} \tag{3.6.8}$$

as $x \to \infty$, for $1 \leqslant r \leqslant m$. Then, as in Lemma 3.3.1, we find that the solutions λ_k of (3.6.3) satisfy

$$\lambda_k \sim \begin{cases} -q_K/p_{K-1} & (k = 2K,\ 1 \leqslant K \leqslant m) \\ -p_K/q_K & (k = 2K+1,\ 0 \leqslant K \leqslant m) \end{cases} \tag{3.6.9}$$

as $x \to \infty$, and hence

$$\lambda_j = o(\lambda_k) \qquad (j > k). \tag{3.6.10}$$

Further, corresponding to (3.3.8), we have

$$M_j \sim \begin{cases} -q_J\lambda_j^{2m-j} & (j = 2J) \\ -p_J\lambda_j^{2m-j} & (j = 2J+1) \end{cases} \tag{3.6.11}$$

and consequently, in place of (3.3.7), we have

$$\lambda_j^{m-K}M_j^{-1/2} = o(\lambda_l^{m-K}M_l^{-1/2}) \tag{3.6.12}$$

for both $j<l\leq 2K$ and $j>l\geq 2K+1$. Finally,

$$\lambda_k^{2m-2K}M_k^{-1}\sim -q_K^{-1} \tag{3.6.13}$$

for both $k=2K$ and $k=2K+1$. These results for M_j follow from (3.6.5), (3.6.8) and (3.6.9).

To give the theorem which corresponds to Theorem 3.3.1, we first define

$$\delta=\max_r\,(|p_rp_{r-1}/q_r^2|,\ |q_rq_{r-1}/p_{r-1}^2|), \tag{3.6.14}$$

as in (3.3.10). Then $\delta=o(1)$ by (3.6.8). We suppose that (3.3.11) holds again, with a similar condition on the q_r as well. Proceeding as in (3.3.17)–(3.3.19), we define

$$M=\mathrm{dg}\,(M_1^{-1/2},\ \dots,\ M_{2m+1}^{-1/2}) \tag{3.6.15}$$

and make the transformation

$$Y=T_1Z$$

with

$$T_1=TM.$$

Then, after a calculation based on (3.6.7) and (3.6.8) but otherwise similar to the one leading to (3.3.31), the Y-system (3.6.2) becomes

$$Z'=(\Lambda+R_2+S)Z, \tag{3.6.16}$$

where

$$\Lambda=\mathrm{dg}\,(\lambda_1,\ \dots,\ \lambda_{2m+1}),$$

S is $L(a,\infty)$, and

$$R_2=\mathrm{dg}\,(0,\rho_1,\ \dots,\ \rho_m).$$

Here the $(1,1)$ entry of R_2 is zero and ρ_j is the 2×2 block

$$\rho_j=\tfrac{1}{2}\mathrm{i}q_j'/q_j\begin{bmatrix}0 & -1\\ 1 & 0\end{bmatrix}.$$

Finally, we make a transformation of the type (3.3.35) in which the 2×2 blocks along the diagonal of $\Lambda+R_2$ are themselves expressed in diagonal form. Under the appropriate conditions on the p_r and q_r, we obtain (3.3.36) and (3.3.37) in the present notation and, at this point in the description of the analysis, we give the full statement of the theorem which corresponds to Theorem 3.3.1.

Theorem 3.6.1 *Let p_r and q_r $(0\leq r\leq m)$ be nowhere zero with locally absolutely continuous first derivatives in $[a,\infty)$. Let (3.6.8) hold and let*

(i) $$\delta^{1/2}p_r'/p_r\in L(a,\infty),\qquad \delta^{1/2}q_r'/q_r\in L(a,\infty), \tag{3.6.17}$$

both for $0 \leq r \leq m$;

(ii) $$q_r'/q_r = o(q_r/p_{r-1}) \qquad (1 \leq r \leq m); \tag{3.6.18}$$

(iii) $$(p_{r-1}q_r'/q_r^2)' \in L(a, \infty) \qquad (1 \leq r \leq m); \tag{3.6.19}$$

(iv) $$p_{r-1}q_r'^2/q_r^3 \in L(a, \infty) \qquad (1 \leq r \leq m). \tag{3.6.20}$$

Let

$$\operatorname{re}(\lambda_j - \lambda_k) \tag{3.6.21}$$

have one sign for each unequal pair j, k *in* $[1, 2m+1]$, *where the* λ_k *are the solutions of* (3.6.3). *Then* (3.6.1) *has solutions* y_k $(1 \leq k \leq 2m+1)$ *such that*

$$y_k^{(m-K)} \sim q_K^{-1/2} \exp\left(\int_a^x \lambda_k(t)\,dt\right) \qquad (k = 2K, 2K+1), \tag{3.6.22}$$

$$y_k^{(m-K)} = o\left\{q_K^{-1/2} \exp\left(\int_a^x \lambda_k(t)\,dt\right)\right\} \qquad (k \neq 2K, 2K+1) \tag{3.6.23}$$

for $0 \leq K \leq m$, *with* $k = 0$ *omitted.*

Here (i)–(iv) are conditions under which the 2×2 blocks in $\Lambda + R_2$ can be expressed in diagonal form, as in Theorems 1.6.1 and 2.2.1. For example, in the case of the jth block, (1.6.2) is

$$q_j'/q_j = o(\lambda_{2j} - \lambda_{2j+1}) = o(q_j/p_{j-1}),$$

by (3.6.8) and (3.6.9), and this condition is imposed here in (3.6.18). Again, (3.6.20) is the simplifying condition (1.6.37) for the jth block. Thus (3.3.39) continues to hold with the present notation, as a consequence of (3.3.37), and the required formulae (3.6.22)–(3.6.23) follow immediately, by (3.6.12) and (3.6.13).

The dichotomy condition (3.6.21) is certainly satisfied when the p_r and q_r are all real. In this case the λ_k are real and, by (3.6.10), $\operatorname{re}(\lambda_j - \lambda_k) \sim -\lambda_k$ if $j > k$. Again, (3.6.12) is satisfied when the p_r are real and the q_r are pure imaginary. In this case the λ_k are pure imaginary, and $\operatorname{re}(\lambda_j - \lambda_k) = 0$. Once again, the theorem is illustrated by the example where the coefficients are powers of x in the form

$$p_r(x) = (\text{const.})x^{\alpha_r}, \qquad q_r(x) = (\text{const.})x^{\beta_r} \tag{3.6.24}$$

for $0 \leq r \leq m$. The conditions (3.6.8) and (3.6.17)–(3.6.20) are satisfied when

$$\left.\begin{aligned} &\beta_0 - \alpha_0 < \alpha_0 - \beta_1 < \beta_1 - \alpha_1 < \cdots < \alpha_{m-1} - \beta_m < \beta_m - \alpha_m \\ \text{and} \quad &\alpha_{m-1} - \beta_m < 1. \end{aligned}\right\} \tag{3.6.25}$$

As a final comment on the theorem, we note that, although the p_r and q_r are assumed to be nowhere zero in $[a, \infty)$, the same ideas continue to apply when some of the coefficients are identically zero, as in the case of (3.3.45).

The discussion up to this point in the chapter can be put into a wider context if we leave aside (3.1.1) and (3.6.1) for the moment and return to the general system (1.1.1). As far as methods based on (3.1.4)–(3.1.7) are concerned, it is a question of identifying matrices A for which

(a) the characteristic equation can be expressed conveniently in terms of the entries in A,

(b) the eigenvalues can be estimated asymptotically,

(c) the product $T^{-1}T'$ can be estimated sufficiently well to allow the theorems in Chapter 1 to be applied to (3.1.6).

These requirements are certainly satisfied in the case of (3.1.1) and (3.6.1). Possible extensions of the methods would therefore be focused on self-adjoint matrix analogues of (3.1.1) and (3.6.1), in which p_r and q_r become matrix coefficients, or yet more generally on the Hamiltonian system

$$JW' = HW \tag{3.6.26}$$

in which J is constant and non-singular, $J^* = -J$ and $H^* = H$. It is unlikely that the analysis could be pursued in full generality but, no doubt, any particular system (3.6.26) that is of interest in this context could be investigated individually.

3.7 EQUATIONS OF GENERALIZED HYPERGEOMETRIC TYPE

In Theorem 3.5.2, we considered an Euler situation for the fourth-order equation (3.5.1). This situation was specified by the condition (3.5.39), and the simplest example is provided by the equation

$$y^{(4)} + a_1(x^{\alpha+2}y')' + a_2x^{\alpha}y = 0, \tag{3.7.1}$$

where a_1 and a_2 are non-zero constants and $\alpha > -4$. The resulting asymptotic formulae (3.5.15) and (3.5.44) show that two solutions of (3.7.1) have an exponential nature and two have an algebraic nature in terms of x.

Here we discuss an equation of the form (3.1.1) which is a higher-order extension of (3.7.1). We consider

$$y^{(2m)} + \sum_{r=\nu}^{m} \{a_r x^{\alpha+2m-2r} y^{(m-r)}\}^{(m-r)} = 0, \tag{3.7.2}$$

where ν is a given integer in $[1, m-1]$, the a_r are constants with $a_\nu \neq 0$, and

$$\alpha > -2m. \tag{3.7.3}$$

This equation is of additional interest because it is one of a class of similar equations which can be expressed in terms of the generalized hypergeometric equation. If the solutions are taken as Meijer G-functions, the asymptotic form can be determined and it is found that, in general, there are 2ν solutions with an exponential nature asymptotically and $2m-2\nu$ solutions with an algebraic nature. There are some exceptional cases, corresponding to $\kappa^2=16$ in Theorem 3.5.2, which arise for certain values of the a_r. We refer to the book of Paris and Wood (1985) for a detailed survey of the asymptotic results obtained in this way for the equation (3.7.2) and for other similar equations. The main purpose here is to establish a connection between the asymptotic formulae arising as just described and those obtained by the general methods of this chapter.

We begin with the details of the way in which (3.7.2) is expressed as a generalized hypergeometric equation. Let $\mathsf{D}=x\,\mathrm{d}/\mathrm{d}x$. Then, considering the general term in (3.7.2), we have

$$\begin{aligned}\frac{\mathrm{d}}{\mathrm{d}x}\{x^{\alpha+2m-2r}y^{(m-r)}\} &= \frac{\mathrm{d}}{\mathrm{d}x}\{x^{\alpha+m-r}(x^{m-r}y^{(m-r)})\}\\ &= x^{\alpha+m-r-1}(\mathsf{D}+\alpha+m-r)(x^{m-r}y^{(m-r)})\\ &= x^{\alpha+m-r-1}(\mathsf{D}+\alpha+m-r)\prod_{s=1}^{m-r}(\mathsf{D}+1-s)y,\end{aligned}$$

and hence

$$\{x^{\alpha+2m-2r}y^{(m-r)}\}^{(m-r)} = x^{\alpha}\prod_{s=1}^{m-r}(\mathsf{D}+\alpha+s)(\mathsf{D}+1-s)y.$$

On multiplying (3.7.2) by x^{2m} and writing

$$\beta=\alpha+2m, \tag{3.7.4}$$

we obtain

$$\left\{\prod_{s=1}^{2m}(\mathsf{D}+1-s)+x^{\beta}\sum_{r=\nu}^{m}a_r\prod_{s=1}^{m-r}(\mathsf{D}+\alpha+s)(\mathsf{D}+1-s)\right\}y=0. \tag{3.7.5}$$

We note that $\beta>0$ by (3.7.3). Now let ρ_j $(1\leqslant j\leqslant 2m-2\nu)$ denote the zeros of the polynomial

$$\sum_{r=\nu}^{m}a_r\prod_{s=1}^{m-r}(\rho+\alpha-s)(\rho+1-s). \tag{3.7.6}$$

Then (3.7.5) is

$$\left\{\prod_{j=1}^{2m}(\mathsf{D}+1-j)+x^{\beta}\prod_{j=1}^{2m-2\nu}(\mathsf{D}-\rho_j)\right\}y=0.$$

Finally, making the change of variable

$$t=\beta^{-2\nu}x^{\beta} \tag{3.7.7}$$

and writing $\mathsf{D}_1 = t\,\mathrm{d}/\mathrm{d}t = \beta^{-1}\mathsf{D}$, we obtain

$$\left\{\prod_{j=1}^{2m}\left(\mathsf{D}_1 - \frac{j-1}{\beta}\right) + t\prod_{j=1}^{2m-2\nu}(\mathsf{D}_1 - \rho_j/\beta)\right\}y=0, \tag{3.7.8}$$

which is an example of the generalized hypergeometric equation. It follows from (3.7.7) and the Meijer G-function solutions of (3.7.8) (Paris and Wood 1981, § 4; 1985, § 5.1.2) that there are solutions y_k of (3.7.2) such that

$$y_k \sim x^{-(2m-1)(\alpha+2m-2\nu)/4\nu}\exp\left\{\frac{2\nu}{\alpha+2m}(-a_\nu)^{1/2\nu}\omega_k x^{(\alpha+2m)/2\nu}\right\}$$

$$(1\leqslant k\leqslant 2\nu), \tag{3.7.9}$$

$$y_k \sim x^{\rho_{k-2\nu}} \qquad (2\nu+1\leqslant k\leqslant 2m), \tag{3.7.10}$$

where $(-a_\nu)^{1/2\nu}$ is any fixed 2νth root and the ω_k are the 2νth roots of unity. These formulae express the exponential and algebraic natures of the solutions.

Turning now to the general methods of this chapter, we consider a slight generalization of (3.7.2) which is

$$y^{(2m)}+\sum_{r=\nu}^{m}\{(a_r+\phi_r)x^{\alpha+2m-2r}y^{(m-r)}\}^{(m-r)}=0, \tag{3.7.11}$$

where the functions ϕ_r satisfy

$$\phi_r = o(1), \tag{3.7.12}$$

$$\phi_r' \in L(a,\infty), \tag{3.7.13}$$

and

$$x^{-1}\phi_r \in L(a,\infty). \tag{3.7.14}$$

The characteristic equation (3.1.8) is now

$$\lambda^{2m}+\sum_{r=\nu}^{m}(a_r+\phi_r)x^{\alpha+2m-2r}\lambda^{2m-2r}=0. \tag{3.7.15}$$

To solve (3.7.15) asymptotically, we first write the equation as

$$\lambda^{2\nu}+x^{\alpha+2m-2\nu}\sum_{r=\nu}^{m}(a_r+\phi_r)(x\lambda)^{-2(r-\nu)}=0.$$

It then follows from (3.7.3) that there are solutions λ_k $(1\leqslant k\leqslant 2\nu)$ such that

$$\lambda_k = x^{(\alpha+2m-2\nu)/2\nu}(-a_\nu)^{1/2\nu}\omega_k \times\{1+O(\phi_\nu)+O(x^{-(\alpha+2m)/\nu})\}. \tag{3.7.16}$$

For these solutions, $x\lambda_k\to\infty$ as $x\to\infty$.

To obtain the other $2m-2v$ solutions λ_k, we write (3.7.15) as

$$\sum_{r=v}^{m} (a_r + \phi_r)(x\lambda)^{2m-2r} + x^{-(\alpha+2m)}(x\lambda)^{2m} = 0.$$

It follows from (3.7.3) again that there are solutions

$$\lambda_k = x^{-1}\mu_{k-2v}\{1 + O(\phi) + O(x^{-(\alpha+2m)})\}, \tag{3.7.17}$$

where $\phi = \max_r |\phi_r|$ and μ_j $(1 \leqslant j \leqslant 2m-2v)$ are the roots of the polynomial

$$g(\mu) = a_v\mu^{2m-2v} + \cdots + a_{m-1}\mu^2 + a_m, \tag{3.7.18}$$

with constant coefficients. We assume that the a_r are such that the μ_j are distinct. Thus the solutions of (3.7.15) fall into the two classes (3.7.16) and (3.7.17), the eigenvalues in the same class having the same size as $x \to \infty$, but those in the first class being large compared to those in the second.

We can now discuss the form which the Z-system (3.3.20) takes in the present situation and, to do this, we estimate the entries σ_{jk} in (3.3.23) in the three cases where λ_j and λ_k are both in the same class (3.7.16), or both in (3.7.17), or in separate classes. First, when λ_j and λ_k both have the form (3.7.16), it follows from (3.1.13) and (3.7.15) that

$$M_j = 2v\lambda_j^{2m-1}\{1 + O(x^{-(\alpha+2m)/v})\}, \tag{3.7.19}$$

and similarly for M_k. Then (3.3.23), (3.7.12), (3.7.13), (3.7.14), and (3.7.16) give

$$\sigma_{jk} = c_{jk}x^{-1} + s_{jk},$$

where

$$c_{jk} = -\{2v(\omega_j - \omega_k)(\omega_j\omega_k)^{v-\frac{1}{2}}\}^{-1} \tag{3.7.20}$$

and $s_{jk} \in L(a, \infty)$.

Next, when λ_i and λ_k have different forms (3.7.16) and (3.7.17), we have (3.7.19) as before, and

$$M_k \sim x^{\alpha+1}g'(\mu_{k-2v})\{1 + O(\phi) + O(x^{-(\alpha+2m)})\}, \tag{3.7.21}$$

by (3.1.13) and (3.7.18). Then (3.3.23), (3.7.19), and (3.7.21) give

$$\sigma_{jk} = O(x^{-1-(\alpha+2m)/4v}) + s_{jk},$$

where s_{jk} contains terms $O(\phi_r')$. Hence $\sigma_{jk} \in L(a, \infty)$ in this case, by (3.7.3) and (3.7.13).

Finally, when λ_j and λ_k both have the form (3.7.17), we have (3.7.21) for both M_j and M_k, and then (3.3.23) gives

$$\sigma_{jk} = (\mu_{j-2\nu} - \mu_{k-2\nu})^{-1}\{g'(\mu_{j-2\nu})g'(\mu_{k-2\nu})\}^{-1/2}$$
$$\times\left(\sum_{r=\nu}^{m} \{\phi_r' + (\alpha + 2m - 2r)(a_r + \phi_r)x^{-1}\}(\mu_{j-2\nu}\mu_{k-2\nu})^{m-r}\right)$$
$$\times\{1 + O(\phi) + O(x^{-(\alpha+2m)})\}$$
$$= x^{-1}d_{jk} + s_{jk},$$

where $s_{jk} \in L(a, \infty)$ and

$$d_{jk} = (\mu_{j-2\nu} - \mu_{k-2\nu})^{-1}\{g'(\mu_{j-2\nu})g'(\mu_{k-2\nu})\}^{-1/2} \times \sum_{r=\nu}^{m} (\alpha + 2m - 2r)a_r(\mu_{j-2\nu}\mu_{k-2\nu})^{m-r}. \tag{3.7.22}$$

Altogether, the system (3.3.20) takes the form

$$Z' = \left\{\begin{pmatrix} \Lambda_1 + x^{-1}C & 0 \\ 0 & x^{-1}(\Lambda_2 + D) \end{pmatrix} + S\right\}Z, \tag{3.7.23}$$

where $S \in L(a, \infty)$, and C and D are the skew-symmetric matrices $C = (c_{jk})$, $D = (d_{jk})$ defined by (3.7.20) and (3.7.22). Also, Λ_1 and Λ_2 are the diagonal matrices formed by the λ_k and $\mu_{k-2\nu}$ in (3.7.16) and (3.7.17).

We can now proceed to the Levinson form by making a transformation

$$Z = \begin{bmatrix} I + Q & 0 \\ 0 & U \end{bmatrix} W$$

in (3.7.23), where $I + Q$ diagonalizes $\Lambda_1 + x^{-1}C$ as in Theorem 1.6.1, and the constant matrix U diagonalizes $\Lambda_2 + D$. The simplifying condition (1.6.35) is satisfied by $\Lambda_1 + x^{-1}C$, and we obtain the Levinson form

$$W' = (\Lambda + S)W$$

with
$$\Lambda = \mathrm{dg}\,(\lambda_1, \ldots, \lambda_{2\nu}, x^{-1}\tau_1, \ldots, x^{-1}\tau_{2m-2\nu}),$$

where the τ_j are the eigenvalues of $\Lambda_2 + D$.

On transforming back to (3.7.11), we meet the same problem as in (3.3.38) because M_j and M_k have different sizes in (3.7.19) and (3.7.21). We deal with the situation as in (3.3.39) by considering derivatives as appropriate. We then find that there are solutions y_k of (3.7.11) such that

$$y_k^{(m-\nu)} \sim \lambda_k^{m-\nu} M_k^{-1/2} \exp\left(\int_a^x \lambda_k(t)\,\mathrm{d}t\right) \quad (1 \leqslant k \leqslant 2\nu), \tag{3.7.24}$$

$$y_k \sim x^{-\frac{1}{2}(1+\alpha)+\tau_{k-2\nu}} \quad (2\nu + 1 \leqslant k \leqslant 2m). \tag{3.7.25}$$

The methods to be given in §§ 3.8 and 3.9 show that, in effect, (3.7.24) can be integrated, and the asymptotic formula for y_k itself has only the factor $M_k^{-1/2}$ multiplying the exponential. By (3.7.16) and (3.7.19), this formula agrees with (3.7.9). The remaining y_k in (3.7.25) are the $2m-2\nu$ solutions of algebraic type. The exponents in (3.7.25) have been obtained by a different method from those in (3.7.10) and, while the two sets of values of these exponents can only tally, it would be an algebraic matter to verify this independently.

3.8 INTEGRATION OF ASYMPTOTIC FORMULAE

We return now to the main question left unanswered by the theory of §§ 3.3–3.4, where the asymptotic formulae appear in terms of derivatives of solutions. The question is whether the asymptotic forms of the solutions themselves can be obtained and, although the matter was resolved for fourth-order equations in § 3.5, it is only in certain specialized, but nonetheless significant, situations that the analysis can be suitably refined for higher-order equations. We begin with the following lemma concerning the integration of an asymptotic formula. The lemma is not confined to solutions of differential equations.

Lemma 3.8.1 *Let the function y be locally absolutely continuous in* $[a,\infty)$, *and let*

$$y' \sim \phi\mu \exp\left(\int_a^x \mu(t)\,\mathrm{d}t\right) \tag{3.8.1}$$

as $x\to\infty$, *where* ϕ *and* $\operatorname{re}\mu$ *are nowhere zero in* $[a,\infty)$, ϕ *is locally absolutely continuous and, as* $x\to\infty$,

(i) $$\mu = O(\operatorname{re}\mu); \tag{3.8.2}$$

(ii) $$\phi'/\phi = o(\mu). \tag{3.8.3}$$

Then, as $x\to\infty$,

$$y \sim \phi \exp\left(\int_a^x \mu(t)\,\mathrm{d}t\right) \tag{3.8.4}$$

in the following specific cases:

(a) $\operatorname{re}\mu>0$, $|\phi|$ *is increasing, and*

$$\int_{\frac{1}{2}x}^x \operatorname{re}\mu(t)\,\mathrm{d}t\to\infty; \tag{3.8.5}$$

(b) $\operatorname{re}\mu>0$, $|\phi|$ *is bounded, and*

$$\phi(x)\exp\left(\int_{\frac{1}{2}x}^x \operatorname{re}\mu(t)\,\mathrm{d}t\right)\to\infty; \tag{3.8.6}$$

(c) re $\mu < 0$, $y(\infty) = 0$, *and*

$$\phi(x) \exp\left(\int_a^x \operatorname{re} \mu(t)\, dt\right) \to 0. \tag{3.8.7}$$

Proof By (3.8.1), we can write

$$y' = \phi\mu(1+\psi) \exp\left(\int_a^x \mu(t)\, dt\right), \tag{3.8.8}$$

where $\psi = o(1)$. Starting with case (a), we integrate (3.8.8) over (a, x). After an integration by parts, we obtain

$$\begin{aligned} y &= \phi \exp\left(\int_a^x \mu(t)\, dt\right) - \int_a^x \phi'(s) \exp\left(\int_a^s \mu(t)\, dt\right) ds \\ &\quad + \int_a^x (\phi\mu\psi)(s) \exp\left(\int_a^s \mu(t)\, dt\right) ds + c \\ &= u_1 + u_2 + u_3 + c, \end{aligned} \tag{3.8.9}$$

say, where c is a constant. By (3.8.2) and (3.8.3),

$$|u_2 + u_3| \leqslant \int_a^x (|\psi|\,|\phi| \operatorname{re} \mu)(s) \exp\left(\int_a^s \operatorname{re} \mu(t)\, dt\right) ds \tag{3.8.10}$$

with a new ψ which is again $o(1)$. Splitting the range of integration (a, x) into $(a, \tfrac{1}{2}x)$ and $(\tfrac{1}{2}x, x)$, and using the fact that $|\phi|$ is increasing, we obtain

$$\begin{aligned} u_2 + u_3 &= O\left\{\phi(\tfrac{1}{2}x) \exp\left(\int_a^{\frac{1}{2}x} \operatorname{re} \mu(t)\, dt\right)\right\} \\ &\quad + o\left\{\phi(x) \exp\left(\int_a^x \operatorname{re} \mu(t)\, dt\right)\right\} \\ &= o\left\{\phi(x) \exp\left(\int_a^x \operatorname{re} \mu(t)\, dt\right)\right\}, \end{aligned} \tag{3.8.11}$$

by (3.8.5). Then (3.8.4) follows immediately from (3.8.9).

In case (b), we argue as before up to (3.8.10). Since $|\phi|$ is bounded now, on splitting the range of integration into $(a, \tfrac{1}{2}x)$ and $(\tfrac{1}{2}x, x)$ again, we obtain

$$\int_a^{\frac{1}{2}x} \cdots ds = O\left(\int_a^{\frac{1}{2}x} \exp \operatorname{re} \mu(t)\, dt\right) \tag{3.8.12}$$

and

$$\int_{\frac{1}{2}x}^x \cdots ds = o\{I(x)\}, \tag{3.8.13}$$

where

$$I(x)=\int_{\frac{1}{2}x}^{x}(|\phi|\,\mathrm{re}\,\mu)(s)\exp\left(\int_a^s \mathrm{re}\,\mu(t)\,\mathrm{d}t\right)\mathrm{d}s.$$

Now

$$I(x)\leq|\phi(x)|\exp\left(\int_a^x \mathrm{re}\,\mu(t)\,\mathrm{d}t\right)$$
$$+\int_{\frac{1}{2}x}^{x}(|\psi|\,|\phi|\,\mathrm{re}\,\mu)(s)\exp\left(\int_a^s \mathrm{re}\,\mu(t)\,\mathrm{d}t\right)\mathrm{d}s$$

when we integrate by parts and use (3.8.3) again. The second term on the right here is $o\{I(x)\}$, and it follows that

$$I(x)=O\left\{\phi(x)\exp\left(\int_a^x \mathrm{re}\,\mu(t)\,\mathrm{d}t\right)\right\}. \tag{3.8.14}$$

Hence the right-hand side of (3.8.13) is

$$o\left\{\phi(x)\exp\left(\int_a^x \mathrm{re}\,\mu(t)\,\mathrm{d}t\right)\right\}$$

and, by (3.8.6), the same is true of the right-hand side of (3.8.12). Thus (3.8.11) holds again, and then (3.8.4) follows from (3.8.9).

Finally, in case (c), we integrate (3.8.8) over (x, ∞). By (3.8.7) and the fact that $y(\infty)=0$, we obtain

$$y(x)=\phi(x)\exp\left(\int_a^x \mu(t)\,\mathrm{d}t\right)-\int_x^\infty \phi'(s)\exp\left(\int_a^s \mu(t)\,\mathrm{d}t\right)\mathrm{d}s$$
$$-\int_x^\infty(\phi\mu\psi)(s)\exp\left(\int_a^s \mu(t)\,\mathrm{d}t\right)\mathrm{d}s. \tag{3.8.15}$$

By (3.8.2) and (3.8.3), the last two terms on the right are

$$o\left\{\int_x^\infty(|\phi|(-\mathrm{re}\,\mu))(s)\exp\left(\int_a^x \mathrm{re}\,\mu(t)\,\mathrm{d}t\right)\mathrm{d}s\right\}$$
$$=o\left\{\phi(x)\exp\left(\int_a^x \mathrm{re}\,\mu(t)\,\mathrm{d}t\right)\right\}$$

after a further integration by parts as in the case of (3.8.14). Then (3.8.4) follows from (3.8.15), and the proof of the lemma is complete. □

The lemma can be applied to the asymptotic formula (3.3.16) when K takes any value such that p_K/p_{K-1} is real and negative or, more generally, when

$$\delta\leq\arg p_K/p_{K-1}\leq 2\pi-\delta \tag{3.8.16}$$

for some K, where $\delta\ (>0)$ is a constant. In the following theorem, we give the result that is obtained for y_{2K-1} and y_{2K} by repeated application of the lemma. The conditions imposed both in the lemma and in the theorem are framed in a way that certainly covers coefficients which are powers of x, as considered in (3.2.20) and (3.3.43).

Theorem 3.8.1 *Let the conditions of Theorem* 3.3.1 *hold and let* (3.8.16) *hold for a given K. Let*

$$p_r'/p_r = o\{(p_K/p_{K-1})^{1/2}\} \qquad (r = K,\, K-1). \tag{3.8.17}$$

Also, for $1 \leqslant \nu \leqslant m-K$, *let*

$$|p_{K-1}^{\nu-\frac{1}{2}}/p_K^{\nu+\frac{1}{2}}| \quad \textit{be monotonic} \tag{3.8.18}$$

and let

$$\Phi \exp\left(\beta \int_{\frac{1}{2}x}^{x} |\mathrm{re}\,(-p_K/p_{K-1})^{1/2}|\,\mathrm{d}t\right) \to \infty \tag{3.8.19}$$

as $x \to \infty$, *for some constant* β $(0 < \beta < 1)$, *where*

$$\Phi = \min\,(1,\, |p_{K-1}^{\frac{1}{2}\nu-\frac{1}{4}}/p_K^{\frac{1}{2}\nu+\frac{1}{4}}|,\, |p_{K+\nu-1}p_{K+\nu}|^{1/4}).$$

Then the two solutions y_k *in* (3.3.16) *also satisfy*

$$y_k^{(m-K-\nu)} \sim (p_{K-1}^{\frac{1}{2}\nu-\frac{1}{4}}/p_K^{\frac{1}{2}\nu+\frac{1}{4}}) \exp\left(\int_a^x \lambda_k(t)\,\mathrm{d}t\right)$$
$$(1 \leqslant \nu \leqslant m-K). \tag{3.8.20}$$

Proof In the lemma, we take

$$\phi = |p_{K-1}^{\frac{1}{2}\nu-\frac{1}{4}}/p_K^{\frac{1}{2}\nu+\frac{1}{4}}| \qquad (\nu = 1, \ldots, m-K)$$

and $\mu = \lambda_k$. Then (3.8.16) and (3.8.17) imply that (3.8.2) and (3.8.3) are satisfied. Also, (3.8.19) and (3.3.2) imply that (3.8.5) and (3.8.6) are satisfied. Further, we have

$$\phi = o\{(p_{K+\nu-1}p_{K+\nu})^{-1/4}\}$$

by (3.3.4) (after a change of notation), and then (3.8.7) follows from (3.8.19) and (3.3.2) again. We also have

$$y_k^{(m-K-\nu)} = o\left\{(p_{K+\nu-1}p_{K+\nu})^{-1/4} \exp\left(\int_a^x \lambda_k(t)\,\mathrm{d}t\right)\right\}$$

for $1 \leqslant \nu \leqslant m-K$ by (3.3.44) and, when $\mathrm{re}\,\lambda_k < 0$, it follows from (3.8.19) and (3.3.2) that

$$y_k^{(m-K-\nu)}(\infty) = 0.$$

Starting therefore with (3.3.16) and $\nu = 1$, we can apply the lemma with

$y = y_k^{(m-K-1)}$ to obtain (3.8.20) in the case $\nu = 1$. The process can be continued with $\nu = 2, \ldots, m-K$ in turn to yield (3.8.20) for the complete range $1 \leqslant \nu \leqslant m-K$. The theorem is now proved. □

We make a special note of the case $\nu = m-K$, which is

$$y_k \sim (p_{K-1}^{\frac{1}{2}(m-K)-\frac{1}{4}}/p_K^{\frac{1}{2}(m-K)+\frac{1}{4}}) \exp\left(\int_a^x \lambda_k(t)\,dt\right). \qquad (3.8.21)$$

This formula does of course tally with (3.5.15) when $m = 2$ and $K = 1$. The condition (3.8.17) is a slight strengthening of (3.3.12) in the case $r = K$, just as occurred in (3.5.11).

Turning to Theorem 3.4.1, the solutions y_k which correspond to $1 \leqslant K \leqslant L$ are also covered by (3.8.20), subject again to the conditions (3.8.16)–(3.8.19). For the remaining solutions y_{2K}, the second formula in (3.4.4) can be integrated immediately and, on taking suitable linear combinations of the y_{2K}, we obtain solutions

$$y_j(x) \sim x^{j-1}/(j-1)! \qquad (1 \leqslant j \leqslant m-L).$$

There still remain, however, the $m-L$ solutions y_{2K-1} in (3.4.4) for which only o-formulae are obtained.

3.9 ESTIMATION OF ERROR TERMS

Next we consider the possibility of improving (3.3.16) to an asymptotic formula for y_k when (3.8.16) does not hold. The integration idea of § 3.8 no longer applies because there is no result like Lemma 3.8.1 in which, for example, μ is pure imaginary. The difficulty is illustrated by the simple function

$$y(x) = e^{ix} + x^{1/2},$$

for which $y'(x) \sim ie^{ix}$, but $y(x)$ is certainly not asymptotic to e^{ix}. An alternative method is to examine the $o(1)$, or error, terms in (3.3.37) more closely and, as a result, improve (3.3.38) directly so that it becomes a true asymptotic formula.

It seems unlikely that a general analysis can be carried out, and therefore we concentrate on the important but relatively simple case in which

$$p_r = P_r \quad (r \neq m), \qquad p_m = P_m + ih, \qquad (3.9.1)$$

where h and the P_r are real-valued and nowhere zero in $[a, \infty)$, with

$$h = o(P_m) \qquad (x \to \infty). \qquad (3.9.2)$$

As in § 3.8, we consider a particular K $(1 \leqslant K \leqslant m)$ and, in contrast to

(3.8.16), we suppose now that

$$P_K/P_{K-1} > 0. \tag{3.9.3}$$

In the analysis which follows, we require the functions F_r defined by

$$F_r = (P_r P_{r-1})^{-1/2}(P_{r-1}/P_r)^{m-r}. \tag{3.9.4}$$

By (3.3.8) and (3.3.2), we have

$$(|M_{2r}| =)|M_{2r-1}| \sim 2\,|F_r|^{-1}. \tag{3.9.5}$$

Also, as in the case of (3.3.9), it follows immediately from (3.9.4) and (3.3.1) that

$$F_r = o(F_{r+1}) \qquad (1 \leqslant r \leqslant m-1). \tag{3.9.6}$$

Now (3.9.6) implies that there is an integer r_0 $(1 \leqslant r_0 \leqslant m+1)$ such that

$$\int_a^\infty |hF_r|\,\mathrm{d}x \tag{3.9.7}$$

converges for $r \leqslant r_0 - 1$ and diverges for $r \geqslant r_0$. In the asymptotic result for y_k to be proved in this section, there are two cases determined by the value of r_0.

We also require some notation relating to the conditions imposed in the earlier result, Therorem 3.3.1. We define the functions P_ν $(\nu = 1, 2, 3)$ by

$$\left.\begin{aligned} \mathsf{P}_1 &= \delta^{3/4} \max |p_r'/p_r|, \\ \mathsf{P}_2 &= \max |\{p_{r-1}^{-1/2} p_r^{-3/2}(p_{r-1}p_r)'\}'|, \\ \mathsf{P}_3 &= \max |p_{r-1}^{-3/2} p_r^{-5/2}(p_{r-1}p_r)'^2|, \end{aligned}\right\} \tag{3.9.8}$$

the maximum being taken over all r with x fixed in each case. In the following theorem, the various conditions are somewhat cumbersome in general, but they are formulated in such a way that certainly covers coefficients which are powers of x. In this respect the conditions have something in common with those in Theorem 3.8.1.

Theorem 3.9.1 *Let the conditions of Theorem* 3.3.1 *hold and let the* p_r *be as in* (3.9.1) *and* (3.9.2). *Let* (3.9.3) *hold for a given K. In addition, let*

(a) $|\mathsf{P}_\nu/hF_J|$ *decrease to zero as* $x \to \infty$ *for* $J \geqslant r_0$;

(b)
$$(\mathsf{P}_\nu/hF_J)(\tfrac{1}{2}x) = O\{(\mathsf{P}_\nu/hF_J)(x)\} \tag{3.9.9}$$

for $J \geqslant r_0$;

(c)
$$\exp\left(-\beta\int_{\frac{1}{2}x}^x |hF_{r_0}|\,\mathrm{d}t\right) = o\{(F_K/F_J)^{1/2}\} \tag{3.9.10}$$

for $J \geqslant r_0$ *and some* β $(0 < \beta < 1)$.

Then the two solutions y_k *in* (3.3.16) *also satisfy*
(I) *if* $K \geqslant r_0$,

$$y_k \sim (p_{K-1}^{\frac{1}{2}(m-K)-\frac{1}{4}}/p_K^{\frac{1}{2}(m-K)+\frac{1}{4}}) \exp\left(\int_a^x \lambda_k(t)\,dt\right); \tag{3.9.11}$$

(II) *if* $K \leqslant r_0 - 1$,

$$y_k = h^{-1/2}\rho_k \exp\left(\int_a^x \lambda_k(t)\,dt\right), \tag{3.9.12}$$

where $\rho_k \in L^2(a, \infty)$.

Proof First we require an estimate for re λ_k. Arguing with (3.3.5) again, we divide (3.1.8) by λ^{2m-2K} and substitute (3.3.5) to obtain

$$\eta_K P_K = \left(\sum_{r=0}^{K-2} + \sum_{r=K+1}^{m}\right) P_r(-P_K/P_{K-1})^{K-r}(1+\eta_K)^{K-r}$$
$$+ ih(-P_K/P_{K-1})^{K-m}(1+\eta_K)^{K-m}.$$

On taking the imaginary part of this equation and using (3.3.4), we obtain

$$(\text{im}\,\eta_K)P_K = o\{(\text{im}\,\eta_K)P_K\} + h(-P_K/P_{K-1})^{K-m}\{1+o(1)\}.$$

Hence

$$\text{im}\,\eta_K \sim hP_K^{-1}(-P_K/P_{K-1})^{K-m}.$$

Then, by (3.3.5) again,

$$\text{re}\,\lambda_k \sim \pm hF_K, \tag{3.9.13}$$

where F_K is defined by (3.9.4).

Next we turn to (3.3.36) and we denote the $o(1)$ vector appearing there by u. If the components of u are denoted by u_j $(1 \leqslant j \leqslant 2m)$, it follows from (3.3.18) and (3.3.35) that (3.3.38) can be written

$$y_k = \left\{M_k^{-1/2}\{1+o(1)\} + O\left(\sum_{J>K} |u_j M_j^{-1/2}|\right)\right\} \exp\left(\int_a^x \lambda_k(t)\,dt\right), \tag{3.9.14}$$

where we use the notation (3.3.2) connecting k, K and j, J, and we have also used (3.3.9) to incorporate the terms with $J < K$ into the $o(1)$ term. We have now to estimate the u_j $(J > K)$ using (1.4.24) and (1.4.25) as applied to the W-formulae (3.3.36). We note that

$$\text{re}\,(\lambda_j - \lambda_k) \sim \text{re}\,\lambda_j \qquad (J > K) \tag{3.9.15}$$

as $x \to \infty$ in both the cases where P_J/P_{J-1} is positive and those where it is negative. In the former case, we use (3.9.13) (with j in place of k) and (3.9.6). In the latter case, we use (3.9.13), (3.9.2), (3.9.6), and (3.3.6) to write

$$hF_K = o(p_m F_m) = o\{(p_m/p_{m-1})^{1/2}\} = o(\text{re}\,\lambda_j). \tag{3.9.16}$$

Let us now deal with part I of the theorem, in which $K \geqslant r_0 - 1$. Then we have

$$\int_a^\infty \text{re}\,\lambda_j\,\mathrm{d}x = \pm\infty \tag{3.9.17}$$

for $J > K$. If the integral diverges to $-\infty$, we use (1.4.24) to write

$$\begin{aligned} u_j(x) &= O\left[\left(\int_a^{\frac{1}{2}x} + \int_{\frac{1}{2}x}^x\right)\left\{\exp\left(\int_t^x \text{re}\,(\lambda_j - \lambda_k)\,\mathrm{d}s\right)\right\}|R(t)|\,\mathrm{d}t\right] \\ &= O\left\{\exp\left(\int_{\frac{1}{2}x}^x (\text{re}\,\lambda_j)\{1 + o(1)\}\,\mathrm{d}s\right)\right\} \\ &\quad + O\left[\int_{\frac{1}{2}x}^x \frac{\mathrm{d}}{\mathrm{d}t}\left\{\exp\left(\int_t^x \text{re}\,(\lambda_j - \lambda_k)\,\mathrm{d}s\right)\right\}\frac{|R(t)|}{|\text{re}\,(\lambda_j - \lambda_k)(t)|}\,\mathrm{d}t\right] \\ &= O\left\{\exp\left(\int_{\frac{1}{2}x}^x (\text{re}\,\lambda_j)\{1 + o(1)\}\,\mathrm{d}s\right)\right\} \\ &\quad + O\left(\sup_{t \geqslant \frac{1}{2}x}\{|R(t)|/|\text{re}\,\lambda_j(t)|\}\right), \end{aligned} \tag{3.9.18}$$

where we have used (3.9.15) and integrated the exponential in the second of the two O-terms. If, on the other hand, the integral in (3.9.17) diverges to $+\infty$, we use (1.4.25) to write, more simply,

$$u_j(x) = O\left(\sup_{t \geqslant x}\{|R(t)|/|\text{re}\,\lambda_j(t)|\}\right). \tag{3.9.19}$$

The matrix R in (3.9.18) and (3.9.19) is associated with the W-system arising from (3.3.35). Hence, by (3.3.27), (3.3.34), and (3.9.8), we have

$$\begin{aligned} |R| &= O\left\{\left(\delta^{1/2}\max_{J>L}|\lambda_j^{1/2}\lambda_l^{-1/2}|\right)\left(\max_r |p_r'/p_r|\right)\right\} + O(\mathsf{P}_2) + O(\mathsf{P}_3) \\ &= O(\mathsf{P}_1 + \mathsf{P}_2 + \mathsf{P}_3), \end{aligned} \tag{3.9.20}$$

where we use (3.3.2), (3.3.6), and (3.3.10) to obtain P_1, and we use the diagonalization of (3.3.34) to obtain P_2 and P_3. We also note that, by (3.9.13) and (3.9.16),

$$(\text{re}\,\lambda_j)^{-1} = O(h^{-1}F_J^{-1}) \tag{3.9.21}$$

whatever the sign of P_J/P_{J-1}.

We can now establish (3.9.11). Let $\mathsf{P} = \mathsf{P}_1 + \mathsf{P}_2 + \mathsf{P}_3$. In the case of (3.9.19), from (3.9.20), (3.9.21), and condition (a) of the theorem, we obtain

$$u_j = O(\mathsf{P}/hF_J).$$

Then, by (3.9.5), (3.9.6), and condition (a).

$$u_j M_j^{-1/2} = O(\mathsf{P}/hF_J^{1/2}) = o(\mathsf{P}/hF_K^{1/2}) = o(F_K^{1/2}) = o(M_k^{-1/2}), \quad (3.9.22)$$

as required. In the case of (3.9.18), the second O-term can be treated similarly because of (3.9.9) and, by (3.9.13) (with j in place of k), the first O-term is

$$O\left\{\exp\left(-\beta\int_{\frac{1}{2}x}^{x} |hF_J|\,\mathrm{d}t\right)\right\} = o\left\{\exp\left(-\beta\int_{\frac{1}{2}x}^{x} |hF_{r_0}|\,\mathrm{d}t\right)\right\} = o\{(M_j/M_k)^{1/2}\}.$$

by (3.9.10) and (3.9.5). Altogether, (3.9.14) can now be written

$$y_k \sim M_k^{-1/2} \exp\left(\int_a^x \lambda_k(t)\,\mathrm{d}t\right),$$

and (3.9.11) follows from (3.9.4) and (3.9.5).

In part II of the theorem, we have $K \leqslant r_0 - 1$ and hence

$$h^{1/2} M_j^{-1/2} \in L^2(a, \infty)$$

for $K \leqslant J \leqslant r_0 - 1$, by (3.9.5) and (3.9.7). Hence, in the summation in (3.9.14), we have only to consider $J \geqslant r_0$. For such J, we can argue as in (3.9.22) to obtain

$$u_j M_j^{-1/2} = O(\mathsf{P}/hF_{r_0}^{1/2}) = o(h^{-1/2}\mathsf{P}^{1/2}),$$

by condition (a). Since $\mathsf{P} \in L(a, \infty)$, we have

$$h^{1/2} u_j M_j^{-1/2} \in L^2(a, \infty),$$

as required. This completes the proof of (3.9.12). □

We note that (3.9.11) agrees with (3.8.21) and therefore this formula for y_k has been established in both the cases (3.8.16) and (3.9.3), subject to the other conditions in Theorem 3.8.1 and part I of Theorem 3.9.1.

In the example where the coefficients are powers of x, that is,

$$P_r(x) = (\text{const.})x^{\alpha_r}, \qquad h(x) = x^{\gamma}, \quad (3.9.23)$$

the conditions of Theorem 3.9.1 are satisfied in a simple way. Taking condition (a) first, we have $hF_J \notin L(a, \infty)$ by (3.9.7) and, by the conditions already imposed in Theorem 3.3.1, $\mathsf{P}_\nu \in L(a, \infty)$. Hence

$$hF_J = k_1 x^{-1+c}, \qquad \mathsf{P}_\nu \sim k_2 x^{-1-d} \quad (3.9.24)$$

with some $c \geqslant 0$, $d > 0$, and constants k_1 and k_2. Then condition (a) is clearly satisfied. The same is true of (3.9.9). In (3.9.10), the left-hand side is exponentially small as compared to the right-hand side, apart from the one exceptional case where $\gamma + c = 0$ with the c which arises when $J = r_0$ in (3.9.24), and this case is not covered.

To discuss the value of r_0 in the example (3.9.23), we re-write (3.3.43) in the equivalent form

$$(m - r - \tfrac{1}{2})\alpha_{r-1} - (m - r + \tfrac{1}{2})\alpha_r$$
$$< (m - r - \tfrac{3}{2})\alpha_r - (m - r - \tfrac{1}{2})\alpha_{r+1} \qquad (1 \leq r \leq m-1),$$

which expresses (3.9.6) in this example. Then $r_0 = m + 1$ if

$$\tfrac{1}{2}(\alpha_{m-1} + \alpha_m) > 1 + \gamma. \tag{3.9.25}$$

Otherwise r_0 is the first value of r such that

$$(m - r - \tfrac{1}{2})\alpha_{r-1} - (m - r + \tfrac{1}{2})\alpha_r \geq -1 - \gamma. \tag{3.9.26}$$

3.10 THE DEFICIENCY INDEX PROBLEM

The main motivating influence on the development of the asymptotic analysis of (3.1.1) over the years has been provided by the deficiency index problem, and here we introduce this problem very briefly in its natural setting in the spectral theory of linear operators in a Hilbert space H. One of the basic aspects of this theory is concerned with the relation between a closed, densely defined, symmetric operator T and its adjoint T^*. Certainly T is a restriction of T^*, and the extent to which T falls short of being self-adjoint is expressed in terms of the domains of T and T^* by the von Neumann formula

$$\mathscr{D}(T^*) = \mathscr{D}(T) \oplus \mathscr{N}_w \oplus \mathscr{N}_{\bar{w}}. \tag{3.10.1}$$

Here w is any non-real number and $\mathscr{N}_w$ is the null space of $T^* - wI$. Thus $\mathscr{N}_w$ is the set of all elements g in H such that

$$T^*g = wg. \tag{3.10.2}$$

As a consequence of the closed, symmetric nature of T, the dimension of $\mathscr{N}_w$ is independent of w provided that w remains in either the upper half-plane or the lower half-plane. Let us suppose that $\operatorname{im} w > 0$. Then the deficiency indices N_+ and N_- of T are defined by

$$N_+ = \dim \mathscr{N}_w, \qquad N_- = \dim \mathscr{N}_{\bar{w}}. \tag{3.10.3}$$

At this stage, N_+ and N_- may be finite or infinite. It follows immediately from (3.10.1) that T is self-adjoint if and only if $N_+ = N_- = 0$. In this case the general spectral theory of self-adjoint operators is applicable to T. The situation where

$$N_+ = N_- \qquad (\neq 0) \tag{3.10.4}$$

is also of particular significance because (3.10.4) is a necessary and sufficient condition for a non-self-adjoint T to possess a self-adjoint extension and, once these extensions have been determined, the spectral theory of self-adjoint operators is again applicable.

Here we are concerned with the situation where H is the function space $L^2(a, \infty)$, with a finite end-point a, and T is associated with a differential operator in such a way that (3.10.2) becomes a differential equation of the type considered in this chapter. The values of N_+ and N_- are then the number of linearly independent solutions of the differential equation (3.10.2) (and of the equation with $\bar{w}$) such that g is $L^2(a, \infty)$. The asymptotic theorems in this chapter provide one means of deciding whether the solutions of the differential equations are $L^2(a, \infty)$ and, in this section and the next, we develop the application of the asymptotic theory to the evaluation of N_+ and N_-.

In the notes at the end of the chapter, we refer elsewhere for the somewhat extensive technical details involved in setting up T and determining T^*, and we give only the briefest indication here. We consider first the case where (3.10.2) emerges as an equation of the type (3.1.1). Accordingly, as in § 3.1 and (1.9.13), let the p_r be given functions such that $1/p_0$ and p_r $(1 \leqslant r \leqslant m)$ are locally Lebesgue integrable in $[a, \infty)$. In order to define the operator T without imposing differentiability conditions on the p_r, we require the quasi-derivatives of an arbitrary function f which were introduced in Definition 1.9.1. We recall that the definition starts with a locally absolutely continuous function f, and then the quasi-derivative $f^{[\nu-1]}$ $(2 \leqslant \nu \leqslant 2m+1)$ is defined provided that $f^{[\nu-2]}$ is locally absolutely continuous. We now introduce the set $\mathscr{D}$ which consists of all f defined in $[a, \infty)$ such that the quasi-derivative $f^{[2m]}$ exists. Also, let $\mathscr{D}_0$ be the subset of $\mathscr{D}$ which consists of functions with compact support in (a, ∞).

Considering $\mathscr{D}_0$ as a subspace of $L^2(a, \infty)$, we define the linear operator T_0 by

$$T_0 f = f^{[2m]} \qquad (f \in \mathscr{D}_0).$$

We are concerned with symmetric operators, and it follows from the definition of quasi-derivatives in Definition 1.9.1 that the usual symmetry relation

$$(T_0 f_1, f_2) = (f_1, T_0 f_2)$$

holds provided that the p_r $(0 \leqslant r \leqslant m)$ are real-valued. Then, passing over the technical details already referred to, we merely state that T_0 is densely defined and closable. The operator T in (3.10.1) then arises as the closure of T_0. Further, T^* is the operator given by

$$T^* g = g^{[2m]} \qquad (g \in \mathscr{D}^*), \tag{3.10.5}$$

where the domain $\mathscr{D}^*$ is the subset of $\mathscr{D}$ which consists of those functions

g such that g and $g^{[2m]}$ are $L^2(a, \infty)$. It follows from (3.10.5) and Definition 1.9.1 that (3.10.2) is just the quasi-derivative formulation of the equation

$$(p_0 y^{(m)})^{(m)} + (p_1 y^{(m-1)})^{(m-1)} + \cdots + p_m y = wy, \qquad (3.10.6)$$

where we now revert to the previous notation y for the solution of a differential equation. Thus we have the equation (3.1.1) but with $p_m - w$ in place of p_m. We repeat that, in the situation under consideration now, the p_r are real-valued and the parameter w is non-real.

If we take the complex conjugate of (3.10.6), w is replaced by $\bar{w}$ and the p_r are unchanged. It then follows from (3.10.3) that

$$N_+ = N_- \qquad (=N, \text{ say}) \qquad (3.10.7)$$

in this case. In the literature cited at the end of the chapter, it is shown that $N \geqslant m$, and therefore N lies in the range

$$m \leqslant N \leqslant 2m. \qquad (3.10.8)$$

Examples of (3.10.6) are known which show that N can take any value in this range. Thus the deficiency index problem for (3.10.6) is the problem of determining general conditions on the p_r which guarantee that N takes a given value in $[m, 2m]$.

The discussion up to this point can be extended to the situation where terms involving odd-order derivatives of y are added to (3.10.6) in the manner described in (1.9.19). The symmetry of T_0 requires that the new coefficients q_r should be pure imaginary. Consequently, the equality (3.10.7) no longer necessarily holds, and (3.10.8) is replaced by

$$m \leqslant N_+ \leqslant 2m, \qquad m \leqslant N_- \leqslant 2m. \qquad (3.10.9)$$

Again, the deficiency index problem in this case is the problem of determining general conditions on the p_r and q_r which guarantee that N_+ and N_- take any given pair of values consistent with (3.10.9). There is also the question whether any pairs of values are never taken, whatever p_r and q_r may be, and one result in this direction is that $N_+ = 2m$ if and only if $N_- = 2m$.

Finally, as in (1.9.22), the odd-order equation corresponding to (3.10.6) is

$$\sum_{j=0}^{m} [\tfrac{1}{2}\{(q_{m-j} y^{(j)})^{(j+1)} + (q_{m-j} y^{(j+1)})^{(j)}\} + (p_{m-j} y^{(j)})^{(j)}] = wy. \qquad (3.10.10)$$

Again, the symmetry of T_0 requires that the q_r should be pure imaginary and the p_r real. The inequalities (3.10.9) are now replaced by

$$\left.\begin{aligned} m \leqslant N_+ \leqslant 2m+1, \quad m+1 \leqslant N_- \leqslant 2m+1 \quad (m \text{ even}) \\ m+1 \leqslant N_+ \leqslant 2m+1, \quad m \leqslant N_- \leqslant 2m+1 \quad (m \text{ odd}) \end{aligned}\right\} \qquad (3.10.11)$$

provided that the sign of the leading coefficient is arranged so that $q_0/\mathrm{i} > 0$. When $m > 0$, we have the result as before that $N_+ = 2m + 1$ if and only if $N_- = 2m + 1$.

3.11 EVALUATION OF DEFICIENCY INDICES

We begin by considering the deficiency index problem for the equation (3.10.6) and we show that the asymptotic theory in this chapter provides various sets of conditions on the p_r under which N takes a given value in the range (3.10.8). There are two stages involved. First the dichotomy condition in, for example, (3.2.7) or (3.3.15) must be verified. Then, with the asymptotic formulae for the y_k thereby validated, the $L^2(a, \infty)$ nature of the solutions must be determined. One possibility which cannot be overlooked is that some linear combination of the solutions may be $L^2(a, \infty)$ even though the solutions themselves may not be.

The value of N in (3.10.8) does not depend on w, provided that w is non-real, and a convenient value of w in the following calculations is $w = \mathrm{i}$. Thus we take (3.10.6) as

$$(p_0 y^{(m)})^{(m)} + (p_1 y^{(m-1)})^{(m-1)} + \cdots + p_m y = \mathrm{i}y, \qquad (3.11.1)$$

where the p_r are real-valued. We take it that $p_0 > 0$ in $[a, \infty)$ and, additionally, we assume that

$$|p_m(x)| \to \infty \qquad \text{as } x \to \infty \qquad (3.11.2)$$

since this is the situation of most interest. In the following theorem, we apply Theorem 3.2.1 to the evaluation of N.

Theorem 3.11.1 *Let the conditions* (*i*)–(*v*) *of Theorem* 3.2.1 *hold with* p_m *as in* (3.11.1) *and* (3.11.2). *Suppose first that*

$$p_m(x) \to \infty \quad \textit{as} \quad x \to \infty. \qquad (3.11.3)$$

Let $2\mathfrak{n}$ $(0 \leq \mathfrak{n} \leq m)$ *denote the number of roots* ξ_k *of* $g(\xi)$ *such that*

$$\mathrm{re}\, \xi_k = 0 \qquad (1 \leq k \leq 2\mathfrak{n}). \qquad (3.11.4)$$

Let the remaining roots ξ_k *have distinct real parts except for the roots which occur in complex conjugate pairs. Suppose also that*

$$g'(\xi_j) \neq g'(\xi_k) \qquad (3.11.5)$$

for distinct j *and* k *in* $[1, 2\mathfrak{n}]$, *and*

$$\mathrm{im}\, g'(\xi_k) \neq 0 \qquad (3.11.6)$$

for $2\mathfrak{n}+1 \leq k \leq 2m$. *Then*

(a) $N = m$ *if the integral*

$$\int_a^{\infty} (p_0 p_m^{2m-1})^{-1/2m}\,\mathrm{d}x \tag{3.11.7}$$

diverges;

(b) $N = m + \mathfrak{n}$ *if the integral* (3.11.7) *converges*.

In the case where

$$p_m(x) \to -\infty \quad as \quad x \to \infty, \tag{3.11.8}$$

the same conclusions hold provided that ξ_k *is replaced by* $\xi_k \exp(\mathrm{i}\pi/2m)$ (*and similarly for* ξ_j) *in* (3.11.4)–(3.11.6).

Proof First we consider the dichotomy condition (3.2.7) where the λ_k are the solutions of

$$p_0\lambda^{2m} + p_1\lambda^{2m-2} + \cdots + p_m - \mathrm{i} = 0, \tag{3.11.9}$$

this equation being (3.1.8) in the present situation. By (3.11.2), we have

$$\lambda_k \sim P\xi_k \tag{3.11.10}$$

as in (3.2.10). Thus (3.2.7) is certainly satisfied for those j and k such that

$$\mathrm{re}\,(\xi_j - \xi_k) \neq 0 \qquad (p_m > 0)$$

or

$$\mathrm{re}\,\{(\xi_j - \xi_k)\exp(\mathrm{i}\pi/2m)\} \neq 0 \qquad (p_m < 0).$$

To deal with the other j and k, we have to take (3.11.10) further and, from this point on, we confine the proof to the case where (3.11.3) holds. The proof when (3.11.8) holds is similar.

In the case of (3.11.3), we have $P > 0$ in (3.2.1). Then, as in (3.2.9), we substitute $\lambda = P\mu$ in (3.11.9) to obtain

$$g(x, \mu) - \mathrm{i}/p_m(x) = 0, \tag{3.11.11}$$

where

$$g(x, \mu) = \sum_{r=0}^{m} (p_r/p_0 P^{2r})\mu^{2m-2r}.$$

Let the roots of $g(x, \mu)$, considered as a polynomial in μ, be denoted by $\xi_k(x)$ where, because of (3.2.2), $\xi_k(x) \to \xi_k$ as $x \to \infty$. We note two properties of the $\xi_k(x)$ which follow from the fact that $g(x, \mu)$ has real coefficients and even powers of μ. The first is that, if $\xi_j = \bar{\xi}_k$, then $\xi_j(x) = \bar{\xi}_k(x)$ in some neighbourhood of ∞. The second is that, if $\mathrm{re}\,\xi_k = 0$, then $\mathrm{re}\,\xi_k(x) = 0$, also in some neighbourhood of ∞.

To solve (3.11.11) asymptotically for μ, we write

$$\mu = \xi_k(x) + \eta(x), \tag{3.11.12}$$

where $\eta(x) = o(1)$. This gives

$$\eta(x)\,\partial g(x, \mu)/\partial \mu + O\{\eta^2(x)\} = \mathrm{i}/p_m(x),$$

with $\mu = \xi_k(x)$ in the partial derivative. Hence

$$\eta(x) \sim \mathrm{i}\{p_m(x)g'(\xi_k)\}^{-1}.$$

Then, by (3.2.9) and (3.11.2), we obtain

$$\lambda_k(x) = P(x)[\xi_k(x) + \mathrm{i}\,\{p_m(x)g'(\xi_k)\}^{-1}\{1 + o(1)\}]. \quad (3.11.13)$$

It now follows immediately that the dichotomy condition (3.2.7) is satisfied in both the cases (3.11.5) and (3.11.6).

We can now use the asymptotic formula (3.2.8) to establish the values of N in parts (a) and (b) of the theorem. By (3.2.8), we have

$$|y_k|^2 \sim (p_0 p_m^{2m-1})^{-1/2m} \exp\left(2\int_a^x \operatorname{re} \lambda_k(t)\,\mathrm{d}t\right), \quad (3.11.14)$$

and we discuss first the $2(m - \mathrm{n})$ values of k for which $\operatorname{re} \xi_k \neq 0$. Then, by (3.11.10) and (3.2.1), (3.11.14) gives

$$|y_k|^2 \sim P p_m^{-1} \exp\left(2(\operatorname{re} \xi_k)\int_a^x P\{1 + o(1)\}\,\mathrm{d}t\right). \quad (3.11.15)$$

To deal with the factor p_m^{-1}, we take $r = m$ in (3.2.4) to obtain

$$p_m'/p_m = o(P),$$

and integration gives

$$p_m = \exp\left\{o\left(\int_a^x P(t)\,\mathrm{d}t\right)\right\}. \quad (3.11.16)$$

Then it also follows from (3.11.3) that

$$\int_a^x P(t)\,\mathrm{d}t \to \infty$$

as $x \to \infty$. We can now write (3.11.15) as

$$|y_k|^2 \sim P \exp\left((\operatorname{re} \xi_k)\{2 + o(1)\}\int_a^x P(t)\,\mathrm{d}t\right). \quad (3.11.17)$$

The change of variable

$$u = \int_a^x P(t)\,\mathrm{d}t$$

now shows that y_k is $L^2(a, \infty)$ or not $L^2(a, \infty)$ according as $\operatorname{re} \xi_k < 0$ or $\operatorname{re} \xi_k > 0$. Since the ξ_k occur in $\pm$ pairs, we therefore obtain $m - \mathrm{n}$ solutions y_k which are $L^2(a, \infty)$ and $m - \mathrm{n}$ which are not.

Next we consider the $2\mathfrak{n}$ values of k such that re $\xi_k = 0$. By (3.11.13) and (3.11.14), we have

$$|y_k|^2 \sim P_1 \exp\left(2b_k \int_a^x P_1\{1 + o(1)\}\, dt\right), \tag{3.11.18}$$

where b_k is the real constant

$$b_k = i/g'(\xi_k)$$

and

$$P_1 = (p_0 p_m^{2m-1})^{-1/2m}.$$

Again, the b_k occur in $\pm$ pairs. As in the case of (3.11.17), it follows from (3.11.18) that, if (3.11.7) diverges, there are $\mathfrak{n}$ solutions y_k which are $L^2(a, \infty)$ and $\mathfrak{n}$ which are not. If, on the other hand, (3.11.7) converges, all $2\mathfrak{n}$ of the solutions (3.11.18) are clearly $L^2(a, \infty)$.

Putting together the results for re $\xi_k \neq 0$ and re $\xi_k = 0$, we find that, when (3.11.7) diverges, there are m solutions y_k which are $L^2(a, \infty)$ and m which are not. When (3.11.7) converges, there are $m - \mathfrak{n} + 2\mathfrak{n} = m + \mathfrak{n}$ solutions y_k which are $L^2(a, \infty)$ and $m - \mathfrak{n}$ which are not. This proves the theorem except that there remains the possibility of extra $L^2(a, \infty)$ solutions arising as linear combinations of non-$L^2(a, \infty)$ solutions. The proof that this possibility does not occur is somewhat lengthy and, since no further development of the asymptotic formulae (3.2.8) is involved, we refer to the literature (Fedorjuk 1966*b*, pp. 359–66; Naimark 1968, pp. 200–2) for the details. □

Returning again to the example (3.2.20), the divergence or convergence of (3.11.7) occurs according as α_0 and α_m, in addition to satisfying (3.2.21), also satisfy

$$\alpha_0 + (2m - 1)\alpha_m \leqslant 2m$$

or

$$\alpha_0 + (2m - 1)\alpha_m > 2m.$$

We move on to the evaluation of N under the conditions imposed in Theorem 3.3.1. Since we are discussing the $L^2(a, \infty)$ nature of the y_k themselves, and not of their derivatives, we also require the refinements given in §§ 3.8 and 3.9. We note that the equation (3.11.1) is covered by the case $h(x) = 1$ of § 3.9. Without mentioning it again in the present discussion, we assume that all the conditions on the p_r in §§ 3.3, 3.8, and 3.9 hold as required.

The dichotomy condition (3.3.15) is easily seen to be satisfied in the present situation because, by (3.3.2) and (3.9.13),

$$\text{re}\,\lambda_k \sim \pm(-p_K/p_{K-1})^{1/2}$$

or

$$\text{re}\,\lambda_k \sim \pm F_K$$

according as $p_K/p_{K-1}<0$ or >0. As k varies, the right-hand sides in these two asymptotic formulae all have different orders of magnitude as $x\to\infty$, apart from $\pm$ pairs, and (3.3.15) follows immediately.

In the following theorem, r_0 is defined as before in terms of the convergence or divergence of (3.9.7), with $h(x)=1$.

Theorem 3.11.2 *Let* $\mathfrak{n}$ *of the ratios* p_r/p_{r-1} $(1\leqslant r\leqslant r_0-1)$ *have positive sign. Then*

$$N=m+\mathfrak{n}.$$

Proof We consider first those ratios p_r/p_{r-1} which are negative, and we apply Theorem 3.8.1 in a typical such case $r=K$. By (3.8.19),

$$\int_a^x (-p_K/p_{K-1})^{1/2}\,\mathrm{d}t\to\infty$$

as $x\to\infty$. Also, by (3.8.17),

$$p_K=\exp\left\{o\left(\int_a^x (-p_K/p_{K-1})^{1/2}\,\mathrm{d}t\right)\right\},$$

with the same for p_{K-1}. Hence (3.8.21) and (3.3.2) give

$$|y_K|^2\sim(-p_K/p_{K-1})^{1/2}\exp\left(\pm\{2+o(1)\}\int_a^x (-p_K/p_{K-1})^{1/2}\,\mathrm{d}t\right).$$

Then, as in (3.11.17), if there are l such values of K, we obtain l solutions y_k which are $L^2(a,\infty)$ and l which are not.

Next we consider those ratios p_r/p_{r-1} which are positive and where $r\geqslant r_0$. This time we apply (3.9.11) to a typical case $r=K$. By (3.9.11), (3.9.4), and (3.9.13), we have

$$|y_k|^2\sim F_K\exp\left(\pm\{2+o(1)\}\int_a^x F_K\,\mathrm{d}t\right). \tag{3.11.19}$$

As before, if there are l_1 such values of K, we obtain l_1 solutions y_k which are $L^2(a,\infty)$ and l_1 which are not. Now $l+l_1+\mathfrak{n}=m$ and therefore, so far, we have $m-\mathfrak{n}$ solutions which are $L^2(a,\infty)$ and $m-\mathfrak{n}$ which are not.

Finally, we consider the ratios p_r/p_{r-1} which are positive and where $r\leqslant r_0-1$. This time (3.9.12) applies (with $h(x)=1$), and we find that all $2\mathfrak{n}$ solutions y_k are $L^2(a,\infty)$ in this case. We also note that no linear combination of non-$L^2(a,\infty)$ solutions y_k can be $L^2(a,\infty)$ because all the non-$L^2(a,\infty)$ solutions have different orders of magnitude as $x\to\infty$. Hence

$$N=m-\mathfrak{n}+2\mathfrak{n}=m+\mathfrak{n},$$

as required. □

In the case $n = m$, N can be evaluated directly from the results of § 3.3 without the need to invoke Theorem 3.9.1, as follows.

Theorem 3.11.3 *Let the conditions of Theorem* 3.3.1 *hold, and let* $p_r/p_{r-1} > 0$ *for all r. Also, let*

$$\int_a^\infty (p_{m-1}p_m)^{-1/2}\,dx < \infty. \tag{3.11.20}$$

Then

$$N = 2m.$$

Proof By (3.9.13) with $h(x) = 1$, we have

$$\operatorname{re}\lambda_k \sim \pm F_K.$$

Since $F_m = (p_{m-1}p_m)^{-1/2}$, it follows from (3.9.6) and (3.11.20) that

$$\int_a^\infty \operatorname{re}\lambda_k(t)\,dt$$

converges. Hence, by (3.3.16) and (3.3.44), both with $K = m$, we have

$$y_k = O\{(p_{m-1}p_m)^{-1/4}\} \qquad (1 \leqslant k \leqslant 2m).$$

Hence all solutions are $L^2(a, \infty)$ by (3.11.20), and $N = 2m$. □

For odd-order equations, the deficiency indices N_+ and N_- may be unequal, and the equation corresponding to (3.11.1) is (3.10.10) with $w = i$ or $w = -i$ respectively. Also, as noted in § 3.10, the p_r are real and the q_r are pure imaginary in the context of symmetric differential operators. Let us write

$$q_r = i\,s_r$$

and we suppose that $s_0 > 0$. Then, subject to (3.11.2) and the conditions in § 3.6, (3.6.9) gives

$$\lambda_k \sim \begin{cases} -i\,s_K/p_{K-1} & (k = 2K,\ 1 \leqslant K \leqslant m) \\ i\,p_K/s_K & (k = 2K+1,\ 0 \leqslant K \leqslant m). \end{cases}$$

The dominant parts of all the λ_k are pure imaginary, and therefore the ideas in § 3.9 are required in order to obtain the asymptotic form of all the y_k (except y_{2K} and y_{2K+1}) in addition to (3.6.22). We indicate briefly the results that are obtained for N_+ and N_-.

Corresponding to (3.9.13), we obtain

$$\operatorname{re}\lambda_k \sim (\operatorname{im} w)G_k, \tag{3.11.21}$$

where

$$G_k = \begin{cases} (-1)^{m-K} s_K^{2m-2K-1}/p_K^{2m-2K} & (k = 2K+1,\ 0 \leqslant K \leqslant m) \\ (-1)^{m-K+1} p_{K-1}^{2m-2K}/s_K^{2m-2K+1} & (k = 2K,\ 1 \leqslant K \leqslant m). \end{cases} \tag{3.11.22}$$

We note that

$$G_{2m+1} = -G_{2m} = s_m^{-1} \tag{3.11.23}$$

and, by (3.6.8),

$$G_r = o(G_{r+1}) \qquad (1 \leqslant r \leqslant 2m-1). \tag{3.11.24}$$

Also, by (3.11.22),

$$G_{2K} \text{ and } G_{2K+1} \text{ have opposite signs } (1 \leqslant K \leqslant m). \tag{3.11.25}$$

As in the case of (3.9.7), it follows from (3.11.23) and (3.11.24) that there is an integer r_0, where

$$1 \leqslant r_0 \leqslant 2m+2, \qquad r_0 \neq 2m+1, \tag{3.11.26}$$

such that

$$\int_a^\infty |G_r| \, dx$$

converges for $r \leqslant r_0 - 1$ and diverges for $r \geqslant r_0$. Corresponding to (3.11.19), analysis similar to that in § 3.9 gives

$$|y_k|^2 \sim G_k \exp\left((\text{im } w)\{2 + o(1)\} \int_a^x G_k(t) \, dt\right) \tag{3.11.27}$$

for $k \geqslant r_0$, where we have used (3.11.21) in the exponential factor.

If r_0 is even, $r_0 = 2\mathfrak{n}$, it follows from (3.11.25) that $m - \mathfrak{n} + 1$ of the solutions in (3.11.27) are $L^2(a, \infty)$ and $m - \mathfrak{n} + 1$ are not. The remaining $2\mathfrak{n} - 1$ solutions y_k, in which $k \leqslant r_0 - 1$, are $L^2(a, \infty)$ as in the case of (3.9.12). Thus

$$N_+ = N_- = (m - \mathfrak{n} + 1) + (2\mathfrak{n} - 1) = m + \mathfrak{n}. \tag{3.11.28}$$

If r_0 is odd, $r_0 = 2\mathfrak{n} + 1$ $(0 \leqslant \mathfrak{n} \leqslant m-1)$, (3.11.27) gives $m - \mathfrak{n}$ solutions which are $L^2(a, \infty)$ and $m - \mathfrak{n}$ which are not. The remaining solution $y_{2\mathfrak{n}+1}$ is $L^2(a, \infty)$ when $(\text{im } w)G_{2\mathfrak{n}+1} < 0$. Again, the $2\mathfrak{n}$ solutions y_k in which $k \leqslant r_0 - 1$ are $L^2(a, \infty)$. Thus we have

$$\begin{aligned} N_- &= N_+ - 1 = m + \mathfrak{n} \qquad (G_{2\mathfrak{n}+1} < 0), \\ N_+ &= N_- - 1 = m + \mathfrak{n} \qquad (G_{2\mathfrak{n}+1} > 0). \end{aligned} \tag{3.11.29}$$

We note that $\mathfrak{n} < m$ by (3.11.26), and this corresponds to the fact that N_+ and N_- can only take the maximum value $2m+1$ together.

Finally, corresponding to Theorem 3.11.3, we have one result which follows directly from the case $K = m$ of (3.6.22) and (3.6.23). This result is that

$$N_+ = N_- = 2n + 1$$

if q_0^{-1} is $L(a, \infty)$.

NOTES AND REFERENCES

3.1 The form of T^{-1} in (3.1.11) and the formulae (3.1.15) and (3.1.16) are due to Fedorjuk (1966*b*, Lemmas 2.1 and 2.2).

3.2 Theorem 3.2.1 is due to Fedorjuk (1965, 1966*b*). Fedorjuk's proof in effect included a proof of Theorem 1.6.1 for the particular system which arises in § 3.2. The case $m = 2$ of Theorem 3.2.1 covers the result of Walker (1971*a*, Theorem 2.3), and a slight modification also covers Theorem 2.5 of the same paper, in which p_2 approaches a non-zero constant as $x \to \infty$.

Fedorjuk (1966*a*) gives corresponding asymptotic results for the non-self-adjoint form (1.9.1). He remarks that, in this case, the analysis continues to apply if one eigenvalue λ_k (and one only) is of smaller magnitude than the rest.

The case $\sigma = [1, m-1]$ of Theorem 3.2.2 is due to Rapoport (1951, 1954); see also Naimark (1968, pp. 185–6), Maksudov (1963), and Walker (1972*b*). In the notes on § 3.4 below, we mention the related papers of Devinatz (1966, 1972*a*).

It is always possible to perturb the p_r in (3.1.1) by terms which are taken by the transformation (3.1.5) into an $L(a, \infty)$ perturbation of the matrix R in (3.1.6). In this context, the asymptotic results of Becker (1981*a*) are covered by Theorem 3.2.1.

3.3 The condition (3.3.1) was identified almost simultaneously by Read (1979) and Eastham (1979). In the latter, different methods are used which lead only to inequalities for solutions. Read obtained the case $K = m$ of (3.3.16) and (3.3.44) subject to additional conditions on the p_r which include $p_r/p_{r-1} > 0$, making λ_k pure imaginary. Theorem 3.3.1 and its proof are due to Eastham and Grudniewicz (1981*b*), as is the extension concerning (3.3.46).

Using a quite different method, Anikeeva (1977) considered the three-term equation (3.3.48) in the case where p_0 and p_m are constant and p_r is a power of x. She obtained the asymptotic form of the solutions y_k themselves, and not just of the derivatives as in (3.3.49).

3.4 Theorem 3.4.1 is due to Eastham and Grudniewicz (1981*b*, § 7). A similar result for the three-term equation (3.3.48) is given by Anikeeva (1977). The transformation (3.4.12), with D diagonal as in (3.4.19), is known as a shearing transformation (Wasow 1985, pp. 22–31) and, as examples of the frequent use of such transformations, we mention the papers of Devinatz (1966, 1972*a,b*) and Kogan and Rofe-Beketov (1975, p. 283). Theorem 3.4.2 and Corollary 3.4.1 appear not to have been stated explicitly before, but the choices (3.4.30) and (3.4.33) were made by Walker (1971*a*, Theorems 2.1 and 2.7) in the case $m = 2$.

It is possible to apply the other general theorems of Chapter 1, besides Theorem 1.3.1, to (3.4.20) and, using Theorem 1.8.3, Devinatz (1966, 1972*a*) gave alternative proofs of Theorems 3.2.1 and 3.2.2. His proofs include a change of independent variable and therefore require that Q_0 (Devinatz 1966, p. 244, 1972*a*) should be real-valued.

3.5 The first papers to investigate the case of large λ_k, with $\lambda_3 = o(\lambda_1)$, were those of Walker (1971*a*, Theorem 2.8, 1972*a*) and Devinatz (1972*b*, (4.2) and (4.3)). The transformations which lead to the Levinson form in these papers are complicated, particularly so in Walker (1972*a*), and no progress to higher-order equations ensued along these lines. Theorem 3.5.1 covers the results of Walker (1972*a*) and Devinatz (1972*b*, (4.3)) in a more precise form.

The case $m = 2$ of Theorem 3.4.1 gives asymptotic formulae for solutions of (3.5.1) in terms of y_1', y_2', y_3, and y_4. However, as in the case of Theorem 3.5.1, the formulae for y_1' and y_2' can be improved to (3.5.15), and then (3.5.15) and (3.4.4) cover the other results of Walker (1971*a*, Theorem 2.8) and Devinatz (1972*b*, (4.2)).

Theorem 3.5.1, together with the standardized methods used in its proof, are due to Grudniewicz (1980*b*); see also Eastham and Grudniewicz (1981*a*, pp. 88–99) (in which, to be correct, it is the case $m = 2$ of Theorem 3.3.1 that is proved). A complete asymptotic analysis of (3.5.1) with the coefficients (3.5.51) is given by Grudniewicz (unpublished thesis, 1980*a*). An asymptotic analysis of the fourth-order equation in the non-self-adjoint form (1.9.1) is given by Miklo (1986).

Theorem 3.5.2 is due to Eastham (1982*b*). When the coefficients p_r are powers of x as in (3.5.51), $\phi = 0$ in (3.5.39) and Theorem 3.5.2 agrees with the result of Paris and Wood (1982, Table 1). These latter results follow from properties of the generalized hypergeometric functions, which are available as solutions of (3.5.1) in this case. Paris and Wood (1982, § 1) also give a description of the alternative method of Anikeeva (1976) for dealing with the same situation. Paris and Wood (1980) give an extensive analysis of corresponding results in the complex domain.

3.6 Theorem 3.6.1 is due to Eastham (1981*b*). The case of (3.6.1) in which all the p_r are zero was investigated by Shirikyan (1967), and he obtained an asymptotic result which corresponds to Theorem 3.2.1. The system formulation used by Shirikyan contained q_r' as well as q_r and therefore he required conditions on the first three derivatives of q_r.

For third-order equations, there is a situation parallel to the one in § 3.5 where greater detail can be added to the general theory. Pfeiffer (1972*a*,*b*) and Unsworth (1973) use a shearing transformation followed by an application of Theorem 1.8.3. Alhammadi (1988) uses the diagonalization method based on (3.1.4). All these papers deal with situations additional to the one covered by Theorem 3.6.1.

For Hamiltonian systems generally, we refer to Atkinson (1964, § 9.1), Yakubovich and Starzhinskii (1975, pp. 109–17) and, especially in relation to the discussion in the text, Kogan and Rofe-Beketov (1976, §§ 1 and 4). Two particular systems which can be expressed in the Hamiltonian form (3.6.26) have been considered by Eastham (1981*a*) and Longani (unpublished thesis, 1982). These systems are both of the form $(PY')' + QY = 0$ with 2×2 symmetric matrices P and Q.

3.7 The recent study of equations of the type (3.7.2) was initiated by Wood (1971) for the fourth-order case and developed to higher-order equations by Paris (1980), Paris and Wood (1981, 1984, 1985), Braaksma (1984), and Adamchik and Lizarev (1986). We refer to the book by Paris and Wood (1986) for a detailed survey of this class of equations in terms of the generalized hypergeometric equation and for an extensive list of references. The approach adopted in § 3.7 is new here.

3.8–3.9 The methods in these sections are developed from those of Eastham and Grudniewicz (1981*b*, pp. 264–5) and Eastham (1981*b*, 1982*a*).

3.10 A general reference for this section is the paper by Everitt and Zettl (1979). For the theory of symmetric extensions and the von Neumann formula, we refer to the books by Akhiezer and Glazman (1981, Chapter 8), Edmunds and Evans (1987, Chapter 3), and Naimark (1968, Chapter 4). We refer to Everitt (1959, 1962, 1963, 1972) for a discussion of the literature on the inequalities (3.10.8), (3.10.9), and (3.10.11), and for proofs of the inequalities. Atkinson (1964, Theorems 9.11.1 and 9.11.2) and Kogan and Rofe-Beketov (1976) give the corresponding inequalities in the wider context of Hamiltonian systems.

Constructions which give all values of N in the range (3.10.8) are due to Glazman (1949), Orlov (1953), Neimark (1962), Fedorjuk (1966*b*), and Devinatz (1972*a*). In the case of (3.10.9) and (3.10.11), Kogan and Rofe-Beketov (1975) show that all possible values such that $|N_+ - N_-| \leqslant 1$ do occur. Gilbert (1978*a,b*) shows that $|N_+ - N_-| = v$ can occur for any v provided that the order of the differential equation is at least $4v - 1$. Surveys of the deficiency index problem are given by Devinatz (1973) and Kauffman *et al.* (1977).

3.11 Theorem 3.11.1 is due to Fedorjuk (1966*b*) and it extends an earlier result of Naimark (1968, pp. 195–6) which gave only the values $N = m$ and $N = m + 1$. Corresponding results for odd-order equations are given by Shirikyan (1967). Theorem 3.11.2 is due to Eastham and Grudniewicz (1981*b*, § 6) and Eastham (1982*a*). Theorem 3.11.3 is due to Read (1979) who gave a different proof which required extra conditions of a subsidiary nature. We refer to Eastham (1981*b*) for the details of the proof of (3.11.28) and (3.11.29) and for a discussion of the relationship of these results to those of Unsworth (1973) in the third-order case.

The deficiency index problem has also been investigated for equations of the generalized hypergeometric type (3.7.2), and we refer to Paris and Wood (1986, pp. 308–21) for a full account.

In connection with the spectral theory of pairs of symmetric differential operators L_1 and L_2, the differential equation $L_1 y = wL_2 y$ is important. The methods of this section are applicable in this situaton as well, and we refer to Braaksma (1984) and Eastham (1982*c*) for details.

4
Resonance and non-resonance

4.1 INTRODUCTION

In this chapter we return to the asymptotically diagonal system

$$Y'(x) = \{\Lambda(x) + R(x)\}Y(x) \tag{4.1.1}$$

and we examine an important class of systems which are not covered by the basic theorems in §§ 1.3–1.7. The situation to be considered is where, first, $\Lambda(x)$ is constant and pure imaginary, the system then being

$$Y'(x) = \{i\Lambda_0 + R(x)\}Y(x) \tag{4.1.2}$$

with a real constant Λ_0. Second, $R(x)$ has the form

$$R(x) = \xi(x)P(x), \tag{4.1.3}$$

where $\xi(x)$ is a scalar factor such that $\xi(x) \to 0$ as $x \to \infty$ and $\xi(x)$ is not $L(a, \infty)$, while $P(x)$ is a non-constant matrix which is either periodic or, more generally, the finite sum of periodic matrices with different periods. The simplest example of $\xi(x)$ is

$$\xi(x) = x^{-\alpha} \qquad (0 < \alpha \leqslant 1). \tag{4.1.4}$$

The form of $R(x)$ in (4.1.3) is not covered by the theorems §§ 1.3, 1.6, and 1.7. Nor is Theorem 1.5.1 applicable because (1.5.3) is not satisfied by (4.1.2).

The aim of this chapter is still to transform (4.1.2) into the Levinson form, but it is no longer always possible to do so with transformations of the type $Y = (I + Q)Z$ with $Q = o(1)$. To discuss this point, we suppose that $P(x)$ has period ω and we write

$$\Lambda_0 = \text{dg}\,(a_1, \ldots, a_n). \tag{4.1.5}$$

Regarding $R(x)$ as a small perturbation when $x \to \infty$, we note that the solutions of the unperturbed system

$$Y'(x) = i\Lambda_0 Y(x)$$

are

$$Y_k(x) = e_k \exp\,(ia_k x),$$

and these solutions have the obvious property of being bounded as $x \to \infty$. Indeed, $|Y_k(x)| = |e_k| = 1$. However, when $R(x)$ is introduced as in (4.1.2), the new feature which appears in the theory is the influence exerted by ω on the size of the solutions $Y(x)$ of (4.1.2). We shall find that it is possible for $Y(x)$ to possess an 'amplitude factor' $\rho(x)$ such that

$$|Y(x)| = \rho(x)\{1 + o(1)\}, \tag{4.1.6}$$

where either $\rho(x) \to \infty$ or $\rho(x) \to 0$ as $x \to \infty$, and the appearance of $\rho(x)$ is caused by ω taking certain specific values which are determined by the a_j in (4.1.5). When such a factor $\rho(x)$ appears, the system (4.1.2) is said to be *resonant,* and otherwise *non-resonant.* It is in non-resonant cases that transformations of the type $Y = (I + Q)Z$ can again be constructed to produce the Levinson form, but in resonant cases a further transformation $Z = CW$, with C constant, is also required.

After these general remarks, it is convenient to express the various values taken by the period ω in the form

$$\omega = 2\pi/\lambda,$$

where λ is a real non-zero parameter. From now on, therefore, we re-write (4.1.3) as

$$R(x) = \xi(x)P(\lambda x), \tag{4.1.7}$$

where $P(s)$ has the fixed period 2π in s, and we discuss resonance and non-resonance in terms of values taken by λ.

The most familiar example covered by (4.1.2) is the second-order equation

$$y''(x) + \{1 + r(x)\}y(x) = 0 \tag{4.1.8}$$

where, corresponding to (4.1.7),

$$r(x) = \xi(x)p(\lambda x) \tag{4.1.9}$$

and $p(s)$ has period 2π. To see this, we first write (4.1.8) in the usual way as the system

$$Y_0' = (C + rB)Y_0,$$

where

$$Y_0 = \begin{bmatrix} y \\ y' \end{bmatrix}, \quad C = \begin{bmatrix} 0 & 1 \\ -1 & 0 \end{bmatrix}, \quad B = \begin{bmatrix} 0 & 0 \\ -1 & 0 \end{bmatrix}.$$

The eigenvalues of C are $\pm \mathrm{i}$, and we proceed to an asymptotically diagonal system as in (1.8.2)–(1.8.4) with the transformation

$$Y_0 = \begin{bmatrix} 1 & 1 \\ \mathrm{i} & -\mathrm{i} \end{bmatrix} Y. \tag{4.1.10}$$

We then obtain (4.1.2) with $\Lambda_0 = \mathrm{dg}\,(1, -1)$ and

$$R(x) = \tfrac{1}{2}\mathrm{i}\xi(x)p(\lambda x)\begin{bmatrix} 1 & 1 \\ -1 & -1 \end{bmatrix}. \tag{4.1.11}$$

A higher-order generalization of (4.1.8) is

$$\left\{\prod_{j=1}^{n}\left(\mathrm{i}\frac{\mathrm{d}}{\mathrm{d}x} + a_j\right) + (-1)^{n-1}r(x)\right\}y(x) = 0, \tag{4.1.12}$$

where the a_j are distinct real constants. This time Y_0 has components $y^{(k-1)}$ $(1 \leqslant k \leqslant n)$ and (4.1.10) is replaced by

$$Y_0 = TY,$$

where T has the (k, j) entry $(\mathrm{i}a_j)^{k-1}$. After a simple calculation, we again obtain (4.1.2) with Λ_0 as in (4.1.5) and

$$R(x) = \mathrm{i}\xi(x)p(\lambda x)D\Omega, \tag{4.1.13}$$

where Ω is the $n \times n$ matrix all of whose entries are unity and D is the constant diagonal matrix

$$D = \mathrm{dg}\,(d_1, \ldots, d_n) \tag{4.1.14}$$

with

$$d_j = 1\Big/\prod_{k \neq j}(a_j - a_k). \tag{4.1.15}$$

We note that (4.1.11) is the special case of (4.1.13) in which $a_1 = -a_2 = 1$ and $d_1 = -d_2 = \frac{1}{2}$.

Returning now to the general situation (4.1.2) and (4.1.5), we add some detail to the above outline introduction. The basic conditions on the function $\xi(x)$ in (4.1.3) are that it is real-valued and locally absolutely continuous in $[a, \infty)$ with

$$\xi(x) \to 0 \quad (x \to \infty), \qquad \xi'(x) \in L(a, \infty). \tag{4.1.16}$$

Also, for some integer M $(\geqslant 1)$,

$$\xi(x) \notin L^M(a, \infty), \qquad \xi(x) \in L^{M+1}(a, \infty). \tag{4.1.17}$$

These conditions are all satisfied by (4.1.4) for example.

Concerning Λ_0, we assume that the a_j in (4.1.5) are distinct, and then we define the set σ of real numbers in terms of Λ_0 by

$$\sigma = \{(a_j - a_k)/N;\, 1 \leqslant j, k \leqslant n, j \neq k, 1 \leqslant N < \infty\}, \tag{4.1.18}$$

N taking integer values. The numbers in σ are the values of λ that will appear in our discussion of resonance and non-resonance, and the way in which σ enters the analysis is indicated by the following lemma.

Lemma 4.1.1 *Let Λ_0 be defined by* (4.1.5) *with distinct a_j and let σ be the set* (4.1.18). *Let $F(x)$ be a given $n \times n$ matrix with period $2\pi/\lambda$ and* $\text{dg}\, F = 0$. *Then, for $\lambda \notin \sigma$, the differential equation*

$$\Phi' + \mathrm{i}(\Phi\Lambda_0 - \Lambda_0\Phi) = F \tag{4.1.19}$$

has a solution $\Phi(x)$ with period $2\pi/\lambda$ and $\text{dg}\, \Phi = 0$.

Proof It is clear that $\text{dg}\, \Phi$ can be chosen to be zero, and we therefore concentrate on the off-diagonal entries. Let

$$E(x) = \exp(\mathrm{i}\Lambda_0 x).$$

Then we can write (4.1.19) as

$$\{E(-x)\Phi(x)E(x)\}' = E(-x)F(x)E(x)$$

and integrate to obtain

$$\Phi(x) = E(x)\left(\int_0^x E(-t)F(t)E(t)\,\mathrm{d}t + C\right)E(-x), \tag{4.1.20}$$

where C is a constant matrix. Then $\Phi(x)$ will have period $2\pi/\lambda$ if C is chosen so that $\Phi(2\pi/\lambda) = \Phi(0)$. Hence we require

$$E(2\pi/\lambda)\left(\int_0^{2\pi/\lambda} E(-t)F(t)E(t)\,\mathrm{d}t\right)E(-2\pi/\lambda)$$
$$= C - E(2\pi/\lambda)CE(-2\pi/\lambda). \tag{4.1.21}$$

A typical (j, k) entry on the right is

$$c_{jk}[1 - \exp\{\mathrm{i}(a_j - a_k)2\pi/\lambda\}],$$

where $C = (c_{jk})$, and there is therefore a solution for c_{jk} if $(a_j - a_k)/\lambda \neq \pm N$ $(N = 1, 2, \ldots)$, that is, if $\lambda \notin \sigma$. This proves the lemma. □

It is also possible to say something when $\lambda \in \sigma$. Suppose that

$$\lambda = (a_j - a_k)/N \tag{4.1.22}$$

for some j, k and N. Then the (j, k) and (k, j) entries on the right of (4.1.21) are both zero for the j and k in question. However, an (arbitrary) choice for both c_{jk} and c_{kj} can still be made if the corresponding (j, k) and (k, j) entries on the left of (4.1.21) are also zero. Thus, if f_{jk} denotes the (j, k) entry of F, we continue to obtain a periodic Φ when (4.1.22) holds provided that both

$$\int_0^{2\pi/\lambda} \mathrm{e}^{-\mathrm{i}(a_j - a_k)t} f_{jk}(t)\,\mathrm{d}t = 0$$

and the same but with j and k interchanged. Hence, by (4.1.22), we require

$$\int_{-\pi}^{\pi} e^{-iNu} f_{jk}(u/\lambda)\, du = 0 \tag{4.1.23}$$

and

$$\int_{-\pi}^{\pi} e^{iNu} f_{kj}(u/\lambda)\, du = 0. \tag{4.1.24}$$

Of course, it may be that (4.1.22) holds simultaneously for more than one choice of j, k, N, in which case (4.1.23) and (4.1.24) are required for all such j, k, N.

4.2 PERTURBATIONS OF HARMONIC OSCILLATION

In the chronological development of the theory of resonance and non-resonance, most attention has been directed to the second-order equation (4.1.8). Since $r(x) = o(1)$, the equation can be described as a perturbation of the harmonic oscillator. The reasons for this emphasis are the unexpected nature of some of the theoretical results and the significance of the equation in applications relating to spectral theory. For the same reasons we also start the main account in this chapter with a discussion of (4.1.8), but with the additional observation that the equation provides a convenient framework within which to introduce both the methods to be developed and the general nature of the asymptotic results.

As in other situations that we have already met, the analysis becomes more difficult as $r(x)$ tends to zero more slowly, that is, as M becomes larger in (4.1.17). We start therefore with the simplest case, $M = 1$. We denote by c_ν $(-\infty < \nu < \infty)$ the complex Fourier coefficients of the function $p(s)$ in (4.1.9), so that $p(s)$ has the Fourier series

$$\sum_{-\infty}^{\infty} c_\nu e^{i\nu s}. \tag{4.2.1}$$

In the following theorem, p is assumed to be real-valued, and then

$$c_\nu = \bar{c}_{-\nu}. \tag{4.2.2}$$

Theorem 4.2.1 *In* (4.1.8), *let* $r(x)$ *have the form* (4.1.9), *where* $\xi(x)$ *is real-valued and non-negative in* $[a, \infty)$, (4.1.16) *holds*, (4.1.17) *holds with* $M = 1$, *and* $p(s)$ *is real-valued with period* 2π *in* s. *Then*

(i) *If* $\lambda \neq \pm 2/N$ $(N = 1, 2, \ldots)$, (4.1.8) *has solutions* $y_k(x)$ $(k = 1, 2)$ *such that, as* $x \to \infty$,

$$y_k(x) \sim \exp\left\{(-1)^{k-1} i\left(x + \tfrac{1}{2}c_0 \int_a^x \xi(t)\, dt\right)\right\}. \tag{4.2.3}$$

(ii) *If* $\lambda = \pm 2/N$ *for some positive integer* N, (4.2.3) *continues to hold when* $c_N = 0$.
(iii) *If* $\lambda = \pm 2/N$ *for some positive integer* N, *with* $c_N \neq 0$ *as well as*

$$c_0 \neq \pm |c_N|, \tag{4.2.4}$$

(4.1.8) *has solutions* $y_k(x)$ $(k = 1, 2)$ *such that*

$$\begin{aligned} y_1(x) &= \{|\zeta| e^{i(x+\alpha)} + |c_N| e^{-i(x+\alpha)} + o(1)\} \exp\left(\gamma \int_a^x \xi(t)\, dt\right), \\ y_2(x) &= \{|c_N| e^{i(x+\beta)} + |\zeta| e^{-i(x+\beta)} + o(1)\} \exp\left(-\gamma \int_a^x \xi(t)\, dt\right), \end{aligned} \tag{4.2.5}$$

where

$$\gamma = \tfrac{1}{2}(|c_N|^2) - c_0^2)^{1/2}, \tag{4.2.6}$$

$$\zeta = 2i\gamma - c_0, \tag{4.2.7}$$

and either

$$\left.\begin{aligned} &\alpha = \tfrac{1}{2}(\arg c_N + \arg \zeta), \qquad \beta = \tfrac{1}{2}(\arg c_N - \arg \zeta), \\ \textit{or} \quad &\alpha = \tfrac{1}{2}(-\arg c_N + \arg \zeta), \quad \beta = -\tfrac{1}{2}(\arg c_N + \arg \zeta) \end{aligned}\right\} \tag{4.2.8}$$

according as $\lambda = 2/N$ *or* $\lambda = -2/N$.

Proof As indicated in § 4.1, we start with the transformation (4.1.10) to write (4.1.8) in the form

$$Y'(x) = \{i\Lambda_0 + \xi(x) P(\lambda x)\} Y(x), \tag{4.2.9}$$

where

$$\Lambda_0 = \mathrm{dg}\,(1, -1) \tag{4.2.10}$$

and, by (4.1.11),

$$P(\lambda x) = \tfrac{1}{2} i p(\lambda x) \begin{bmatrix} 1 & 1 \\ -1 & -1 \end{bmatrix}. \tag{4.2.11}$$

The set σ in (4.1.18) is now

$$\sigma = \{\pm 2/N;\ 1 \leqslant N < \infty\}. \tag{4.2.12}$$

Next we make the '$I + Q$' type of transformation

$$Y = (I + \xi P_1) Z, \tag{4.2.13}$$

where the 2×2 matrix $P_1(x)$ has period $2\pi/\lambda$ and is otherwise still to be chosen. Then (4.2.9) becomes

$$\begin{aligned} Z' &= (1 + \xi P_1)^{-1} \{(i\Lambda_0 + \xi P)(I + \xi P_1) - \xi P_1' - \xi' P_1\} Z \\ &= (I - \xi P_1 + S)\{i\Lambda_0 + \xi(i\Lambda_0 P_1 - P_1' + P) + S\} Z \end{aligned}$$

where, as usual, S denotes any $L(a, \infty)$ matrix. Here S contains terms involving ξ^2 and ξ', and it arises from the conditions (4.1.16) and

(4.1.17) (with $M = 1$). Hence we obtain the Z-system

$$Z' = \{\mathrm{i}\Lambda_0 + \xi(\mathrm{i}\Lambda_0 P_1 + \mathrm{i}P_1\Lambda_0 - P_1' + P) + S\}Z. \qquad (4.2.14)$$

By hypothesis, ξ is not $L(a, \infty)$ and therefore we eliminate, as far as possible, the off-diagonal terms in the coefficient of ξ here.

Dealing with part (i) of the theorem first, we have $\lambda \neq \pm 2/N$ and thus $\lambda \notin \sigma$, by (4.2.12). We can therefore apply Lemma 4.1.1 with $F = P - \mathrm{dg}\, P$ and $P_1 = \Phi$. Then (4.2.14) reduces to

$$\begin{aligned} Z' &= (\mathrm{i}\Lambda_0 + \xi\, \mathrm{dg}\, P + S)Z \\ &= [\mathrm{i}\{1 + \tfrac{1}{2}\xi p(\lambda x)\}\Lambda_0 + S]Z, \end{aligned} \qquad (4.2.15)$$

by (4.2.11). This system has the Levinson form and, since ξ and p are real-valued, the dichotomy condition L in § 1.3 is satisfied trivially. Hence there are solutions

$$Z_k(x) = \{e_k + o(1)\} \exp\left((-1)^{k-1}\mathrm{i}\left(x + \frac{1}{2}\int_a^x \xi(t)p(\lambda t)\, \mathrm{d}t\right)\right\} \qquad (4.2.16)$$

for $k = 1, 2$. To simplify the exponential term, we write

$$\int_a^x \xi(t)p(\lambda t)\, \mathrm{d}t = c_0\int_a^x \xi(t)\, \mathrm{d}t + \int_a^x \xi(t)\{p(\lambda t) - c_0\}\, \mathrm{d}t.$$

Since $\xi' \in L(a, \infty)$, the second integral on the right converges as $x \to \infty$ and therefore, after adjusting Z_k by a constant multiple, we can write (4.2.16) as

$$Z_k(x) = \{e_k + o(1)\} \exp\left((-1)^{k-1}\mathrm{i}\left(x + \tfrac{1}{2}c_0\int_a^x \xi(t)\, \mathrm{d}t\right)\right\}.$$

Then (4.2.3) follows when we transform back to Y_0 via (4.2.13) and (4.1.10).

Moving on to part (ii) of the theorem, we consider $\lambda = 2/N$ for example. Thus (4.1.22) holds with $a_1 = 1$, $a_2 = -1$, $j = 1$, $k = 2$, by (4.2.10). Then, with $F(x) = P(\lambda x)$, (4.1.23) and (4.1.24) are just $c_N = 0$ and c_{-N} $(= \bar{c}_N) = 0$, by (4.2.11). Hence P_1 can be chosen as in part (i), and (4.2.3) follows again.

To prove part (iii), we return to (4.2.14) but we can no longer afford to take $\mathrm{dg}\, P_1 = 0$ in our choice of P_1. We write

$$\tilde{P} = P - \mathrm{dg}\, P, \qquad \tilde{P}_1 = P_1 - \mathrm{dg}\, P_1,$$

and then the coefficient of ξ in (4.2.14) is

$$\{\mathrm{i}(\Lambda_0\tilde{P}_1 - \tilde{P}_1\Lambda_0) - \tilde{P}' + \tilde{P}\} + (\mathrm{dg}\, P - \mathrm{dg}\, P_1'). \qquad (4.2.17)$$

We deal with the off-diagonal terms $\{\cdots\}$ first. Taking $\lambda = 2/N$ again, for example, we apply the modification of Lemma 4.1.1 based on (4.1.22),

(4.1.23) and (4.1.24) where, as before, we have $a_1=1$, $a_2=-1$, $j=1$, $k=2$. By (4.2.11), we can apply the lemma with

$$F(x)=\tfrac{1}{2}\mathrm{i}\begin{bmatrix} 0 & p(\lambda x)-c_N \mathrm{e}^{\mathrm{i}N\lambda x} \\ -p(\lambda x)+\bar{c}_N\mathrm{e}^{-\mathrm{i}N\lambda x} & 0 \end{bmatrix}$$

$$=\tfrac{1}{2}\mathrm{i}\begin{bmatrix} 0 & p(\lambda x)-c_N \mathrm{e}^{2\mathrm{i}x} \\ -p(\lambda x)+\bar{c}_N\mathrm{e}^{-2\mathrm{i}x} & 0 \end{bmatrix}.$$

We therefore obtain a periodic $\tilde{P}_1$ so that the term $\{\cdots\}$ in (4.2.17) reduces to

$$\tfrac{1}{2}\mathrm{i}\begin{bmatrix} 0 & c_N\mathrm{e}^{2\mathrm{i}x} \\ -\bar{c}_N\mathrm{e}^{-2\mathrm{i}x} & 0 \end{bmatrix}.$$

Turning to the diagonal terms in (4.2.17), we can define a periodic dg P_1 so that

$$\mathrm{dg}\, P_1'=\mathrm{dg}\, P-\tfrac{1}{2}\mathrm{i}c_0\Lambda_0,$$

by (4.2.11) again. Altogether, we have now reduced the expression (4.2.17) as far as we can and, with our choice of P_1, the system (4.2.14) becomes

$$Z'=\left\{\begin{bmatrix} \mathrm{i} & 0 \\ 0 & -\mathrm{i} \end{bmatrix}+\tfrac{1}{2}\mathrm{i}\xi\begin{bmatrix} c_0 & c_N\mathrm{e}^{2\mathrm{i}x} \\ -\bar{c}_N\mathrm{e}^{-2\mathrm{i}x} & -c_0 \end{bmatrix}+S\right\}Z. \tag{4.2.18}$$

The further transformation

$$Z=(\exp \mathrm{i}\Lambda_0 x)W \tag{4.2.19}$$

simplifies the system to

$$W'=(\xi\Gamma+S)W, \tag{4.2.20}$$

where Γ is the constant matrix

$$\Gamma=\tfrac{1}{2}\mathrm{i}\begin{bmatrix} c_0 & c_N \\ -\bar{c}_N & -c_0 \end{bmatrix}. \tag{4.2.21}$$

The eigenvalues of Γ are $\pm\gamma$, where γ is given by (4.2.6), and these eigenvalues are distinct by (4.2.4). Eigenvectors corresponding to γ and $-\gamma$ respectively are

$$v_1=\begin{pmatrix} \zeta \\ \bar{c}_N \end{pmatrix},\qquad v_2=\begin{pmatrix} c_N \\ \zeta \end{pmatrix}, \tag{4.2.22}$$

where ζ is given by (4.2.7). The final transformation

$$W=(v_1\; v_2)V \tag{4.2.23}$$

takes (4.2.20) into the Levinson form

$$V'=(\xi\Lambda_1+S)V \tag{4.2.24}$$

with $\Lambda_1 = \mathrm{dg}\,(\gamma, -\gamma)$. The dichotomy condition L in § 1.3 is certainly satisfied since ξ is real-valued and non-negative. Then (4.2.24) has solutions

$$V_k(x) = \{e_k + o(1)\} \exp\left((-1)^{k-1}\gamma \int_a^x \xi(t)\, \mathrm{d}t\right)$$

for $k = 1, 2$. On transforming back to Y_0 via (4.2.23), (4.2.19), (4.2.13) and (4.1.10), we obtain solutions $y_1(x)$ and $y_2(x)$ of (4.1.8) such that

$$y_1(x) = \{\zeta \mathrm{e}^{\mathrm{i}x} + \bar{c}_N \mathrm{e}^{-\mathrm{i}x} + o(1)\} \exp\left(\gamma \int_a^x \xi(t)\, \mathrm{d}t\right),$$

$$y_2(x) = \{c_N \mathrm{e}^{\mathrm{i}x} + \zeta \mathrm{e}^{-\mathrm{i}x} + o(1)\} \exp\left(-\gamma \int_a^x \xi(t)\, \mathrm{d}t\right).$$

We then obtain (4.2.5) for $\lambda = 2/N$ by adjusting $y_1(x)$ and $y_2(x)$ by constant multiples $\mathrm{e}^{\mathrm{i}\beta}$ and $\mathrm{e}^{-\mathrm{i}\alpha}$ respectively. When $\lambda = -2/N$, the only difference in the above working is that c_N is replaced by c_{-N} $(=\bar{c}_N)$. The proof of the theorem is now complete. □

We note that, in addition to (4.2.3) and (4.2.5), there are also asymptotic formulae for $y_1'(x)$ and $y_2'(x)$ which are obtained from the above proof by examining the second component of Y_0. A further point which arises from the $I + o(1)$ nature of the transformation (4.2.13) is that the theorem continues to hold if (4.1.9) is extended to

$$r(x) = \xi(x)p(\lambda x) + \xi_1(x), \tag{4.2.25}$$

where $\xi_1 \in L(a, \infty)$. The term ξ_1 simply contributes to the matrix S in (4.2.14). A similar point was made at the end of § 1.6.

4.3 RESONANCE AND EMBEDDED EIGENVALUES

In (4.2.3), we have $|y_k(x)| \to 1$ as $x \to \infty$ and, since there is no amplitude factor of the kind discussed in (4.1.6), we refer to the equation (4.1.8), with $r(x)$ given by (4.1.9), as being non-resonant when $\lambda \neq \pm 2/N$ $(N = 1, 2, \ldots)$. The same remark applies when $\lambda = \pm 2/N$ and $c_N = 0$, by part (ii) of Theorem 4.2.1.

In part (iii), however, when $|c_N| > |c_0|$, the value of γ in (4.2.6) is real and, since $\xi(x)$ is not $L(a, \infty)$ by (4.1.17) (with $M = 1$), (4.2.5) shows that $y_1(x)$ has the large amplitude factor

$$\exp\left(\gamma \int_a^x \xi(t)\, \mathrm{d}t\right),$$

while $y_2(x)$ has the reciprocal amplitude factor. Thus resonance occurs for

(4.1.8) when

$$\lambda = \pm 2/N \qquad \text{and} \qquad |c_N| > |c_0| \tag{4.3.1}$$

for some N.

In the case

$$c_0 = 0, \tag{4.3.2}$$

(4.2.5) simplifies somewhat since we then have

$$\zeta = \mathrm{i}\,|c_N|, \qquad \alpha = \pm\tfrac{1}{2}\arg c_N + \tfrac{1}{4}\pi, \qquad \beta = \pm\tfrac{1}{2}\arg c_N - \tfrac{1}{4}\pi$$

in (4.2.7) and (4.2.8) and, after adjusting the solutions by a constant multiple $\frac{1}{2}|c_N|^{-1}$, we obtain

$$\left.\begin{aligned} y_1(x) &= \{\cos(x + \tfrac{1}{2}\phi + \tfrac{1}{4}\pi) + o(1)\}\exp\left(\tfrac{1}{2}|c_N|\int_a^x \xi(t)\,\mathrm{d}t\right), \\ y_2(x) &= \{\sin(x + \tfrac{1}{2}\phi + \tfrac{1}{4}\pi) + o(1)\}\exp\left(-\tfrac{1}{2}|c_N|\int_a^x \xi(t)\,\mathrm{d}t\right), \end{aligned}\right\} \tag{4.3.3}$$

where

$$\phi = \pm\arg c_N. \tag{4.3.4}$$

Hence, subject to (4.3.2), resonance occurs for (4.1.8) when $\lambda = \pm 2/N$ and $c_N \neq 0$.

A simple example where (4.3.2) holds is

$$p(s) = \sin s. \tag{4.3.5}$$

In this case, the only non-zero Fourier coefficients are $c_1 = \bar{c}_{-1} = -\frac{1}{2}\mathrm{i}$, and therefore resonance occurs only when

$$\lambda = \pm 2. \tag{4.3.6}$$

Further, taking $\lambda = 2$, we have $\phi = -\frac{1}{2}\pi$ in (4.3.4), and then (4.3.3) reduces to

$$\left.\begin{aligned} y_1(x) &= \{\cos x + o(1)\}\exp\left(\frac{1}{4}\int_a^x \xi(t)\,\mathrm{d}t\right), \\ y_2(x) &= \{\sin x + o(1)\}\exp\left(-\frac{1}{4}\int_a^x \xi(t)\,\mathrm{d}t\right). \end{aligned}\right\} \tag{4.3.7}$$

We also mention here the example where $p(s)$ is the 2π-periodic extension of the step-function

$$h(s) = \begin{cases} 1 & (0 \leq s < \pi), \\ -1 & (\pi \leq s < 2\pi). \end{cases} \tag{4.3.8}$$

Here the Fourier coefficients are

$$c_\nu = \begin{cases} 0 & (\nu \text{ even}), \\ -2\mathrm{i}/\nu\pi & (\nu \text{ odd}). \end{cases} \tag{4.3.9}$$

This time resonance occurs when

$$\lambda = \pm 2/N \qquad (N = 1, 3, 5, \ldots) \tag{4.3.10}$$

and correspondingly, in (4.3.3), we have $\phi = \mp\frac{1}{2}\pi$.

The discussion of (4.1.8) has been limited so far to the case $M = 1$ in (4.1.17). Thus, in the example (4.1.4), only the range

$$\tfrac{1}{2} < \alpha \leqslant 1 \tag{4.3.11}$$

has been covered. Later, in §§ 4.6–4.10, we go on to general M both for (4.1.8) and for the wider class of system (4.1.2). Before we do so, however, there are some further points to be made about the case $M = 1$.

The first point concerns an application of the asymptotic formulae (4.3.3) to the spectral theory of the equation

$$y''(x) + \{\mu + q(x)\}y(x) = 0 \tag{4.3.12}$$

in the Hilbert space $L^2(a, \infty)$. Here μ is the spectral parameter, $q(x)$ is real-valued, and a boundary condition

$$a_1 y(a) + a_2 y'(a) = 0$$

is imposed at $x = a$ with real constants a_1 and a_2. The constants may be finite or one of them may be infinite. In the case where

$$q(x) \to 0 \tag{4.3.13}$$

as $x \to \infty$, it is a long-standing result going back go Weyl (1910) that (4.3.12) has a discrete spectrum for $\mu < 0$ and a continuous spectrum for $\mu \geqslant 0$. However, it is also known that there may be positive eigenvalues μ embedded in the continuous spectrum, and a matter for investigation is the possible location of such eigenvalues under suitable conditions on $q(x)$ additional to (4.3.13).

There are four main conditions each of which guarantees that (4.3.12) has no (non-trivial) $L^2(a, \infty)$ solutions when μ is in some half-range $\mu > \Lambda$. These conditions are

(i) $$\limsup x\, |q(x)| = K \tag{4.3.14}$$

(ii) $$\limsup\, (\log x)^{-1} \int_a^x |q(t)|\, \mathrm{d}t = K \tag{4.3.15}$$

(iii) $$\limsup x\, |q'(x)| = L \tag{4.3.16}$$

(iv) $$\limsup\, (\log x)^{-1} \int_a^x |q'(t)|\, \mathrm{d}t = L, \tag{4.3.17}$$

the lim sup being as $x \to \infty$, and K and L are finite. The corresponding values of Λ are

$$4K^2/\pi^2, \quad K^2, \quad L/\pi, \quad \tfrac{1}{2}L \tag{4.3.18}$$

respectively. References for the proofs of these results are given in the notes at the end of the chapter, but here we show that (4.3.3) provides examples which demonstrate the best-possible nature of these values of Λ.

Example 4.3.1 In the case of (4.3.14), we have to define a function $q(x)$ such that (4.3.12) has an $L^2(a, \infty)$ solution for values of μ arbitrarily close to $4K^2/\pi^2$. In (4.1.8) we have $\mu = 1$ and we can deal with the situation by taking $q(x) = r(x)$ and showing that (4.1.8) has an $L^2(a, \infty)$ solution for values of K arbitrarily close to $\frac{1}{2}\pi$. We take the step-function example based on (4.3.8) and we choose $N = 1$ and

$$\xi(x) = kx^{-1}, \tag{4.3.19}$$

where k is a positive constant. By (4.3.9), we obtain from (4.3.3)

$$|y_2(x)| \leqslant (\text{const.})x^{-k/\pi}$$

and hence y_2 is $L^2(a, \infty)$ if $k > \frac{1}{2}\pi$. Then, with $\lambda = 2$ in (4.3.10), we have

$$q(x) = \xi(x)p(2x) = kx^{-1}p(2x), \tag{4.3.20}$$

giving

$$|q(x)| = kx^{-1}.$$

Thus (4.3.14) holds with $K = k > \frac{1}{2}\pi$, as required.

Example 4.3.2 In the case of (4.3.15), we have to construct an example in which K is arbitrarily close to 1, by (4.3.18). Here we take $p(s)$ to be the 2π-periodic extension of the step-function

$$h_1(s) = \begin{cases} 1 & (0 \leqslant s \leqslant \epsilon) \\ -1 & (\pi \leqslant s \leqslant \pi + \epsilon) \\ 0 & (\text{otherwise}), \end{cases} \tag{4.3.21}$$

where ϵ is fixed and $0 < \epsilon < \pi$. Then (4.3.2) holds and we again consider $N = 1$ in (4.3.3). We have

$$c_1 = (2\pi)^{-1} \int_0^{2\pi} h_1(s) \mathrm{e}^{-\mathrm{i}s} \, \mathrm{d}s = (1 - \mathrm{e}^{-\mathrm{i}\epsilon})/\mathrm{i}\pi. \tag{4.3.22}$$

With $\xi(x)$ as in (4.3.19), we have

$$|y_2(x)| \leqslant (\text{const.})x^{-\frac{1}{2}k|c_1|},$$

by (4.3.3), and hence y_2 is $L^2(a, \infty)$ if

$$k\,|c_1| > 1. \tag{4.3.23}$$

With $\lambda = 2$ in (4.3.1), we again have (4.3.20), but now

$$\int_0^x |q(t)| \, dt = k \int_0^x t^{-1} |p(2t)| \, dt = kC_0 \log x + O(1)$$

as $x \to \infty$, where

$$C_0 = \pi^{-1} \int_0^{\pi} |p(2t)| \, dt = (2\pi)^{-1} \int_0^{2\pi} |h_1(s)| \, ds = \epsilon/\pi.$$

Thus (4.3.15) holds with

$$K = kC_0 = k \, |c_1| \, \epsilon / |1 - e^{-i\epsilon}|,$$

by (4.3.22). Then, by (4.3.23) and by choice of ϵ, this value of K can be made arbitrarily close to 1, as required.

Example 4.3.3 In the case of (4.3.16), we require an example in which L is arbitrarily close to π, by (4.3.18). Let $h(s)$ be as in (4.3.8). Then we define $p(s)$ to be the 2π-periodic extension of the function

$$\int_0^s h(t) \, dt - \kappa, \tag{4.3.24}$$

where the constant κ is chosen so that $p(s)$ has mean value zero, as required by (4.3.2). Thus $p(s)$ is a 'saw-tooth' function whose c_1 Fourier coefficient is $2/\pi$, by (4.3.9). As in Example 4.3.1, y_2 is $L^2(a, \infty)$ if $k > \frac{1}{2}\pi$ in (4.3.19). With the present definition of $p(s)$, (4.3.20) gives

$$\limsup x \, |q'(x)| = 2k.$$

Thus (4.3.16) holds with $L = 2k > \pi$, as required.

Example 4.3.4 In the case of (4.3.17), we require an example in which L is arbitrarily close to 2. In the same way as (4.3.24) represents an integrated version of Example 4.3.1, so here we define $p(s)$ to be the 2π-periodic extension of the function

$$\int_0^s h_1(t) \, dt - \kappa,$$

where $h_1(s)$ is as in (4.3.21). The remaining details follow the same lines as in the previous cases.

4.4 TWO EXAMPLES

In this section we give two separate examples of (4.1.8) in which $r(x) \to 0$ as $x \to \infty$ and the equations can be solved explicitly. The examples can be adjusted to illustrate both resonant and non-resonant situations, but we concentrate here on the case of resonance. The amplitude factor will then be exhibited in the explicit form of the solutions.

Example 4.4.1 The first example is obtained from a general construction in which a solution of one differential equation gives rise to a solution of another. Let $\phi(x)$ be a solution of

$$y''(x) + q_0(x)y(x) = 0$$

and define

$$\psi(x) = \phi(x) \exp\left(\int_a^x g(t)\phi(t)\,\mathrm{d}t\right), \tag{4.4.1}$$

where $g(x)$ is a given differentiable function. Then it is easily verified that $\psi(x)$ is a solution of the equation

$$y''(x) + q(x)y(x) = 0, \tag{4.4.2}$$

where

$$q = q_0 - 3g\phi' - g'\phi - g^2\phi^2. \tag{4.4.3}$$

To obtain a resonant equation (4.4.2), we start with $q_0(x) = 1$ and $\phi(x) = \cos x$. Next we choose $g(x)$ so that the exponential term in (4.4.1) becomes the large amplitude factor, and a convenient choice is

$$g(x) = x^{-\alpha} \cos x \qquad (0 < \alpha \leqslant 1).$$

Then (4.4.1) is

$$\psi(x) = (\cos x) \exp\left(\int_a^x t^{-\alpha} \cos^2 t\,\mathrm{d}t\right) \tag{4.4.4}$$

and, by (4.4.3), (4.4.2) takes the form (4.1.8) with

$$r(x) = 2x^{-\alpha} \sin 2x + O(x^{-2\alpha}).$$

If we now suppose further that $\frac{1}{2} < \alpha \leqslant 1$, as in (4.3.11), the O-term here is $L(a, \infty)$ and is therefore of secondary importance. Thus $r(x)$ has the form (4.2.25) with $\xi(x) = 2x^{-\alpha}$, $p(s) = \sin s$ and $\lambda = 2$. Also, (4.4.4) shows that, apart from a constant multiple, $\psi(x)$ has the form

$$\{\cos x + o(1)\} \exp\left(\frac{1}{2}\int_a^x t^{-\alpha}\,\mathrm{d}t\right),$$

in line with the first formula in (4.3.7).

Example 4.4.2 The second example is also obtained from a general construction involving (4.4.2) with $q(x)$ now defined, not by (4.4.3), but as a step-function. Suppose then that

$$q(x) = q_n^2 \qquad (x_n \leqslant x < x_{n+1},\ n = 1, 2, \ldots), \tag{4.4.5}$$

where $q_n > 0$, $x_1 = 0$ and $x_n \to \infty$ as $n \to \infty$. We define

$$w_n = x_{n+1} - x_n$$

and we consider the special situation in which

$$w_n q_n = (L + \tfrac{1}{2})\pi, \tag{4.4.6}$$

where L is a non-negative integer. It follows from (4.4.5) and (4.4.6) that continuously differentiable solutions of (4.4.2) are given by

$$y_1(x) = \begin{cases} A_m \eta_m \sin\{q_{2m-2}(x - x_{2m-2})\} & (x_{2m-2} \leqslant x \leqslant x_{2m-1}) \\ B_m \eta_m \cos\{q_{2m-1}(x - x_{2m-1})\} & (x_{2m-1} \leqslant x \leqslant x_{2m}) \end{cases}$$

$$y_2(x) = \begin{cases} C_m \zeta_m \sin\{q_{2m-1}(x - x_{2m-1})\} & (x_{2m-1} \leqslant x \leqslant x_{2m}) \\ D_m \zeta_m \cos\{q_{2m}(x - x_{2m})\} & (x_{2m} \leqslant x \leqslant x_{2m+1}) \end{cases}$$

for $m \geqslant 1$, where A_m, B_m, C_m, D_m are all ± 1, x_0 is taken to be zero, and

$$\left.\begin{aligned} \eta_m &= q_1 q_3 \cdots q_{2m-3}/q_2 q_4 \cdots q_{2m-2} \\ \zeta_m &= q_2 q_4 \cdots q_{2m-2}/q_3 q_5 \cdots q_{2m-1} \end{aligned}\right\} \tag{4.4.7}$$

with $\eta_1 = \zeta_1 = 1$.

We now make the choice

$$q_n = \begin{cases} \{1 + (n+1)^{-\alpha}\}^{-1} & (n \text{ odd}), \\ (1 - n^{-\alpha})^{-1} & (n \text{ even}), \end{cases} \tag{4.4.8}$$

where $0 < \alpha \leqslant 1$. Then (4.4.6) gives

$$x_{2m} = (L + \tfrac{1}{2})\{2m - 1 + (2m)^{-\alpha}\}\pi$$

$$x_{2m+1} = (L + \tfrac{1}{2})2m\pi,$$

and (4.4.2) takes the form (4.1.8), where $r(x)$ is a step-function with positive and negative values, and $r(x) = O(x^{-\alpha})$ as $x \to \infty$. It is also clear from (4.4.7) and (4.4.8) that $\eta_m \to 0$ and $\zeta_m \to \infty$ as $m \to \infty$. Thus $y_1(x)$ and $y_2(x)$ have the type of amplitude factors η_m and ζ_m which imply that resonance occurs.

As in the previous example, if we suppose that $\frac{1}{2} < \alpha \leqslant 1$ in (4.4.8), it is easy to check that the step-function $r(x)$ has the form (4.2.25) where $\xi(x) = \{(L + \frac{1}{2})\pi/x\}^{\alpha}$, $p(s)$ is the 2π-periodic extension of the step-function (4.3.8), and $\lambda = 2/(2L+1)$. These values of λ, with $L = 1, 2, \ldots$, are in line with (4.3.10).

4.5 HIGHER-ORDER EQUATIONS

The ideas in § 4.2 also apply to the nth order equation

$$\left\{\prod_{j=1}^{n} \left(\mathrm{i}\frac{\mathrm{d}}{\mathrm{d}x} + a_j\right) + (-1)^{n-1} r(x)\right\} y(x) = 0 \tag{4.5.1}$$

which was introduced in (4.1.12), and here we discuss briefly the corresponding results, still with $M=1$ in (4.1.17). As indicated in (4.1.13), the equation (4.5.1) can be written in the form

$$Y'(x) = \{\mathrm{i}\Lambda_0 + \xi(x)P(\lambda x)\}Y(x), \tag{4.5.2}$$

where Λ_0 is given by (4.1.5),

$$P(s) = \mathrm{i}p(s)D\Omega, \tag{4.5.3}$$

and D is given by (4.1.14) and (4.1.15). We assume as before that the periodic function $p(s)$ is real-valued, so that (4.2.1) and (4.2.2) continue to apply. In the transformation (4.2.13), P_1 is now an $n \times n$ matrix and, with the present notation, we obtain (4.2.14) once again. If $\lambda \notin \sigma$, where σ is the set (4.1.18), we can again apply Lemma 4.1.1 to choose P_1 so that

$$Z' = \mathrm{i}\{\Lambda_0 + \xi p(\lambda x)D + S\}Z,$$

corresponding to (4.2.15). This system has the Levinson form and then, on applying the Levinson Theorem and transforming back, we obtain solutions $y_k(x)$ $(1 \leqslant k \leqslant n)$ of (4.5.1) such that

$$y_k(x) \sim \exp\left\{\mathrm{i}\left(a_k x + c_0 d_k \int_a^x \xi(t)\,\mathrm{d}t\right)\right\}. \tag{4.5.4}$$

This formula corresponds to (4.2.3).

Turning next to $\lambda \in \sigma$, we suppose that

$$\lambda = (a_j - a_k)/N \tag{4.5.5}$$

for some j, k, and N, as in (4.1.22). We also recall the point made at the end of § 4.1 that there may be more than one value of N associated with the same λ in this way. Then, as in part (ii) of Theorem 4.2.1, the formula (4.5.4) continues to hold provided that $c_N = 0$ for all the relevant values of N.

If however $c_N \neq 0$ for some or all of the relevant values of N, we proceed as in the working leading to (4.2.20) and now we obtain

$$W' = (\xi\Gamma + S)W$$

with

$$\Gamma = \mathrm{i}D(c_0 I + C_1),$$

where C_1 has zero entries apart from (j, k) entries c_N and (k, j) entries $\bar{c}_N$ for all possible j, k, N in (4.5.5) (with a fixed λ). The simplest situation is where

$$(a_j - a_k)/(a_{j'} - a_{k'})$$

is irrational for all distinct pairs (j, k) and (j', k'). There is then only one possible N in (4.5.5) and C_1 has only two non-zero entries c_N and $\bar{c}_N$. The diagonalization of Γ is straightforward in this case and we can continue as in the proof of Theorem 4.2.1. To give the final asymptotic result, let us suppose that (4.5.5) holds for $j=J$ and $k=K$. Then we find that (4.1.12) has solutions $y_k(x)$ $(1 \leq k \leq n)$ such that (4.5.4) holds for $k \neq J$ and $k \neq K$, while

$$y_J(x) = \{|\zeta|\,\mathrm{e}^{\mathrm{i}(a_Jx+\alpha)} - 2d_K\mathrm{e}^{\mathrm{i}(a_Kx-\alpha)} + o(1)\}$$

$$\times \exp\left(\{\tfrac{1}{2}\mathrm{i}c_0(d_J+d_K)+\gamma\}\int_a^x \xi(t)\,\mathrm{d}t\right),$$

$$y_K(x) = \{2d_J\,|c_N|\,\mathrm{e}^{\mathrm{i}(a_Jx+\beta)} + |\zeta|\,\mathrm{e}^{\mathrm{i}(a_Kx-\beta)} + o(1)\}$$

$$\times \exp\left(\{\tfrac{1}{2}\mathrm{i}c_0(d_J+d_K)-\gamma\}\int_a^x \xi(t)\,\mathrm{d}t\right),$$

where

$$\gamma = \tfrac{1}{2}\sqrt{\{-4d_Jd_K\,|c_N|^2 - c_0^2(d_J-d_K)^2\}},$$
$$\zeta = 2\mathrm{i}\gamma - c_0(d_J-d_K),$$

and α, β are as in the first pair (4.2.8). Also, corresponding to (4.2.4), we assume that $\gamma \neq 0$. Finally, as in the case of (4.3.1), γ is real, and therefore resonance occurs, when

$$\lambda = \pm(a_J - a_K)/N, \tag{4.5.6}$$

$$d_Jd_K < 0, \tag{4.5.7}$$

and

$$|c_N| > \tfrac{1}{2}(-d_Jd_K)^{-1/2}\,|d_J-d_K|\,|c_0|. \tag{4.5.8}$$

A particular case of (4.5.1) is the even-order equation

$$\left\{\prod_{j=1}^{n}\left(\frac{\mathrm{d}^2}{\mathrm{d}x^2}+b_j^2\right)+r(x)\right\}y(x)=0, \tag{4.5.9}$$

where the constants b_j are distinct, real and positive. The previous a_j are now alternately b_j and $-b_j$ and, corresponding to (4.1.14) and (4.1.15), we have now

$$D = \mathrm{dg}\,(d_1, \ldots, d_{2n})$$

with

$$d_{2l-1} = -d_{2l} = \left\{2b_l\prod_{k\neq l}(b_l^2-b_k^2)\right\}^{-1} \qquad (1 \leq l \leq n), \tag{4.5.10}$$

If we consider $a_J = b_l$ and $a_K = -b_l$ in (4.5.6), the condition (4.5.7) becomes $d_{2l-1}d_{2l} < 0$, which is certainly satisfied, by (4.5.10). Then, in the case where $c_0 = 0$, it follows from (4.5.8) that resonance occurs for (4.5.9) when $\lambda = \pm 2b_l/N$ and $c_N \neq 0$.

4.6 NON-RESONANCE FOR SYSTEMS

The discussion of (4.1.2) has been limited so far to the special cases which are associated with the equations (4.1.8) and (4.1.12). We move on now to the general system (4.1.2) and therefore, recalling (4.1.5) and (4.1.7), we consider the form

$$Y'(x) = \{\mathrm{i}\Lambda_0 + \xi(x)P(\lambda x)\}Y(x), \tag{4.6.1}$$

where

$$\Lambda_0 = \mathrm{dg}\,(a_1, \ldots, a_n) \tag{4.6.2}$$

with distinct real a_j, and $P(s)$ has period 2π in s. Further, we suppose that ξ satisfies (4.1.16) and (4.1.17) with the restriction to $M = 1$ now lifted. The set σ is as defined in (4.1.18). In this section we deal with the situation where $\lambda \notin \sigma$ and we aim to establish non-resonance. There are two steps in this process. The first is to extend the transformation (4.2.13) to include higher powers of ξ in such a way that the transformed system contains no off-diagonal terms involving the non-$L(a, \infty)$ powers $\xi, \ldots, \xi^M$ and consequently takes the Levinson form. The second step is to determine the circumstances in which the diagonal matrix in this Levinson form is pure imaginary, with the implication that the solutions of the transformed system have no amplitude factor of the kind in (4.1.6).

There are more than one way of extending (4.2.13) and the following lemma gives the extension which is appropriate for our purpose here.

Lemma 4.6.1 *Let ξ satisfy* (4.1.16) *and* (4.1.17), *and let $\lambda \notin \sigma$. Then there are matrices $P_m(x)$ with period $2\pi/\lambda$ and*

$$\mathrm{dg}\, P_m = 0 \tag{4.6.3}$$

such that the transformation

$$Y = \{\exp(\xi P_1 + \xi^2 P_2 + \cdots + \xi^M P_M)\}Z \tag{4.6.4}$$

takes (4.6.1) *into the Levinson form*

$$Z'(x) = \{\Lambda(x) + S(x)\}Z(x), \tag{4.6.5}$$

where Λ is diagonal and $S \in L(a, \infty)$.

Proof Let

$$\mathscr{P} = \xi P_1 + \xi^2 P_2 + \cdots + \xi^M P_M.$$

Then substitution of (4.6.4) into (4.6.1) gives

$$Z' = \{\mathrm{i}e^{-\mathscr{P}}\Lambda_0 e^{\mathscr{P}} + \xi e^{-\mathscr{P}} P e^{\mathscr{P}} - e^{-\mathscr{P}}(e^{\mathscr{P}})'\}Z. \tag{4.6.6}$$

We wish to choose the P_m satisfying the conditions of the lemma so that

(4.6.6) becomes

$$Z' = \{i\Lambda_0 + \xi\Lambda_1 + \cdots + \xi^M\Lambda_M + S\}Z, \qquad (4.6.7)$$

where $S \in L(a, \infty)$ and the Λ_m are diagonal with period $2\pi/\lambda$. Because of (4.1.16), terms involving ξ' in (4.6.6) go into S and take no further part in the analysis. Accordingly, from now on, we treat ξ simply as a parameter independent of x and require that

$$\begin{aligned} &ie^{-\mathcal{P}}\Lambda_0 e^{\mathcal{P}} + \xi e^{-\mathcal{P}}Pe^{\mathcal{P}} - e^{-\mathcal{P}}(e^{\mathcal{P}})' \\ &\qquad = i\Lambda_0 + \xi\Lambda_1 + \cdots + \xi^M\Lambda_M + O(\xi^{M+1}). \qquad (4.6.8) \end{aligned}$$

The P_m and Λ_m will be defined inductively by equating the terms in ξ^m on each side of (4.6.8); in this process $e^{\mathcal{P}}$ will be replaced by its power series in $\mathcal{P}$. To start the process wih $m = 1$, we require that

$$i(\Lambda_0 P_1 - P_1\Lambda_0) - P_1' + P = \Lambda_1.$$

Hence, with $\operatorname{dg} P_1 = 0$, we define Λ_1 by

$$\Lambda_1 = \operatorname{dg} P. \qquad (4.6.9)$$

This leaves

$$P_1' + i(P_1\Lambda_0 - \Lambda_0 P_1) = P - \operatorname{dg} P, \qquad (4.6.10)$$

and the existence of a periodic P_1 follows from Lemma 4.1.1. At this stage we are at the same position as we were with (4.2.14) and (4.2.15).

Suppose now that $P_1, \ldots, P_m, \Lambda_1, \ldots, \Lambda_m$ have been determined. Then we write

$$Q_m = \xi P_1 + \xi^2 P_2 + \cdots + \xi^m P_m \qquad (4.6.11)$$

and $E_m = \exp Q_m$. We introduce the power series expansions

$$\left.\begin{aligned} iE_m^{-1}\Lambda_0 E_m &= i\Lambda_0 + \sum_1^\infty \xi^r A_r^{(m)} \\ E_m^{-1}PE_m &= P + \sum_1^\infty \xi^r B_r^{(m)} \\ E_m^{-1}E_m' &= \sum_1^\infty \xi^r C_r^{(m)}, \end{aligned}\right\} \qquad (4.6.12)$$

where the $A_r^{(m)}$, $B_r^{(m)}$ and $C_r^{(m)}$ have period $2\pi/\lambda$. To define P_{m+1} and Λ_{m+1}, we first write

$$E_{m+1} = E_m + \xi^{m+1}P_{m+1} + O(\xi^{m+2})$$

and

$$E_{m+1}^{-1} = E_m^{-1} - \xi^{m+1}P_{m+1} + O(\xi^{m+2})$$

as consequences of the power series representation of the exponentials.

Then, replacing M by $m+1$ and $\mathscr{P}$ by Q_{m+1} in (4.6.8), the $(m+1)$th stage of the process is to arrange that

$$\begin{aligned}
&\mathrm{i}\{E_m^{-1} - \xi^{m+1}P_{m+1} + O(\xi^{m+2})\}\Lambda_0\{E_m + \xi^{m+1}P_{m+1} + O(\xi^{m+2})\} \\
&\quad + \xi E_m^{-1}PE_m \\
&\quad - \{E_m^{-1} - \xi^{m+1}P_{m+1} + O(\xi^{m+2})\}\{E'_m + \xi^{m+1}P'_{m+1} + O(\xi^{m+2})\} \\
&\qquad\qquad = \mathrm{i}\Lambda_0 + \xi\Lambda_1 + \cdots + \xi^{m+1}\Lambda_{m+1} + O(\xi^{m+2}),
\end{aligned}$$

giving

$$\begin{aligned}
&\mathrm{i}E_m^{-1}\Lambda_0 E_m - \xi^{m+1}\{P'_{m+1} + \mathrm{i}(P_{m+1}\Lambda_0 - \Lambda_0 P_{m+1})\} + \xi E_m^{-1}PE_m - E_m^{-1}E'_m \\
&\qquad = \mathrm{i}\Lambda_0 + \xi\Lambda_1 + \cdots + \xi^{m+1}\Lambda_{m+1} + O(\xi^{m+2}). \quad (4.6.13)
\end{aligned}$$

Hence, substituting the power series (4.6.12), we wish to arrange that

$$\begin{aligned}
&-\xi^{m+1}\{P'_{m+1} + \mathrm{i}(P_{m+1}\Lambda_0 - \Lambda_0 P_{m+1})\} + \sum_1^\infty \xi^r(A_r^{(m)} + B_{r-1}^{(m)} - C_r^{(m)}) \\
&\qquad = \xi\Lambda_1 + \xi^2\Lambda_2 + \cdots + \xi^{m+1}\Lambda_{m+1} + O(\xi^{m+2}),
\end{aligned}$$

where $B_0^{(m)} = P$. Hence, writing $A_{m+1}^{(m)} + B_m^{(m)} - C_{m+1}^{(m)} = D_{m+1}$, we define

$$\Lambda_{m+1} = \mathrm{dg}\, D_{m+1} \qquad (4.6.14)$$

and then

$$P'_{m+1} + \mathrm{i}(P_{m+1}\Lambda_0 - \Lambda_0 P_{m+1}) = D_{m+1} - \mathrm{dg}\, D_{m+1}. \qquad (4.6.15)$$

We note that D_{m+1} involves only P and $P_1, \ldots, P_m$, and then the required definition of P_{m+1} follows from Lemma 4.1.1. This completes the proof of Lemma 4.6.1. □

We can now prove the main results on the nature of the transformed system (4.6.5) and on non-resonance for (4.6.1). We write

$$\tilde{P} = P - \mathrm{dg}\, P.$$

Theorem 4.6.1 *Let the conditions of Lemma* 4.6.1 *hold and let* ξ *be real-valued. Let* P *have the form given by*

(i) $\quad \mathrm{dg}\, P$ *is pure imaginary* (4.6.16)

(ii) $\quad \tilde{P} = DU,$ (4.6.17)

where D *is a real constant diagonal matrix and* U *is skew-hermitian. Then the diagonal matrix* $\Lambda(x)$ *in* (4.6.5) *is pure imaginary.*

We recall the definition of the term skew-hermitian, which is that $U^* = -U$, where U^* denotes the complex conjugate transpose of U. The proof of the theorem is based on the following lemma in which we show that the form (4.6.17) carries through to all the P_m $(1 \le m \le M)$. In the sequel, we use the same symbol U to denote any skew-hermitian matrix.

Lemma 4.6.2 *Let* (4.6.16) *and* (4.6.17) *hold. Then* P_m *also has the form* DU *for some skew-hermitian* U.

Proof We first note that, if F has the DU form in (4.1.19), it follows from (4.1.20) and (4.1.21) that Φ also has this form. In particular, by (4.6.10) and (4.6.17), P_1 has the form DU. We now proceed by induction and suppose that $P_1, \ldots, P_m$ have the form DU. Then the same is true of Q_m in (4.6.11) and, in turn, it is easy to verify from the power series for $\exp Q_m$ that

$$(\exp Q_m)^* = D^{-1} \exp(-Q_m) D. \tag{4.6.18}$$

At this point we are making the assumption that D is non-singular. This assumption will be removed later in the proof. We also note that, in the case of a non-singular D, any matrix A has the form DU if and only if

$$A^* = -D^{-1}AD. \tag{4.6.19}$$

The definition of P_{m+1} in (4.6.15) involves the three formulae (4.6.12) and we now examine each term on the left-hand side of (4.6.12). First, writing $E_m = \exp Q_m$ again, we have

$$\begin{aligned}(E_m^{-1}\tilde{P}E_m)^* &= E_m^*\tilde{P}^*(E_m^{-1})^* \\ &= -D^{-1}E_m^{-1}\tilde{P}E_mD,\end{aligned}$$

by (4.6.17) and (4.6.18). Hence, by (4.6.19), $E_m^{-1}\tilde{P}E_m$ has the form DU. Similar reasoning applies to both

$$E_m^{-1}(\mathrm{dg}\, P)E_m - \mathrm{dg}\, P,$$

where we use (4.6.16) (to complete the consideration of the second equation in (4.6.12)), and

$$\mathrm{i}E_m^{-1}\Lambda_0 E_m - \mathrm{i}\Lambda_0,$$

which occurs in the first equation in (4.6.12). Turning finally to the last equation in (4.6.12), we use $E_m^{-1}E_m = I$ and differentiate to obtain

$$(E_m^{-1}E_m')^* = -\{(E_m^{-1})'E_m\}^* = -E_m^*\{(E_m^{-1})'\}^* = -D^{-1}E_m^{-1}E_m'D,$$

by (4.6.18). Hence, by (4.6.19), $E_m^{-1}E_m'$ also has the form DU. It now follows from (4.6.12) that each $A_r^{(m)}$, $B_r^{(m)}$ and $C_r^{(m)}$ has the form DU and therefore, by (4.6.15), so does P_{m+1}. This completes the proof of the lemma in the case where D is non-singular.

If now D is singular, we replace the zero diagonal entries by a non-zero real constant ϵ and call the resulting matrix D_ϵ. We then define

$$P_\epsilon = D_\epsilon U$$

with the same U as in (4.6.17). Then, starting with $P_\epsilon + \mathrm{dg}\, P$ in place of P, the process in Lemma 4.6.1 now leads to an equation (4.6.15) which defines a new P_{m+1}. By what we have just proved, this P_{m+1} has the form $D_\epsilon U_\epsilon$, where U_ϵ is skew-hermitian and, as is easy to check, U_ϵ is analytic in ϵ. Hence, letting $\epsilon \to 0$, we finally obtain

$$P_{m+1} = DU$$

in the case where D is singular. This completes the proof of the lemma. □

Proof of Theorem 4.6.1 By Lemma 4.6.2, (4.6.9) and (4.6.14), we now have

$$\Lambda_m = \mathrm{dg}\, D_m \qquad (1 \leqslant m \leqslant M),$$

where each D_m has the form DU. Hence $\Lambda_m = D\,\mathrm{dg}\, U$ and, since U is skew-hermitian, it follows that each Λ_m is pure imaginary. The same is therefore true of Λ, by (4.6.7), and the theorem is proved. □

Theorem 4.6.2 *Let the real-valued function* ξ *satisfy* (4.1.16) *and* (4.1.17), *and let* $\lambda \notin \sigma$. *Let* P *have the form* (4.6.16)–(4.6.17). *Then all solutions* $Y(x)$ *of* (4.6.1) *are bounded as* $x \to \infty$ *and there is no resonance.*

Proof In (4.6.5), we write

$$\Lambda(x) = \mathrm{i}\,\mathrm{dg}\,\{\mu_1(x), \ldots, \mu_n(x)\},$$

where the μ_k are real. The dichotomy condition in Theorem 1.3.1 is trivially satisfied by $\Lambda(x)$, and hence there are solutions

$$Z_k(x) = \{e_k + o(1)\} \exp\left(\mathrm{i}\int_a^x \mu_k(t)\,\mathrm{d}t\right) \tag{4.6.20}$$

for $1 \leqslant k \leqslant n$. Since the transformation (4.6.4) has the form

$$Y = \{I + o(1)\}Z,$$

it follows from (4.6.20) that the system (4.6.1) has a fundamental set of solutions $Y_k(x)$ such that

$$|Y_k(x)| = 1 + o(1).$$

Thus, for $\lambda \notin \sigma$, no solution $Y(x)$ has an amplitude factor of the type (4.1.6), and there is therefore no resonance. □

We note that, when $p(s)$ is real-valued, (4.5.3) provides a simple example of (4.6.16)–(4.6.17), and therefore Theorem 4.6.2 applies in particular to the nth order equation (4.5.1), with the second-order equation (4.1.8) as a yet more special case.

It follows from Theorem 4.6.2 that any value of λ for which (4.6.1) is resonant must lie in σ. Whether resonance actually occurs for some given value in σ depends on the nature of $P(s)$ but certainly, as we have seen in §§ 4.3 and 4.5, any value in σ can produce resonance for a suitably chosen $P(s)$. In view of these properties of σ, we refer to σ as the *resonance set* for (4.6.1).

Although we have emphasized the form (4.6.17) up to now, Theorems 4.6.1 and 4.6.2 continue to hold if (4.6.17) is replaced by

$$\tilde{P} = UD. \tag{4.6.21}$$

In the proof of Lemma 4.6.2 we simply replace DU by UD throughout. In fact, when D is non-singular, these two forms for $\tilde{P}$ are equivalent since, for example, (4.6.17) can be written

$$\tilde{P} = (DUD)D^{-1} = U_1 D_1.$$

The proofs of Lemma 4.6.2 and Theorem 4.6.1 also hold if (4.6.17) is generalized to either the 'lower-triangular' form

$$\tilde{P} = \begin{bmatrix} D_1 U_1 & & \cdots & 0 \\ & D_2 U_2 & & \vdots \\ & & \ddots & \\ & & & \end{bmatrix} \tag{4.6.22}$$

or the corresponding 'upper-triangular' form. Here there are square blocks $D_j U_j$ along the diagonal, with zeros above, and the other entries are arbitrary. The D_j and U_j have the same properties as D and U in (4.6.17). Since the product of two 'lower-triangular' matrices is again a matrix of the same type, the previous argument now applies to each block along the diagonal. Then Theorems 4.6.1 and 4.6.2 continue to hold with (4.6.22) in place of (4.6.17).

For future reference, we note a convenient way of writing (4.6.22) in the case where some of the D_j may be singular. In any particular block $D_j U_j$ where there are rows of zeros, these rows can be moved to the top of the block by re-numbering the components of Y in the original system (4.6.1). We can then write $\tilde{P}$ in the new block form

$$\tilde{P} = \begin{bmatrix} O_1 & & & \cdots & 0 \\ & D_2 U_1 & & & \vdots \\ & & O_2 & & \\ & & & \ddots & \end{bmatrix}, \tag{4.6.23}$$

where the O_j (possibly empty) are lower-triangular and the new D_j are non-singular.

4.7 THE MATRICES $\Lambda_1, \Lambda_2, \Lambda_3$

The main feature of Theorem 4.6.1 is the class of matrices P defined by (4.6.16) and (4.6.17) (or the variants (4.6.21) and (4.6.22)). In this section and the next we discuss the extent to which these types of matrix P necessarily arise from the methods in § 4.6 if the final diagonal matrix Λ in (4.6.5) is always to be pure imaginary. It turns out that the form of P in fact arises from the pure imaginary nature of Λ_1, Λ_2 and Λ_3 only. To put the matter more precisely, we suppose that $P(s)$ is a given matrix with period 2π in s and that, for all Λ_0 with distinct real entries, the process in Lemma 4.6.1 leads to pure imaginary Λ_1, Λ_2 and Λ_3 when $\lambda \notin \sigma$, where σ is defined in terms of Λ_0 by (4.1.18). We wish to draw out the consequences for P. The analysis is lengthy and therefore we divide it into several stages. The final results for P are given in § 4.8. We begin with the following lemma.

Lemma 4.7.1 *Let P_1, P_2, Λ_1, Λ_2, Λ_3 be defined as in* § 4.6. *Then*

(i) $$\Lambda_1 = \mathrm{dg}\, P \tag{4.7.1}$$

(ii) $$\Lambda_2 = \tfrac{1}{2}\, \mathrm{dg}\,(PP_1 - P_1P) \tag{4.7.2}$$

(iii) $$\Lambda_3 = \tfrac{1}{2}\, \mathrm{dg}(PP_2 - P_2P) + \tfrac{1}{6}\, \mathrm{dg}\,(\tfrac{1}{2}PP_1^2 - P_1PP_1 + \tfrac{1}{2}P_1^2P) + \tfrac{1}{3}\, \mathrm{dg}\,(\tfrac{1}{2}\Lambda_1P_1^2 - P_1\Lambda_1P_1 + \tfrac{1}{2}P_1^2\Lambda_1). \tag{4.7.3}$$

Proof As in § 4.6, we express the left-hand side of (4.6.8) as a power series in ξ, but now we have to be more explicit about the terms in ξ^2 and ξ^3. We write (4.6.8) as

$$\begin{aligned} &\mathrm{i}\{I - \xi P_1 - \xi^2P_2 - \xi^3P_3 + \tfrac{1}{2}(\xi^2P_1^2 + \xi^3P_1P_2 + \xi^3P_2P_1) - \tfrac{1}{6}\xi^3P_1^3\}\Lambda_0 \\ &\quad \times \{I + \xi P_1 + \xi^2P_2 + \xi^3P_3 + \tfrac{1}{2}(\xi^2P_1^2 + \xi^3P_1P_2 + \xi^3P_2P_1) + \tfrac{1}{6}\xi^3P_1^3\} \\ &\quad + \xi(I - \xi P_1 - \xi^2P_2 + \tfrac{1}{2}\xi^2P_1^2)P(I + \xi P_1 + \xi^2P_2 + \tfrac{1}{2}\xi^2P_1^2) \\ &\quad - (I - \xi P_1 - \xi^2P_2 + \tfrac{1}{2}\xi^2P_1^2)[\xi P_1' + \xi^2P_2' + \xi^3P_3' \\ &\quad + \tfrac{1}{2}\{\xi^2(P_1P_1' + P_1'P_1) + \xi^3(P_1'P_2 + P_1P_2' + P_2'P_1 + P_2P_1') + \tfrac{1}{6}\xi^3(P_1^3)'\}] \\ &\quad = \mathrm{i}\Lambda_0 + \xi\Lambda_1 + \xi^2\Lambda_2 + \xi^3\Lambda_3 + O(\xi^4). \end{aligned} \tag{4.7.4}$$

The terms in ξ on each side of (4.7.4) give

$$\Lambda_1 = P + \mathrm{i}(\Lambda_0P_1 - P_1\Lambda_0) - P_1', \tag{4.7.5}$$

as we have already noted in the working leading to (4.6.9) and (4.6.10); indeed, (4.7.1) is just a repeat of (4.6.9).

Next, the terms in ξ^2 in (4.7.4) give

$$\begin{aligned} \Lambda_2 = \mathrm{i}\{&\Lambda_0(P_2 + \tfrac{1}{2}P_1^2) - P_1\Lambda_0P_1 - (P_2 - \tfrac{1}{2}P_1^2)\Lambda_0\} \\ &+ PP_1 - P_1P - P_2' + \tfrac{1}{2}(P_1P_1' - P_1'P_1). \end{aligned} \tag{4.7.6}$$

On substituting for P_1' from (4.7.5), we obtain

$$\Lambda_2 = \tfrac{1}{2}(PP_1 - P_1P) + \mathrm{i}(\Lambda_0P_2 - P_2\Lambda_0) - P_2'. \tag{4.7.7}$$

Since $\mathrm{dg}\, P_2 = 0$, (4.7.2) follows.

Moving on to Λ_3, the terms in ξ^3 in (4.7.4) give

$$\begin{aligned}\Lambda_3 = &\, \mathrm{i}\{\Lambda_0(P_3 + \tfrac{1}{2}P_1P_2 + \tfrac{1}{2}P_2P_1 + \tfrac{1}{6}P_1^3) - P_1\Lambda_0(P_2 + \tfrac{1}{2}P_1^2) \\ & - (P_2 - \tfrac{1}{2}P_1^2)\Lambda_0 P_1 - (P_3 - \tfrac{1}{2}P_1P_2 - \tfrac{1}{2}P_2P_1 + \tfrac{1}{6}P_1^3)\Lambda_0\} \\ & + P(P_2 + \tfrac{1}{2}P_1^2) - P_1PP_1 - (P_2 - \tfrac{1}{2}P_1^2)P \\ & - \{P_3' + \tfrac{1}{2}(P_1'P_2 + P_1P_2' + P_2'P_1 + P_2P_1') + \tfrac{1}{6}(P_1^3)'\} \\ & + P_1\{P_2' + \tfrac{1}{2}(P_1P_1' + P_1'P_1)\} + (P_2 - \tfrac{1}{2}P_1^2)P_1'.\end{aligned}$$

We now substitute for P_1' and P_2' from (4.7.5) and (4.7.7) and, after some simplification, we obtain

$$\begin{aligned}\Lambda_3 = &\, \tfrac{1}{2}(PP_2 - P_2P) + \tfrac{1}{12}PP_1^2 - \tfrac{1}{6}P_1PP_1 + \tfrac{1}{12}P_1^2P \\ & + \tfrac{1}{6}\Lambda_1P_1^2 - \tfrac{1}{3}P_1\Lambda_1P_1 + \tfrac{1}{6}P_1^2\Lambda_1 + \mathrm{i}(\Lambda_0P_3 - P_3\Lambda_0) - P_3'.\end{aligned}$$

Since $\operatorname{dg} P_3 = 0$, (4.7.3) follows, and the lemma is proved. □

In the rest of this section, without stating it again, we assume that each off-diagonal entry p_{jk} of P is locally absolutely continuous and non-zero almost everywhere. Some relaxation of these conditions is possible at the cost of extra analysis.

Lemma 4.7.2 *Let Λ_2, as given by* (4.7.2), *be pure imaginary for all Λ_0. Then*

$$p_{jk}(s) = d_{jk}\bar{p}_{kj}(s) \qquad (j \neq k), \tag{4.7.8}$$

where d_{jk} is a real constant.

Proof Taking a typical jth diagonal entry in Λ_2, we obtain from (4.7.2) that

$$\sum_{k=1}^{n} (p_{jk}q_{kj} - q_{jk}p_{kj}) = (\text{pure imaginary}), \tag{4.7.9}$$

where $P_1 = (q_{jk})$. By (4.6.10) and Lemma 4.1.1, we can express q_{jk} in terms of p_{jk} by the formula

$$\begin{aligned}q_{jk}(x) = (1 - e_{jk}^2)^{-1}\mathrm{e}^{\mathrm{i}(a_j - a_k)x}\bigg\{ & \int_0^x \mathrm{e}^{-\mathrm{i}(a_j - a_k)\tau} p_{jk}(\lambda\tau)\,\mathrm{d}\tau \\ & + e_{jk}^2 \int_x^{2\pi/\lambda} \mathrm{e}^{-\mathrm{i}(a_j - a_k)\tau} p_{jk}(\lambda\tau)\,\mathrm{d}\tau\bigg\},\end{aligned} \tag{4.7.10}$$

where

$$e_{jk} = \exp\{\mathrm{i}(a_j - a_k)\pi/\lambda\}.$$

In the first integral on the right of (4.7.10), we make the change of variable $\tau = u + x - \pi/\lambda$ and, in the second integral, $\tau = u + x + \pi/\lambda$.

Then, bearing in mind that p_{jk} has period 2π, we obtain

$$q_{jk}(x) = \tfrac{1}{2}\mathrm{i} \operatorname{cosec}\{(a_j - a_k)\pi/\lambda\} \int_{-\pi/\lambda}^{\pi/\lambda} \mathrm{e}^{-\mathrm{i}(a_j - a_k)u}\, p_{jk}(\lambda u + \lambda x - \pi)\, \mathrm{d}u. \tag{4.7.11}$$

We now fix $a_1 - a_2$ (let us say) and let every other $a_j - a_k \to \pm\infty$. The integral in (4.7.11) tends to zero by the Riemann–Lebesgue Lemma and, if we also arrange that $(a_j - a_k)/\lambda$ avoids integer values, (4.7.9) reduces to

$$p_{12}q_{21} - q_{12}p_{21} = (\text{pure imaginary}).$$

Hence also

$$p_{12}q_{21} - \bar{q}_{12}\bar{p}_{21} = (\text{pure imaginary}). \tag{4.7.12}$$

On substituting for q_{21} and q_{12} from (4.7.11), we obtain

$$\int_{-\pi/\lambda}^{\pi/\lambda} \mathrm{e}^{-\mathrm{i}\mu u}\phi(x, u)\, \mathrm{d}u = (\text{real}), \tag{4.7.13}$$

where $\mu = a_2 - a_1$ and

$$\phi(x, u) = p_{12}(\lambda x)p_{21}(\lambda u + \lambda x - \pi) - \bar{p}_{21}(\lambda x)\bar{p}_{12}(\lambda u + \lambda x - \pi). \tag{4.7.14}$$

Here, (4.7.13) is true in the first place for μ/λ not an integer (since $\lambda \notin \sigma$) and then for all real μ by continuity.

Next we take the complex conjugate of (4.7.13) and change u to $-u$ in the integral. The right-hand side of (4.7.13) is unaffected, and hence we obtain

$$\phi(x, u) = \bar{\phi}(x, -u). \tag{4.7.15}$$

Taking $u = \pi/\lambda$, we obtain

$$p_{12}(\lambda x)p_{21}(\lambda x) = (\text{real})$$

by (4.7.14) and (4.7.15). Hence

$$p_{12}(\lambda x) = d(x)\bar{p}_{21}(\lambda x),$$

where $d(x)$ is real. To show that $d(x)$ is constant, we substitute for p_{12} in (4.7.15) to obtain

$$\begin{aligned}\bar{p}_{21}(\lambda x)p_{21}(\lambda u + \lambda x - \pi)\{d(x) - d(u + x - \pi/\lambda)\} \\ = p_{21}(\lambda x)\bar{p}_{21}(-\lambda u + \lambda x - \pi)\{d(x) - d(-u + x - \pi/\lambda)\}.\end{aligned}$$

On differentiating with respect to u and taking $u = \pi/\lambda$ again, we obtain

$$d'(x)\,|p_{21}(\lambda x)|^2 = 0.$$

Hence $d'(x) = 0$ almost everywhere, and (4.7.8) follows. □

The lemma just proved dealt specifically with the consequences of Λ_2 being pure imaginary. The next lemma brings in Λ_1 and Λ_3 as well and establishes a multiplication property of the d_{jk} in (4.7.8).

Lemma 4.7.3 *Let* Λ_1, Λ_2, *and* Λ_3, *as given by* (4.7.1)–(4.7.3), *be pure imaginary for all* Λ_0. *Then, for any distinct j, k, l, we have either*

(i)
$$d_{jk}d_{kl}d_{lj} = -1 \tag{4.7.16}$$
or

(ii)
$$p_{jk}(s) = k_1 e^{iLs}, \qquad p_{kl}(s) = k_2 e^{iMs}, \quad p_{lj}(s) = k_3 e^{iNs}, \tag{4.7.17$_1$}$$
$$p_{kj}(s) = K_1 e^{-iLs}, \qquad p_{lk}(s) = K_2 e^{-iMs}, \qquad p_{jl}(s) = K_3 e^{-iNs}, \tag{4.7.17$_2$}$$
where L, M, and N are integers with
$$L + M + N = 0 \tag{4.7.18}$$
and the k_ν *and* K_ν *are constants such that the products*
$$k_1k_2k_3 \qquad \text{and} \qquad K_1K_2K_3 \tag{4.7.19}$$
are both pure imaginary.

Proof We consider $(j, k, l) = (1, 2, 3)$ for example. We begin by noting that the hypothesis on Λ_1 and (4.7.1) give simply
$$\mathrm{dg}\, P = (\text{p.i.}), \tag{4.7.20}$$
where the right-hand side denotes a pure imaginary expression. We also note that, by (4.7.11), the results in Lemma 4.7.2 concerning p_{jk} and p_{kj} lead to the same properties for q_{jk} and q_{kj}. It is then simple to check that the last term $\frac{1}{3}\,\mathrm{dg}\,(\cdots)$ on the right of (4.7.3) is pure imaginary, as is that part of the middle term $\frac{1}{6}\,\mathrm{dg}\,(\cdots)$ on the right of (4.7.3) which involves $\mathrm{dg}\, P$, by (4.7.20). Hence (4.7.3) gives
$$\tfrac{1}{2}\,\mathrm{dg}\,(PP_2 - P_2P) + \tfrac{1}{6}\,\mathrm{dg}\,(\tfrac{1}{2}\tilde{P}P_1^2 - P_1\tilde{P}P_1 + \tfrac{1}{2}P_1^2\tilde{P}) = (\text{p.i.}), \tag{4.7.21}$$
where $\tilde{P} = P - \mathrm{dg}\, P$.

We now write
$$PP_1 - P_1P = (r_{jk}), \qquad P_2 = (t_{jk}).$$
Then, as in the case of (4.7.11), it follows from (4.7.7) that
$$t_{jk}(x) = \tfrac{1}{4}\,i\,\mathrm{cosec}\,\{(a_j - a_k)\pi/\lambda\}\int_{-\pi/\lambda}^{\pi/\lambda} e^{-i(a_j-a_k)u}\, r_{jk}(u + x - \pi/\lambda)\, du \tag{4.7.22}$$
where, in terms of P and P_1
$$r_{jk} = \sum_{\nu=1}^{n} (p_{j\nu}q_{\nu k} - q_{j\nu}p_{\nu k}). \tag{4.7.23}$$

Next, we hold $a_1 - a_2$ and $a_2 - a_3$ fixed and let every other $a_j - a_k \to \pm\infty$ in the same way as we did to obtain (4.7.12). We now find that the part of (4.7.21) which contains only p_{jk}, q_{jk}, and t_{jk}, with j, $k = 1, 2, 3$, is pure imaginary. In particular, the (1, 1) diagonal term in (4.7.21) gives

$$\begin{aligned} &p_{12}t_{21} + p_{13}t_{31} - t_{12}p_{21} - t_{13}p_{31} \\ &\quad + \tfrac{1}{6}(p_{12}q_{23}q_{31} + p_{13}q_{32}q_{21}) - \tfrac{1}{3}(q_{12}p_{23}q_{31} + q_{13}p_{32}q_{21}) \\ &\quad + \tfrac{1}{6}(q_{12}q_{23}p_{31} + q_{13}q_{32}p_{21}) = (\text{p.i.}). \end{aligned} \tag{4.7.24}$$

Also, further to (4.7.23), we write

$$r_{jk} = (p_{jj} - p_{kk})q_{jk} + \tilde{r}_{jk}, \tag{4.7.25}$$

thus defining $\tilde{r}_{jk}$, and then we define $\tilde{t}_{jk}$ by (4.7.22) but with $\tilde{r}_{jk}$ in place of r_{jk}.

Let us now suppose that, contrary to (i),

$$d_{12}d_{23}d_{31} \neq -1. \tag{4.7.26}$$

The lemma will be proved if we show that (ii) must hold. Using the obvious fact that $\alpha + \beta = (\text{p.i.})$ implies that $\alpha + \bar{\beta} = (\text{p.i.})$ (as we did in (4.7.12)), we obtain from (4.7.24)

$$\begin{aligned} &\bar{p}_{12}\bar{t}_{21} + p_{13}t_{31} - t_{12}p_{21} - \bar{t}_{13}\bar{p}_{31} \\ &\quad + \tfrac{1}{6}(\bar{p}_{12}\bar{q}_{23}\bar{q}_{31} + p_{13}q_{32}q_{21}) - \tfrac{1}{3}(\bar{q}_{12}\bar{p}_{23}\bar{q}_{31} + q_{13}p_{32}q_{21}) \\ &\quad + \tfrac{1}{6}(\bar{q}_{12}\bar{q}_{23}\bar{p}_{31} + q_{13}q_{32}p_{21}) = (\text{p.i.}). \end{aligned}$$

In the case of (4.7.26), we substitute for the t_{jk} from (4.7.22) and (4.7.23), and then use (4.7.8) for both the p_{jk} and q_{jk} in the terms with conjugate signs. By (4.7.8), those expressions which arise from $(p_{jj} - p_{kk})q_{jk}$ in (4.7.25) cancel out, and we are left with

$$\begin{aligned} (d_{12}d_{23}d_{31} + 1)(-\tilde{t}_{12}p_{21} + p_{13}\tilde{t}_{31} + \tfrac{1}{6}p_{13}q_{32}q_{21} \\ - \tfrac{1}{3}q_{13}p_{32}q_{21} + \tfrac{1}{6}q_{13}q_{32}p_{21}) = (\text{p.i.}). \end{aligned}$$

By (4.7.26) we have therefore

$$-\tilde{t}_{12}p_{21} + p_{13}\tilde{t}_{31} + \tfrac{1}{6}p_{13}q_{32}q_{21} - \tfrac{1}{3}q_{13}p_{32}q_{21} + \tfrac{1}{6}q_{13}q_{32}p_{21} = (\text{p.i.}). \tag{4.7.27}$$

Apart from the terms which have cancelled out, the expression on the left here consists of just half of the terms in (4.7.24). We now define

$$S_1 = \tilde{t}_{31}p_{13} + \tfrac{1}{6}(p_{21}q_{13}q_{32} - p_{32}q_{21}q_{13}) \tag{4.7.28}$$

and similarly for S_2 and S_3 by cyclic change of the subscripts. Then (4.7.27) is

$$S_1 - S_2 = (\text{p.i.}). \tag{4.7.29}$$

Similarly, the (2, 2) and (3, 3) diagonal terms in (4.7.21) give

$$S_2 - S_3 = (\text{p.i.}), \qquad S_3 - S_1 = (\text{p.i.}). \tag{4.7.30}$$

To examine the consequences of (4.7.29), we consider S_1 first and express (4.7.28) in terms of the p_{jk}, using (4.7.22), (4.7.23), and (4.7.11). We write

$$\alpha = a_3 - a_2, \quad \beta = a_1 - a_3, \quad \gamma = a_2 - a_1, \tag{4.7.31}$$

noting that

$$\alpha + \beta + \gamma = 0. \tag{4.7.32}$$

Then (4.7.28) is

$$\begin{aligned}
&24 \sin(\alpha\pi/\lambda) \sin(\beta\pi/\lambda) \sin(\gamma\pi/\lambda) S_1 \\
&= 3p_{13}(\lambda x)\Big\{\sin(\alpha\pi/\lambda)\Big[\int e^{i\beta u} p_{32}(\lambda u + \lambda x - \pi) \\
&\quad \times \int e^{-i\gamma v} p_{21}(\lambda v + \lambda u + \lambda x)\, dv\, du\Big] \\
&\quad - \sin(\gamma\pi/\lambda)\Big[\int e^{i\beta u} p_{21}(\lambda u + \lambda x - \pi) \\
&\quad \times \int e^{-i\alpha v} p_{32}(\lambda v + \lambda u + \lambda x)\, dv\, du\Big]\Big\} \\
&\quad - \sin(\gamma\pi/\lambda) p_{21}(\lambda x)\Big[\int e^{-i\beta u} p_{13}(\lambda u + \lambda x - \pi)\, du \\
&\quad \times \int e^{-i\alpha u} p_{32}(\lambda u + \lambda x - \pi)\, du\Big] \\
&\quad + \sin(\alpha\pi/\lambda) p_{32}(\lambda x)\Big[\int e^{-i\gamma u} p_{21}(\lambda u + \lambda x - \pi)\, du \\
&\quad \times \int e^{-i\beta u} p_{13}(\lambda u + \lambda x - \pi)\, du\Big] \\
&= \tilde{S}_1,
\end{aligned} \tag{4.7.33}$$

say, where all the integrations are over $(-\pi/\lambda, \pi/\lambda)$. There is a similar expression $\tilde{S}_2$ arising from S_2, in which the cyclic change $(1, 2, 3) \to (2, 3, 1)$, $(\alpha, \beta, \gamma) \to (\beta, \gamma, \alpha)$ is made. Then (4.7.29) becomes

$$\tilde{S}_1 - \tilde{S}_2 = (\text{p.i.}). \tag{4.7.34}$$

We are assuming that the p_{jk} are not identically zero and, considering the complex Fourier coefficients of p_{32}, we choose an integer N such that

$$(2\pi/\lambda) c_N = \int_{-\pi/\lambda}^{\pi/\lambda} e^{-iN\lambda s} p_{32}(\lambda s)\, ds \neq 0. \tag{4.7.35}$$

Although we cannot take $\alpha = N\lambda$ in (4.7.34), such a choice placing λ in the set σ by (4.1.18) and (4.7.31), we can however let $\alpha \to N\lambda$. Then, with γ held fixed, (4.7.32) gives $\beta \to -\gamma - N\lambda$. In the limit, it follows from (4.7.33) and the corresponding expression for $\tilde{S}_2$ that (4.7.34) becomes

$$-3p_{13}(\lambda x)\left[\int e^{-i(\gamma+N\lambda)u} p_{21}(\lambda u + \lambda x - \pi) \times \int e^{-iN\lambda v} p_{32}(\lambda v + \lambda u + \lambda x)\, dv\, du\right]$$
$$- p_{21}(\lambda x)\left[\int e^{i(\gamma+N\lambda)u} p_{13}(\lambda u + \lambda x - \pi)\, du \times \int e^{-iN\lambda u} p_{32}(\lambda u + \lambda x - \pi)\, du\right]$$
$$+ 3(-1)^N p_{21}(\lambda x)\left[\int e^{i\gamma u} p_{13}(\lambda u + \lambda x - \pi) \times \int e^{-iN\lambda v} p_{32}(\lambda v + \lambda u + \lambda x)\, dv\, du\right]$$
$$+ (-1)^N p_{13}(\lambda x)\left[\int e^{-iN\lambda u} p_{32}(\lambda u + \lambda x - \pi)\, du \times \int e^{-i\gamma u} p_{21}(\lambda u + \lambda x - \pi)\, du\right]$$
$$= (\text{p.i.}).$$

We next express the four integrals involving p_{32} in terms of c_N (see (4.7.35)) by making the obvious change of integration variable and then restoring the limits of integration to $-\pi/\lambda$, π/λ by periodicity. Also, in the integrals involving $e^{i\gamma u}$, we change the integration variable to $-u$, so that the factor becomes $e^{-i\gamma u}$ in all integrals. The result is that

$$\int_{-\pi/\lambda}^{\pi/\lambda} e^{-i\gamma u} \psi(x, u)\, du = (\text{p.i.}), \tag{4.7.36}$$

where

$$\psi(x, u) = c_N e^{iN\lambda x}\{p_{13}(\lambda x) p_{21}(\lambda u + \lambda x - \pi) - (-1)^N e^{-iN\lambda u} p_{21}(\lambda x) p_{13}(\lambda x - \lambda u - \pi)\}. \tag{4.7.37}$$

Now (4.7.36) is similar to (4.7.13) (in fact $\gamma = \mu$) and, as for (4.7.15), we conclude that

$$\psi(x, u) = -\bar{\psi}(x, -u).$$

On differentiating with respect to u and taking $u = \pi/\lambda$, we obtain

$$c_N(e^{iNs}p_{13}p_{21})' = \bar{c}_N(e^{-iNs}\bar{p}_{13}\bar{p}_{21})',$$

where p_{13} and p_{21} are evaluated at s. On integration, we obtain

$$c_N e^{iNs}p_{13}(s)p_{21}(s) = \rho_N(s) + i\kappa_N, \tag{4.7.38}$$

where ρ_N is real-valued and κ_N is a real constant.

We now assert that (4.7.38) is possible for only one value of N because, if another value M could be taken, we would have

$$\rho_N + i\kappa_N = (c_N/c_M)e^{i(N-M)s}(\rho_M + i\kappa_M).$$

The imaginary part of this equation is

$$\kappa_N = |c_N/c_M|[\rho_M \sin\{(N-M)s + \theta\} + \kappa_M \cos\{(N-M)s + \theta\}],$$

where $\theta = \arg(c_N/c_M)$. On choosing s so that $\sin\{\cdots\} = 0$ and $\cos\{\cdots\} = \pm 1$, we obtain $\kappa_N = \kappa_M = 0$ and hence also $\rho_N = \rho_M = 0$. This is a contradiction, and therefore p_{32} has only one non-zero Fourier coefficient c_N in (4.7.35). Hence

$$p_{32}(s) = c_N e^{iNs}.$$

A similar argument applies to p_{21} and p_{13}, and we have therefore established (4.7.17$_1$). Then (4.7.17$_2$) also follows for the other triplet p_{12}, p_{31}, p_{23} by (4.7.8). To establish (4.7.18), we note that (4.7.38) is now

$$k_1k_2k_3 \exp\{i(L+M+N)s\} = \rho(s) + i\kappa,$$

from which it follows immediately that $L+M+N=0$. To establish (4.7.19), we substitute (4.7.17$_1$) back into (4.7.33) and (4.7.34) with general α, β, and γ. The integrals are now simple to evaluate and, after a calculation involving (4.7.32) and (4.7.18), the left-hand side of (4.7.34) becomes a real constant multiple of $k_1k_2k_3$, and (4.7.19) follows. The proof of Lemma 4.7.3 is now complete. □

4.8 THE FORM OF P

The lemmas in § 4.7 provide the information on the form of P which we need to discuss the necessity of the conditions (i) and (ii) in Theorem 4.6.1. By (4.7.1), the consequence of Λ_1 being pure imaginary is that (4.6.16) necessarily holds. Moving on to the necessity of the DU form for $\tilde{P}$ in (4.6.17), we deal separately with the cases $n=2$, $n=3$, and $n>3$. As in § 4.7, we assume throughout that each off-diagonal entry of P is locally absolutely continuous and non-zero almost everywhere.

Theorem 4.8.1 *Let $n=2$ and let Λ_2 be pure imaginary for all Λ_0. Then $\tilde{P}$ has the form DU.*

Proof By Lemma 4.7.2, we must have

$$\tilde{P}=\begin{bmatrix} 0 & p_{12} \\ d\bar{p}_{12} & 0 \end{bmatrix},$$

where d is a non-zero real constant. Hence $\tilde{P}=DU$ with $D=\mathrm{dg}\,(1,-d)$. □

Theorem 4.8.2 *Let $n=3$ and let Λ_1, Λ_2, and Λ_3 be pure imaginary for all Λ_0. Then, either*
(a) *$\tilde{P}$ has the form DU, or*
(b)

$$\tilde{P}=ECE^{-1}, \tag{4.8.1}$$

where

$$E(s)=\mathrm{dg}\,(\mathrm{e}^{\mathrm{i}\zeta s}, \mathrm{e}^{\mathrm{i}\eta s}, \mathrm{e}^{\mathrm{i}\theta s}) \tag{4.8.2}$$

with real constants ζ, η, and θ, and C is a constant matrix,

$$C=\begin{bmatrix} 0 & k_1 & K_3 \\ K_1 & 0 & k_2 \\ k_3 & K_2 & 0 \end{bmatrix}, \tag{4.8.3}$$

where the k_ν and K_ν have the same properties as in Lemma 4.7.3 (ii).

Proof We consider the two cases in the statement of Lemma 4.7.3. In case (i) we have immediately

$$\tilde{P}=DU \tag{4.8.4}$$

with

$$D=\mathrm{dg}\,(-1, d_{21}, d_{31}), \tag{4.8.5}$$

and the skew-hermitian nature of U follows from (4.7.16).

In case (ii), we have (4.7.17) with $(j, k, l)=(1, 2, 3)$. By (4.7.18), we can define ζ, η, and θ so that

$$L=\zeta-\eta, \qquad M=\eta-\theta, \qquad N=\theta-\zeta, \tag{4.8.6}$$

and then (4.8.1)–(4.8.3) follow immediately from (4.7.17). □

For general $n>3$, we have the following result which corresponds to part (a) of the last theorem.

Theorem 4.8.3 *Let Λ_1, Λ_2 and Λ_3 be pure imaginary for all Λ_0. Suppose also that case* (ii) *of Lemma* 4.7.3 *does not occur for any triplet (j, k, l). Then $\tilde{P}$ has the form DU.*

Proof Let us consider, for example, the real non-zero constants d_{j1} provided by Lemma 4.7.2. As in (4.8.4) and (4.8.5), we write

$$\tilde{P}=DU$$

with

$$D=\mathrm{dg}\,(-1, d_{21}, \ldots, d_{n1}).$$

Then the entries u_{jk} in U are given by

$$u_{jk} = d_{j1}^{-1} p_{jk} = d_{1j} p_{jk}.$$

Then

$$\bar{u}_{jk} = d_{1j}\bar{p}_{jk} = d_{1j} d_{jk} p_{kj} = d_{1j} d_{jk} d_{k1} u_{kj} = -u_{kj}$$

by (4.7.16). Hence U is skew-hermitian, as required. □

In Lemma (4.7.3), the hypothesis is that Λ_1, Λ_2, and Λ_3 are pure imaginary. It is natural to ask whether (4.7.17) (which leads to part (b) of Theorem 4.8.2) can be excluded if further Λ_m ($m \geqslant 4$) are also taken to be pure imaginary. That this is not so can be shown by considering (4.8.1)–(4.8.3) directly and establishing the asymptotic form of the solutions Y of (4.6.1) in terms of pure imaginary exponentials. Let us suppose for example that $\operatorname{dg} P = 0$ and that $P = ECE^{-1}$ as in (4.8.1)–(4.8.3). The transformation $Y = EZ$ takes (4.6.1) into

$$Z' = (\mathrm{i}\Delta + \xi C)Z, \tag{4.8.7}$$

where

$$\begin{aligned} \Delta &= \operatorname{dg}(a_1 - \lambda\zeta,\ a_2 - \lambda\eta,\ a_3 - \lambda\theta) \\ &= \operatorname{dg}(b_1, b_2, b_3), \end{aligned} \tag{4.8.8}$$

say. The diagonal entries in Δ are distinct since, for example, $\zeta - \eta = L$ by (4.8.6), and $\lambda \notin \sigma$. By (4.1.16), the conditions of Theorem 1.6.1 are satisfied by the system (4.8.7), and therefore there are solutions Z_k ($1 \leqslant k \leqslant 3$) such that

$$Z_k(x) = \{e_k + o(1)\} \exp\left(\int_a^x \mu_k(t)\,\mathrm{d}t\right),$$

where the μ_k are the eigenvalues of the matrix $i\Delta + \xi C$. We now show that the μ_k are pure imaginary, thereby establishing non-resonance. By (4.8.3) and (4.8.8), the μ_k are the solutions of

$$\begin{vmatrix} ib_1 - \mu & \xi k_1 & \xi K_3 \\ \xi K_1 & ib_2 - \mu & \xi k_2 \\ \xi k_3 & \xi K_2 & ib_3 - \mu \end{vmatrix} = 0,$$

which is

$$\begin{aligned} &(ib_1 - \mu)(ib_2 - \mu)(ib_3 - \mu) \\ &\qquad - \xi^2\{(\mathrm{i}b_1 - \mu)k_2K_2 + (\mathrm{i}b_2 - \mu)k_3K_3 + (\mathrm{i}b_3 - \mu)k_1K_1\} \\ &\qquad\qquad + \xi^3(k_1k_2k_3 + K_1K_2K_3) = 0. \end{aligned}$$

It follows from (4.7.17) and Lemma 4.7.2 that the products k_1K_1, k_2K_2, k_3K_3 are all real. Hence, writing $\mu = \mathrm{i}\nu$ and using (4.7.18), we obtain

$$\begin{aligned} &(b_1 - \nu)(b_2 - \nu)(b_3 - \nu) \\ &\qquad + \xi^2\{(b_1 - \nu)c_1 + (b_2 - \nu)c_2 + (b_3 - \nu)c_3\} + \kappa\xi^3 = 0, \end{aligned}$$

with real constants c_1, c_2, c_3, κ. When ξ is sufficiently small, that is, when x is in some neighbourhood of ∞, v can be expressed as a power series in ξ with real coefficients which are determined recursively in terms of the b_r, c_r, and κ. Thus v is real for large x, leading to pure imaginary μ_k, as required.

4.9 VALUES OF λ IN THE RESONANT SET

We return to the general system (4.6.1) with a given Λ_0 as in (4.6.2) and we consider the effect of the transformation (4.6.4) when λ takes values in the resonant set σ, that is,

$$\lambda = (a_J - a_K)/N \tag{4.9.1}$$

for some J, K and N. As in § 4.6, the aim is to choose the periodic matrices P_m so that the transformed system (4.6.6) takes as simple a form as can be arranged. It is no longer possible to guarantee diagonal matrices $\Lambda_1, \ldots, \Lambda_M$ in (4.6.7), and the result corresponding to Lemma 4.6.1 is as follows.

Lemma 4.9.1 *Let ξ satisfy* (4.1.16) *and* (4.1.17), *and let $\lambda \in \sigma$ as in* (4.9.1). *Then there are matrices $P_m(x)$ with period $2\pi/\lambda$ such that the transformation*

$$Y = \{\exp(\xi P_1 + \xi^2 P_2 + \cdots + \xi^M P_M)\}Z \tag{4.9.2}$$

takes (4.6.1) *into the form*

$$Z' = (\mathrm{i}\Lambda_0 + \xi\Delta_1 + \cdots + \xi^M \Delta_M + S)Z \tag{4.9.3}$$

with $S \in L(a, \infty)$ and

$$\Delta_m = E(x)\Gamma_m E(-x) \qquad (1 \le m \le M), \tag{4.9.4}$$

where $E(x) = \exp \mathrm{i}\Lambda_0 x$ and Γ_m is constant.

Proof We follow a similar procedure to the one used for (4.6.8), and now we require that

$$\mathrm{i}e^{-\mathscr{P}}\Lambda_0 e^{\mathscr{P}} + \xi e^{-\mathscr{P}} P e^{\mathscr{P}} - e^{-\mathscr{P}}(e^{\mathscr{P}})'$$
$$= \mathrm{i}\Lambda_0 + \xi\Delta_1 + \cdots + \xi^M \Delta_M + O(\xi^{M+1}). \tag{4.9.5}$$

Starting with the term in ξ on each side, we aim to define P_1 and Δ_1 so that

$$\mathrm{i}(\Lambda_0 P_1 - P_1 \Lambda_0) - P_1' + P = \Delta_1. \tag{4.9.6}$$

We define Δ_1 by taking its (J, K) and (K, J) entries to be

$$c_N^{(J,K)} e^{\mathrm{i}N\lambda x} \qquad \text{and} \qquad c_{-N}^{(K,J)} e^{-\mathrm{i}N\lambda x}, \tag{4.9.7}$$

where $c_N^{(J,K)}$ denotes the Nth complex Fourier coefficient of the (J, K) entry in P, and likewise for $c_{-N}^{(K,J)}$. We do this for all possible values of J, K and N corresponding to the same λ in (4.9.1). The remaining off-diagonal entries in Δ_1 are then defined to be zero. Thus, so far, Δ_1 has the form (4.9.4). We complete the definition of Δ_1 by considering the diagonal entries in (4.9.6) and taking

$$\mathrm{dg}\,\Delta_1 = (2\pi)^{-1}\int_{-\pi}^{\pi} \mathrm{dg}\,P(s)\,\mathrm{d}s, \tag{4.9.8}$$

and then we can define a periodic dg P_1 by

$$(\mathrm{dg}\,P_1)' = \mathrm{dg}\,P - (2\pi)^{-1}\int_{-\pi}^{\pi} \mathrm{dg}\,P(s)\,\mathrm{d}s.$$

Altogether, by (4.9.1), (4.9.7) and (4.9.8), Δ_1 has the form (4.9.4) and, indeed, the form can be given explicitly as

$$\Delta_1 = (\lambda/2\pi)E(x)\left(\int_{-\pi/\lambda}^{\pi/\lambda} E(-t)P(\lambda t)E(t)\,\mathrm{d}t\right)E(-x). \tag{4.9.9}$$

Finally, the off-diagonal elements in P_1 can be defined as in Lemma 4.1.1 (with $\Phi = P_1 - \mathrm{dg}\,P_1$) since the choice of Δ_1 means that (4.9.6) falls under the case considered in (4.1.22)–(4.1.24). At this point we have a similar position to the one in (4.2.18). We can now go on to the higher powers of ξ and, as in the proof of Lemma 4.6.1, the method of defining P_1 and Δ_1 can be repeated for each power of ξ in (4.9.5) to determine the P_m and Δ_m inductively. This establishes (4.9.3) with Δ_m having the required form (4.9.4). □

As in (4.2.19), the transformation

$$Z = (\exp \mathrm{i}\Lambda_0 x)W \tag{4.9.10}$$

simplifies (4.9.3) to

$$W' = (\xi\Gamma_1 + \cdots + \xi^M\Gamma_M + S)W. \tag{4.9.11}$$

This system does not necessarily have the Levinson form but, before proceeding further, we obtain two general properties of the constant matrices Γ_m.

Lemma 4.9.2 (i) *The trace of Γ_m is zero for $m \geq 2$.*

(ii) *Let*

$$P = DU, \tag{4.9.12}$$

where D is a real constant diagonal matrix and U is skew-hermitian. Then

$$\Gamma_m = DU_m, \tag{4.9.13}$$

where U_m is constant and skew-hermitian.

Proof (i) Since the trace of a matrix is invariant under similarity transformations, it follows from (4.9.4) and (4.9.5) that

$$\xi \operatorname{tr} P - \operatorname{tr} e^{-\mathscr{P}}(e^{\mathscr{P}})' = \sum_{1}^{M} \xi^m \operatorname{tr} \Gamma_m + O(\xi^{M+1}). \tag{4.9.14}$$

By the Liouville–Jacobi formulae (1.2.15) and (1.2.16), we have

$$-\operatorname{tr} e^{-\mathscr{P}}(e^{\mathscr{P}})' = (\log \det e^{\mathscr{P}})' = (\operatorname{tr} \mathscr{P})' = \sum_{1}^{M} \xi^m \operatorname{tr} P'_m.$$

On substituting into (4.9.14) and equating the terms in ξ^m, we obtain $\operatorname{tr} \Gamma_m = -\operatorname{tr} P'_m$ $(m \geqslant 2)$. Since P_m has period $2\pi/\lambda$, it follows that $\operatorname{tr} \Gamma_m = 0$.

(ii) The proof is almost the same as for Lemma 4.6.2. First, the definition of P_1 and Δ_1 in (4.9.6) and (4.9.9) imply that P_1 and Δ_1 have the form DU for some skew-hermitian U. The induction argument of Lemma 4.6.2 can then be applied again to show that P_m and Δ_m have the form DU for $m \geqslant 1$, and then (4.9.13) follows from (4.9.4). □

We note that (4.9.12) is equivalent to (4.6.16) and (4.6.17) when D is non-singular, but places more restriction on dg P when D is singular.

We have now to consider the transformation of (4.9.11) into the Levinson form. In the case where the Γ_m $(1 \leqslant m \leqslant M)$ are all zero, (4.9.11) is already in the Levinson form and the transformation back to Y via (4.9.10) and (4.9.2) gives solutions Y_k of (4.6.1) such that

$$Y_k(x) = \{e_k + o(1)\} \exp(\mathrm{i}a_k x).$$

There is no resonance in this case. The situation of interest therefore is when the Γ_m are not all zero, and we denote by Γ_L the first non-zero Γ_m. Thus (4.9.11) is now

$$W' = (\xi^L \Gamma_L + \cdots + \xi^M \Gamma_M + S)W. \tag{4.9.15}$$

As a general condition on Γ_L, we assume that

$$\Gamma_L \text{ has } n \text{ distinct eigenvalues } \rho_k \qquad (1 \leqslant k \leqslant n), \tag{4.9.16}$$

leaving this as a condition to be verified for individual systems (4.6.1). We denote the corresponding eigenvectors of Γ_L by v_k. The asymptotic form of the solutions of (4.6.1) is now given by the following theorem.

Theorem 4.9.1 *Let ξ satisfy* (4.1.16) *and* (4.1.17), *and let ξ be real-valued and positive. Let $\lambda \in \sigma$ as in* (4.9.1) *and let* (4.9.16) *hold. Then* (4.6.1) *has solutions*

$$Y_k(x) = \{(\exp \mathrm{i}\Lambda_0 x)v_k + o(1)\} \exp\left(\int_a^x \xi^L(t)\mu_k(t)\, \mathrm{d}t\right), \tag{4.9.17}$$

where the μ_k *are the eigenvalues of the matrix*

$$\Gamma_L + \cdots + \xi^{M-L}\Gamma_M. \tag{4.9.18}$$

Proof First we make the transformation

$$W = (v_1 \cdots v_n)V$$

to replace Γ_L by its diagonal form. By (4.9.16) and (4.1.16), we can apply Theorem 1.6.1 to the resulting V-system provided only that the $\xi^L \mu_k$ $(1 \leqslant k \leqslant n)$ satisfy the dichotomy condition L in § 1.3. To see that this is so, we note that the μ_k are analytic in ξ in some x-interval $[X, \infty)$, by (4.9.16) and (4.9.18). Then, since ξ is real-valued, re $\xi^L(\mu_j - \mu_k)$ is either identically zero in $[X, \infty)$ or has the form $\xi^{L+s}\{\alpha + o(1)\}$ for some $s \geqslant 0$ and a non-zero constant α. Then again, since $\xi > 0$, either (1.3.1) or (1.3.2) is satisfied by $\lambda_k = \xi^L \mu_k$. Thus (4.9.15) has solutions

$$W_k = \{v_k + o(1)\} \exp\left(\int_a^x \xi^L(t)\mu_k(t)\, dt\right),$$

and (4.9.17) follows when we transform back to Y. □

The theory of §§ 4.6–4.7 has focused attention on matrices P with the DU form in (4.9.12). For such matrices, the circumstances in which resonance may occur when $\lambda \in \sigma$ can be narrowed down, and the following theorem gives a further non-resonant situation.

Theorem 4.9.2 *Let the conditions of Theorem* 4.9.1 *hold. Also, let* $P = DU$ *as in* (4.9.12), *and let all the diagonal entries in* D *be non-zero with the same sign. Then* (4.6.1) *is non-resonant.*

Proof By (4.9.13), the matrix (4.9.18) has the form DU_0 for some skew-hermitian U_0. Let μ denote any one of the eigenvalues μ_k of this matrix with corresponding eigenvector v. Then

$$U_0 v = \mu D^{-1} v$$

and, taking adjoints,

$$-v^* U_0 = \bar{\mu} v^* D^{-1}.$$

On pre-multiplying the first equation by v^*, post-multiplying the second by v and adding, we obtain

$$(\mu + \bar{\mu}) v^* D^{-1} v = 0.$$

Since the diagonal entries in D are either all positive or all negative, we have $v^* D^{-1} v \neq 0$, and hence $\mu + \bar{\mu} = 0$. Thus the μ_k are pure imaginary in (4.9.17), giving

$$|Y_k| = |v_k|\{1 + o(1)\},$$

and there is no amplitude factor of the type ρ in (4.1.6). □

It follows from this theorem that resonance can occur only in situations where D either has both positive and negative entries or has some zero entries. We have already met examples of the first alternative in (4.5.6)–(4.5.8) and (4.5.10). In this case, D is given by (4.1.14) and, for the second-order equation (4.1.8), we have $d_1 = \frac{1}{2}$ and $d_2 = -\frac{1}{2}$. Thus the resonance results of § 4.3 are based upon the opposite signs of d_1 and d_2. An example where D has a zero entry will be considered at the end of the next section.

4.10 RESONANCE FOR $M \geqslant 2$

In Theorem 4.9.1, the size $|Y_k|$ of the solutions is giverned by re μ_k and, in order to discuss the resonance situation in more detail than in Theorem 4.9.2, it is necessary to calculate at least some of the matrices Γ_m. Now Γ_1 is easily obtained from (4.9.4) and (4.9.9), and consequently resonance can be discussed readily when $M = 1$, as was done in §§ 4.2–4.5. However, the calculations for Γ_2, $\Gamma_3, \ldots$ rapidly become complicated and therefore we shall confine the detailed account here to the case of 2×2 matrices.

The general formula for Δ_2 can be obtained from (4.9.5) in the same way as (4.7.7) was obtained from (4.6.8). On equating the terms in ξ^2 on each side of (4.9.5), we have (4.7.6) as before and, substituting for P_1' from (4.9.6), we find that

$$\Delta_2 = \tfrac{1}{2}(PP_1 - P_1 P) + \tfrac{1}{2}(\Delta_1 P_1 - P_1 \Delta_1) + \mathrm{i}(\Lambda_0 P_2 - P_2 \Lambda_0) - P_2'.$$

We shall be dealing with situations where

$$\Delta_1 = 0 \tag{4.10.1}$$

and it then follows from (4.9.4) and (4.9.9), with $\frac{1}{2}(PP_1 - P_1 P)$ in place of P, that

$$\Gamma_2 = (\lambda/4\pi)\int_{-\pi/\lambda}^{\pi/\lambda} E(-t)\{P(\lambda t)P_1(t) - P_1(t)P(\lambda t)\}E(t)\,\mathrm{d}t. \tag{4.10.2}$$

Turning now to $n = 2$, we have the following useful lemma concerning the eigenvalues μ_k of the matrix (4.9.18).

Lemma 4.10.1 *Let* $n = 2$ *and let* $P = DU$ *as in* (4.9.12). *Further, let* $\operatorname{tr} \Gamma_1 = 0$. *Then*

$$\mu_k^2 \text{ is real.} \tag{4.10.3}$$

Proof By Lemma 4.9.2, the matrix (4.9.18) has zero trace and it also has the form DU_1 for some skew-hermitian U_1. Hence the characteristic equation for the μ_k has the form

$$\begin{vmatrix} i\eta-\mu & d_1\theta \\ -d_2\bar{\theta} & -i\eta-\mu \end{vmatrix} = 0$$

with real d_1, d_2 and η. Thus

$$\mu^2 = -\eta^2 - d_1 d_2 |\theta|^2,$$

and the right-hand side is real, as required. □

We continue now with the case $n=2$ and

$$P = DU = \begin{bmatrix} d_1 & 0 \\ 0 & d_2 \end{bmatrix} \begin{bmatrix} ip & r \\ -\bar{r} & iq \end{bmatrix}, \tag{4.10.4}$$

where d_1, d_2, p and q are real. The situation of interest is where the value (4.9.1) of λ does not produce resonance when $M=1$, but may do so when $M \geqslant 2$. We therefore suppose that (4.10.1) holds when

$$\lambda = (a_1 - a_2)/N \tag{4.10.5}$$

and we go on to calculate Γ_2 from the formula (4.10.2). We denote the complex Fourier coefficients of p, q and r by

$$\alpha_\nu, \quad \beta_\nu, \quad \gamma_\nu \qquad (-\infty < \nu < \infty) \tag{4.10.6}$$

respectively. By (4.9.9), (4.10.1) holds when

$$\alpha_0 = 0, \quad \beta_0 = 0, \quad \gamma_N = 0. \tag{4.10.7}$$

It then follows from (4.9.6) that

$$P_1 = DU_1, \tag{4.10.8}$$

where U_1 has the skew-hermitian form

$$U_1 = \begin{bmatrix} ip_1 & r_1 \\ -\bar{r}_1 & iq_1 \end{bmatrix} \tag{4.10.9}$$

and p_1, q_1, r_1 have the complex Fourier coefficients

$$\left.\begin{array}{ll} \alpha_\nu / i\nu\lambda, \quad \beta_\nu / i\nu\lambda & (\nu \neq 0) \\ \gamma_\nu / \{i(\nu - N)\lambda\} & (\nu \neq N) \end{array}\right\} \tag{4.10.10}$$

respectively. Here p_1 and q_1 are real-valued, and the omitted Fourier coefficients are zero.

In the integrand in (4.10.2), we have

$$PP_1 - P_1 P = \begin{bmatrix} v & d_1 w \\ -d_2 \bar{w} & -v \end{bmatrix} \tag{4.10.11}$$

where

$$v = d_1 d_2 (\bar{r} r_1 - r \bar{r}_1) = 2i d_1 d_2 \operatorname{im}(\bar{r} r_1) \tag{4.10.12}$$

and

$$w = \mathrm{i}\{(d_1 p - d_2 q) r_1 - (d_1 p_1 - d_2 q_1) r\}, \tag{4.10.13}$$

by (4.10.4), (4.10.8) and (4.10.9). Correspondingly, we write (4.10.2) as

$$\Gamma_2 = \begin{bmatrix} \mathrm{i}\alpha & d_1\beta \\ -d_2\bar{\beta} & -\mathrm{i}\alpha \end{bmatrix}. \tag{4.10.14}$$

Considering the diagonal terms in (4.10.11) first, the only contribution made by these terms to Γ_2 in (4.10.2) comes from the zeroth Fourier coefficient of v. Hence, by (4.10.6), (4.10.10) and (4.10.12),

$$\begin{aligned} \alpha &= d_1 d_2 \,\mathrm{im}\left(\sum_{\nu=-\infty}^{\infty} \bar{\gamma}_\nu \gamma_\nu / \{\mathrm{i}(\nu - N)\lambda\} \right) \\ &= d_1 d_2 \lambda^{-1} \left\{ N^{-1} |\gamma_0|^2 - \sum_1^\infty |\gamma_\nu|^2 \left(\frac{1}{\nu - N} - \frac{1}{\nu + N} \right) \right\} \\ &= d_1 d_2 (2N/\lambda) \left\{ (2N^2)^{-1} |\gamma_0|^2 - \sum_1^\infty |\gamma_\nu|^2 / (\nu^2 - N^2) \right\}. \end{aligned} \tag{4.10.15}$$

In these summations the terms with $\nu = N$ are omitted because of (4.10.7).

To deal with the terms involving w in (4.10.11) and (4.10.13), we write

$$s = d_1 p - d_2 q \tag{4.10.16}$$

and we denote the Fourier coefficients of s by c_ν. Then

$$c_\nu = d_1 \alpha_\nu - d_2 \beta_\nu \tag{4.10.17}$$

by (4.10.6). It follows from (4.10.11) and (4.10.5) that the only contribution made by w to (4.10.2) and (4.10.14) comes from the Nth Fourier coefficient of w. Hence, by (4.10.13), (4.10.6), (4.10.10), (4.10.16), and (4.10.17), we have

$$\beta = \lambda^{-1} \sum_{-\infty}^{\infty} c_{N-\nu} \gamma_\nu / (\nu - N). \tag{4.10.18}$$

Again, the term with $\nu = N$ is omitted from the summation.

We can now deal with the general condition (4.9.16) in the case $L = 2$. By (4.10.14), the eigenvalues ρ_k of Γ_2 are given by

$$\rho_K^2 = -\alpha^2 - d_1 d_2 |\beta|^2 \tag{4.10.19}$$

$$= -(d_1 d_2/\lambda^2) \left\{ 4N^2 d_1 d_2 \left((2N^2)^{-1} |\gamma_0|^2 - \sum_1^\infty |\gamma_\nu|^2/(\nu^2 - N^2) \right)^2 + \left| \sum_{-\infty}^{\infty} c_{N-\nu} \gamma_\nu / (\nu - N) \right|^2 \right\}, \tag{4.10.20}$$

by (4.10.15) and (4.10.18). The ρ_k are distinct provided that the right-hand side of (4.10.20) is not zero. Now, assuming that $d_1d_2 \neq 0$, we note that the vanishing of the expression $\{\cdots\}$ here would imply some special relationship between the Fourier coefficients c_ν and γ_ν as a whole. The expression is therefore normally non-zero, although of course verification would still be necessary in individual cases. Thus, in the situation $n = 2$ which we are considering in (4.10.4), Theorem 4.9.1 is normally applicable with $L = 2$ provided that $d_1d_2 \neq 0$.

The eigenvalues μ_k of (4.9.18) are given by

$$\mu_k^2 = \rho_k^2 + O(\xi) \tag{4.10.21}$$

and, since μ_k^2 is real as noted in (4.10.3), it follows from (4.9.17) that resonance does or does not occur according as $\rho_k^2 > 0$ or $\rho_k^2 < 0$. We note that, when $d_1d_2 > 0$, (4.10.19) gives $\rho_k^2 < 0$ and there is then non-resonance, in agreement with Theorem 4.9.2.

In the case of the second-order equation (4.1.8), we have the system formulation (4.2.9)–(4.2.11). Then, in (4.10.4), we have

$$d_1 = \tfrac{1}{2}, \qquad d_2 = -\tfrac{1}{2}, \qquad q = p, \qquad r = ip \tag{4.10.22}$$

and, by (4.10.16),

$$s = p.$$

The c_ν in (4.10.17) are now the Fourier coefficients of p, and

$$\gamma_\nu = ic_\nu. \tag{4.10.23}$$

Then (4.10.20) becomes

$$\rho_k^2 = \tfrac{1}{4}\lambda^{-2}\Big\{ -N^2\Big((2N^2)^{-1}|c_0|^2 - \sum_1^\infty |c_\nu|^2/(\nu^2 - N^2)\Big)^2 + \Big|\sum_{-\infty}^\infty c_{N-\nu}c_\nu/(\nu - N)\Big|^2\Big\}, \tag{4.10.24}$$

and the value of λ under consideration in (4.10.5) becomes

$$\lambda = 2/N. \tag{4.10.25}$$

In § 4.3, we dealt with resonance and non-resonance for (4.1.8) in the case $M = 1$, and we considered the two examples (4.3.5) and (4.3.8). As an application of the theory leading to (4.10.24), we can now complete the analysis of these examples by dealing with $M \geqslant 2$.

Example 4.10.1 In (4.10.4), let (4.10.22) hold with

$$p(s) = \sin s. \tag{4.10.26}$$

Then $c_1 = \bar{c}_{-1} = -\tfrac{1}{2}i$ and the other c_ν are zero. We are concerned here with the values (4.10.25) of λ which do not produce resonance when

$M = 1$, that is, with

$$\lambda = 2/N \qquad (N \neq \pm 1). \tag{4.10.27}$$

First, we suppose further that $N \neq \pm 2$. Then we have $c_{N-\nu}c_\nu = 0$ for all ν, and (4.10.24) reduces to

$$\rho_k^2 = -\tfrac{1}{16}(\lambda^2 - 4)^{-2}. \tag{4.10.28}$$

The right-hand side is negative, and therefore there is no resonance when

$$\lambda = 2/N \qquad (N \neq \pm 1, \pm 2). \tag{4.10.29}$$

To obtain the asymptotic form of the solutions of (4.1.8) in this case, we require the eigenvectors v_k which appear in (4.9.17). In (4.10.15) and (4.10.18), we have

$$\alpha = \tfrac{1}{4}(\lambda^2 - 4)^{-1}, \qquad \beta = 0,$$

and the matrix Γ_2 in (4.10.14) is simply

$$\Gamma_2 = \mathrm{i}\tfrac{1}{4}(\lambda^2 - 4)^{-1}\,\mathrm{dg}\,(1, -1).$$

Thus $v_k = e_k$, We can now transform back to the original equation (4.1.8), starting from (4.9.17) (with $L = 2$) and using (4.10.21) and (4.10.28). The final step is (4.1.10) and, taking the first component of Y_0, we obtain solutions y_k $(k = 1, 2)$ of (4.1.8) such that

$$y_k(x) \sim \exp\left\{(-1)^{k-1}\mathrm{i}\left(x + \tfrac{1}{4}(\lambda^2 - 4)^{-1}\int_a^x \xi^2(t)\chi(t)\,\mathrm{d}t\right)\right\}, \tag{4.10.30}$$

where

$$\chi(x) = 1 + O\{\xi(x)\} \quad (x \to \infty).$$

This formula for y_k is valid when λ has the values (4.10.29).

We turn now to the remaining values of λ,

$$\lambda = \pm 1.$$

In (4.10.15), we have $\alpha = -\frac{1}{12}$ as before, but the summation in (4.10.18) now contains one non-zero term arising from either $\nu = 1$ or $\nu = -1$ and, recalling (4.10.23), we obtain $\beta = \frac{1}{4}\mathrm{i}$. Then (4.10.19) gives

$$\rho_k^2 = -\tfrac{1}{144} + \tfrac{1}{64} = 5/2^6 \cdot 3^2.$$

This time $\rho_k^2 > 0$ and therefore resonance does occur. The matrix Γ_2 in (4.10.14) is

$$\Gamma_2 = \frac{\mathrm{i}}{24}\begin{bmatrix} -2 & 3 \\ -3 & 2 \end{bmatrix}.$$

The eigenvectors corresponding to the eigenvalues $\sqrt{5}/24$ and $-\sqrt{5}/24$

are

$$v_1 = \begin{bmatrix} 3\mathrm{i} \\ 2\mathrm{i} + \sqrt{5} \end{bmatrix}, \qquad v_2 = \begin{bmatrix} 2\mathrm{i} + \sqrt{5} \\ 3\mathrm{i} \end{bmatrix}$$

respectively. The transformation from (4.9.17) back to (4.1.8) now gives solutions y_1 and y_2 of (4.1.8) such that

$$y_1(x) = \{3\mathrm{i}\mathrm{e}^{\mathrm{i}x} + (2\mathrm{i} + \sqrt{5})\mathrm{e}^{-\mathrm{i}x} + o(1)\} \exp\left(\frac{\sqrt{5}}{24}\int_a^x \xi^2(t)\chi(t)\,\mathrm{d}t\right),$$

$$y_2(x) = \{(2\mathrm{i} + \sqrt{5})\mathrm{e}^{\mathrm{i}x} + 3\mathrm{i}\mathrm{e}^{-\mathrm{i}x} + o(1)\} \exp\left(-\frac{\sqrt{5}}{24}\int_a^x \xi^2(t)\chi(t)\,\mathrm{d}t\right),$$

where $\chi(x)$ is as in (4.10.30). To obtain real-valued solutions, we can for example take the imaginary parts of y_1 and y_2 and adjust them by a multiple $1/\sqrt{5}$ to arrive at new solutions

$$y_1(x) = \{\sqrt{5}\cos x - \sin x + o(1)\} \exp\left(\frac{\sqrt{5}}{24}\int_a^x \xi^2(t)\chi(t)\,\mathrm{d}t\right),$$

$$y_2(x) = \{\sqrt{5}\cos x + \sin x + o(1)\} \exp\left(-\frac{\sqrt{5}}{24}\int_a^x \xi^2(t)\chi(t)\,\mathrm{d}t\right).$$

In the second example we require the sums of the two infinite series in the following lemma.

Lemma 4.10.2 *For all non-zero* $z \neq 1, 3, 5, \ldots$,

(i) $$\sum_{\tau=0}^{\infty} \frac{1}{(2\tau+1)^2\{(2\tau+1)^2 - z^2\}} = -\frac{\pi^2}{8z^2} + \frac{\pi}{4z^3}\tan\tfrac{1}{2}\pi z.$$

(ii) $$\sum_{\tau=0}^{\infty} \frac{1}{\{(2\tau+1)^2 - z^2\}^2} = \frac{1}{16}\frac{\pi^2}{z^2}\sec^2\tfrac{1}{2}\pi z - \frac{1}{8}\frac{\pi}{z^3}\tan\tfrac{1}{2}\pi z.$$

Proof (i) We use the standard summation

$$\sum_0^{\infty} \frac{1}{(2\tau+1)^2 - z^2} = \frac{1}{4}\frac{\pi}{z}\tan\tfrac{1}{2}\pi z$$

(Copson 1935, p. 157). Letting $z \to 0$, we also have

$$\sum_0^{\infty} \frac{1}{(2\tau+1)^2} = \tfrac{1}{8}\pi^2,$$

and the result follows when we subtract these two series.

(ii) Here we simply differentiate the above standard summation with respect to z and then divide by z. □

Example 4.10.2 Let $p(s)$ be the 2π-periodic extension of the step-function (4.3.8). Then, as in (4.3.9),

$$c_\nu = 0 \ (\nu \text{ even}), \qquad = -2\mathrm{i}/\nu\pi \ (\nu \text{ odd}). \tag{4.10.31}$$

Again, we are concerned with values of λ in (4.10.25) which do not produce resonance when $M = 1$, and these values correspond to even N. Then, writing $N = 2K$, we consider

$$\lambda = 1/K \qquad (K = \pm 1, \pm 2, \ldots). \tag{4.10.32}$$

By (4.10.31) and (4.10.23), (4.10.15) is now

$$\begin{aligned}\alpha &= (4K^2/\pi^2)\sum_0^\infty \frac{1}{(2\tau+1)^2\{(2\tau+1)^2 - 4K^2\}} \\ &= (4K^2/\pi^2)(-\pi^2/32K^2) = -1/8 \end{aligned} \tag{4.10.33}$$

by Lemma 4.10.2(i) with $z = 2K$. Also, by (4.10.31) and (4.10.23), (4.10.18) is now

$$\begin{aligned}\beta &= \mathrm{i}(4K/\pi^2)\sum_{-\infty}^\infty \frac{1}{(2\tau+1)\{(2\tau+1) - 2K\}^2} \\ &= \mathrm{i}(4K/\pi^2)\left(\sum_0^\infty + \sum_{-\infty}^{-1}\right).\end{aligned}$$

In the second summation we write $\tau = -\tau' - 1$ and then combine the two summations to obtain

$$\begin{aligned}\beta &= \mathrm{i}(4K/\pi^2)\sum_0^\infty \left\{\frac{1}{2\tau+1}\left(\frac{1}{\{(2\tau+1)-2K\}^2} - \frac{1}{\{(2\tau+1)+2K\}^2}\right)\right\} \\ &= \mathrm{i}(32K^2/\pi^2)\sum_0^\infty \frac{1}{\{(2\tau+1)^2 - 4K^2\}^2} \\ &= \mathrm{i}(32K^2/\pi^2)(\pi^2/64K^2) = \tfrac{1}{2}\mathrm{i}\end{aligned} \tag{4.10.34}$$

by Lemma 4.10.2(ii) with $z = 2K$.

By (4.10.19), (4.10.33) and (4.10.34), we have

$$\rho_k^2 = -\tfrac{1}{64} + \tfrac{1}{16} = \tfrac{3}{64}.$$

Thus $\rho_k^2 > 0$ and therefore resonance occurs for every K in (4.10.32). The matrix Γ_2 in (4.10.14) is

$$\Gamma_2 = \frac{\mathrm{i}}{8}\begin{bmatrix} -1 & 2 \\ -2 & 1 \end{bmatrix}.$$

The eigenvectors corresponding to the eigenvalues $\sqrt{3}/8$ and $-\sqrt{3}/8$ are

$$v_1 = \begin{bmatrix} 2\mathrm{i} \\ \mathrm{i} + \sqrt{3} \end{bmatrix}, \qquad v_2 = \begin{bmatrix} \mathrm{i} + \sqrt{3} \\ 2\mathrm{i} \end{bmatrix}.$$

As in the previous example, the transformation back to (4.1.8) gives real-valued solutions

$$y_1(x) = \{\sqrt{3}\cos x - \sin x + o(1)\}\exp\left(\frac{\sqrt{3}}{8}\int_a^x \xi^2(t)\chi(t)\,\mathrm{d}t\right),$$

$$y_2(x) = \{\sqrt{3}\cos x + \sin x + o(1)\}\exp\left(-\frac{\sqrt{3}}{8}\int_a^x \xi^2(t)\chi(t)\,\mathrm{d}t\right).$$

These formulae are valid when λ has the values (4.10.32). The formulae are independent of K to the accuracy shown.

We consider one more example which touches on a question raised by Theorem 4.9.2 and not so far mentioned in this section, and that is whether resonance can occur when D has a zero diagonal entry in (4.10.4).

Example 4.10.3 Let $\Lambda_0 = \mathrm{dg}\,(1, -1)$ and let

$$P(s) = \sin s\begin{bmatrix} 1 & 0 \\ 0 & 0 \end{bmatrix}\begin{bmatrix} \mathrm{i} & \mathrm{i} \\ \mathrm{i} & \mathrm{i} \end{bmatrix} = \sin s\begin{bmatrix} \mathrm{i} & \mathrm{i} \\ 0 & 0 \end{bmatrix},$$

in which $d_2 = 0$. Then, with $Y = \begin{bmatrix} u \\ v \end{bmatrix}$, the system (4.6.1) is

$$\left.\begin{aligned} u' &= \mathrm{i}\{1 + \xi(x)\sin\lambda x\}u + \mathrm{i}v\xi(x)\sin\lambda x \\ v' &= -\mathrm{i}v. \end{aligned}\right\} \tag{4.10.35}$$

Let

$$\begin{aligned} f(x) &= x - \int_x^\infty \xi(t)\sin\lambda t\,\mathrm{d}t \\ &= x - \lambda^{-1}\xi(x)\cos\lambda x + O\{\xi'(x)\} \end{aligned} \tag{4.10.36}$$

if, say, $\xi(x)$ is a negative power of x. Then, corresponding to $v = 0$, (4.10.35) certainly has a solution

$$Y_1(x) = \{e_1 + o(1)\}\mathrm{e}^{\mathrm{i}x}.$$

Next, with $v(x) = \mathrm{e}^{-\mathrm{i}x}$, (4.10.35) and (4.10.36) give

$$\begin{aligned} (u\mathrm{e}^{-\mathrm{i}f})' &= \mathrm{i}\xi(x)\mathrm{e}^{-\mathrm{i}f}\mathrm{e}^{-\mathrm{i}x}\sin\lambda x \\ &= \mathrm{i}\mathrm{e}^{-2\mathrm{i}x}\sin\lambda x\sum_{m=1}^{M}(\mathrm{i}\lambda^{-1}\cos\lambda x)^{m-1}\xi^m(x)/(m-1)! \\ &\quad + O\{\xi^{M+1}(x)\} + O\{\xi'(x)\}. \end{aligned}$$

Now the product $\sin\lambda x\,(\cos\lambda x)^{m-1}$ contains terms $\mathrm{e}^{\pm\mathrm{i}m\lambda x}$ which cancel the factor $\mathrm{e}^{-2\mathrm{i}x}$ when $\lambda = \pm 2/m$. This leaves a term (const.)$\xi^m(x)$ which, when integrated, produces an amplitude factor for u. Thus, when (4.1.17) holds, resonance occurs for $\lambda = \pm 2/m$ $(1 \leqslant m \leqslant M)$. We note that the amplitude factor is attached to the component u but not to v in this second solution of (4.10.35).

4.11 PERTURBATIONS WITHOUT PERIODICITY

When the perturbation $R(x)$ in (4.1.2) is no longer restricted to the specific type (4.1.7), it is still possible to develop the ideas of resonance and non-resonance for (4.1.2) although, of course, only to a more limited degree. To discuss what can be done, we begin with the transformation

$$Y(x) = E(x)Z(x), \tag{4.11.1}$$

where

$$E(x) = \exp \mathrm{i}\Lambda_0 x, \tag{4.11.2}$$

and we express (4.1.2) in the form

$$Z' = E^{-1}REZ. \tag{4.11.3}$$

At the next stage, transformations such as (4.2.13) and (4.6.4) are no longer available since they involve ξ, and we proceed with

$$Z = (I + Q)W, \tag{4.11.4}$$

where the $o(1)$ matrix Q is still to be chosen. Then (4.11.3) becomes

$$(I + Q)W' = (E^{-1}RE - Q' + E^{-1}REQ)W. \tag{4.11.5}$$

The simplest choice of Q is the one first introduced in (1.11.7) which, in the present case, we modify slightly to

$$Q(x) = -\int_x^{\infty} E(-t)\tilde{R}(t)E(t)\,\mathrm{d}t, \tag{4.11.6}$$

where

$$\tilde{R} = R - \mathrm{dg}\, R,$$

on the assumption that the infinite integral exists. This choice leads to the following non-resonance result which corresponds to the case $M = 1$ of Theorem 4.6.2. In the statement of the theorem, we also include a condition on $\mathrm{dg}\, R$ which corresponds to (4.6.16) and, although this condition is not strictly necessary as far as the asymptotic form of solutions is concerned, it is a natural one to impose in the context of this chapter.

Theorem 4.11.1 *Let* $\mathrm{dg}\, R$ *be pure imaginary and let the infinite integral in* (4.11.6) *converge. Also let*

$$REQ \in L(a, \infty), \qquad Q\, \mathrm{dg}\, R \in L(a, \infty). \tag{4.11.7}$$

Then (4.1.2) *has solutions* $Y_k(x)$ $(1 \leqslant k \leqslant n)$ *such that*

$$Y_k(x) = \{e_k + o(1)\} \exp\left(\mathrm{i}a_k x + \int_a^x r_{kk}(t)\,\mathrm{d}t\right). \tag{4.11.8}$$

Proof With the choice (4.11.6), the system (4.11.5) reduces to

$$\begin{aligned} W' &= (I+Q)^{-1}(\mathrm{dg}\,R + E^{-1}REQ)W \\ &= (\mathrm{dg}\,R + S)W, \end{aligned}$$

where $S \in L(a, \infty)$ by (4.11.7), and we have expressed $(I+Q)^{-1}$ as $I + O(Q)$. Since $\mathrm{dg}\,R$ is pure imaginary, the W-system satisfies the conditions of Theorem 1.3.1. Then (4.11.8) follows when we transform back to Y via (4.11.4) and (4.11.1). □

We note the form which Theorem 4.11.1 takes in the case of the nth order equation (4.1.12). Here we have

$$\Lambda_0 = \mathrm{dg}\,(a_1, \ldots, a_n) \tag{4.11.9}$$

and, as in (4.1.13),

$$R(x) = \mathrm{i}r(x)D\Omega, \tag{4.11.10}$$

where D is given by (4.1.14) and (4.1.15). The special form of $r(x)$ in (4.1.9) is not now assumed.

Theorem 4.11.2 *Let* $r(x)$ *be real-valued. For unequal* j *and* k *in* $[1, n]$, *let the integrals*

$$q_{jk}(x) = -\int_x^\infty r(t) \exp\{i(a_k - a_j)t\}\,\mathrm{d}t \tag{4.11.11}$$

converge, and let

$$rq_{jk} \in L(a, \infty). \tag{4.11.12}$$

Then (4.1.12) *has solutions*

$$y_k(x) \sim \exp\left\{\mathrm{i}\left(a_k x + d_k \int_a^x r(t)\,\mathrm{d}t\right)\right\}. \tag{4.11.13}$$

Proof By (4.11.2) and (4.11.9), the q_{jk} defined in (4.11.11) are the off-diagonal entries of the matrix Q in (4.11.6). Then again, by (4.11.10), the conditions (4.11.7) reduce to (4.11.12), and the theorem follows from (4.11.8) when we take the first component of Y_k. □

We note that (4.11.13) reduces to (4.5.4) when $r(x)$ has the form (4.1.9). Turning now to the situation where the integral in (4.11.6) does not converge, we can still simplify (4.11.5) somewhat by writing

$$\tilde{R} = R_1 + R_2,$$

where R_1 makes a divergent contribution to the integral and R_2 a convergent contribution. Then we define Q by (4.11.6) but with R_2 in place of $\tilde{R}$. However, this procedure leaves a term

$$\mathrm{dg}\, R + E^{-1}R_1E$$

on the right-hand side of (4.11.5) and, in order to make a further transformation which produces the Levinson form, it is necessary to know more about the detailed form of R_1.

We consider first the case of the second-order equation

$$y''(x) + \{1 + r(x)\}y(x) = 0. \tag{4.11.14}$$

Here $\Lambda_0 = \mathrm{dg}\,(1, -1)$ and, by (4.11.10), the Z-system (4.11.3) becomes

$$Z' = \tfrac{1}{2}\mathrm{i}r\begin{pmatrix} 1 & \mathrm{e}^{-2\mathrm{i}x} \\ -\mathrm{e}^{2\mathrm{i}x} & -1 \end{pmatrix}Z. \tag{4.11.15}$$

The connection between y and Z is given by (4.1.10) and (4.11.1), that is,

$$\begin{pmatrix} y \\ y' \end{pmatrix} = \begin{pmatrix} 1 & 1 \\ \mathrm{i} & -\mathrm{i} \end{pmatrix}\begin{pmatrix} \mathrm{e}^{\mathrm{i}x} & 0 \\ 0 & \mathrm{e}^{-\mathrm{i}x} \end{pmatrix}Z. \tag{4.11.16}$$

Then we have the following theorem which generalizes the resonance situation considered in (4.3.2) and (4.3.3).

Theorem 4.11.3 *Let $r(x)$ be real-valued, and suppose that*

(i) *the infinite integral*

$$q_1(x) = -\int_x^{\infty} r(t)\,\mathrm{d}t \tag{4.11.17}$$

converges;

(ii) *there are real constants α and β, not both zero, such that the integral*

$$q_2(x) = -\int_x^{\infty} r(t)(\alpha \sin 2t + \beta \cos 2t)\,\mathrm{d}t \tag{4.11.18}$$

also converges;

(iii) $$rq_j \in L(a, \infty) \qquad (j = 1, 2); \tag{4.11.19}$$

(iv) *for $a \leq t \leq x < \infty$, there is a constant κ such that*

$$\int_t^x r(s)(\alpha \cos 2s - \beta \sin 2s)\,\mathrm{d}s \tag{4.11.20}$$

is either bounded below by κ or bounded above by κ.

Then (4.11.14) *has solutions* $y_1(x)$ *and* $y_2(x)$ *such that*

$$\left.\begin{aligned} y_1(x) &= \{\cos(x + \tfrac{1}{2}\phi + \tfrac{1}{4}\pi) + o(1)\} \exp \tau(x) \\ y_2(x) &= \{\sin(x + \tfrac{1}{2}\phi + \tfrac{1}{4}\pi) + o(1)\} \exp\{-\tau(x)\}, \end{aligned}\right\} \tag{4.11.21}$$

$$\left.\begin{aligned} y_1'(x) &= \{-\sin(x + \tfrac{1}{2}\phi + \tfrac{1}{4}\pi) + o(1)\} \exp \tau(x) \\ y_2'(x) &= \{\cos(x + \tfrac{1}{2}\phi + \tfrac{1}{4}\pi) + o(1)\} \exp\{-\tau(x)\}, \end{aligned}\right\} \tag{4.11.22}$$

where

$$\tau(x) = \tfrac{1}{2}(\alpha^2 + \beta^2)^{-1/2} \int_a^x r(t)(\alpha \cos 2t - \beta \sin 2t)\, \mathrm{d}t \tag{4.11.23}$$

and

$$\phi = \arg(\alpha + \mathrm{i}\beta). \tag{4.11.24}$$

Proof In (4.11.15), we write

$$\mathrm{e}^{2\mathrm{i}x} = (\alpha + \mathrm{i}\beta)^{-1}\{\alpha \cos 2x - \beta \sin 2x + \mathrm{i}(\alpha \sin 2x + \beta \cos 2x)\}, \tag{4.11.25}$$

and similarly for $\mathrm{e}^{-2\mathrm{i}x}$ in terms of the complex conjugate. Then, in the transformation (4.11.4), we take

$$Q = \tfrac{1}{2}\begin{bmatrix} \mathrm{i}q_1 & (\alpha - \mathrm{i}\beta)^{-1}q_2 \\ (\alpha + \mathrm{i}\beta)^{-1}q_2 & -\mathrm{i}q_1 \end{bmatrix}, \tag{4.11.26}$$

where q_1 and q_2 are defined by (4.11.17) and (4.11.18). By (4.11.19), the transformed system (4.11.5) is

$$W' = \left\{\tfrac{1}{2}r(\alpha \cos 2x - \beta \sin 2x)\begin{pmatrix} 0 & \mathrm{i}(\alpha - \mathrm{i}\beta)^{-1} \\ -\mathrm{i}(\alpha + \mathrm{i}\beta)^{-1} & 0 \end{pmatrix} + S\right\} W,$$

where $S \in L(a, \infty)$. The constant matrix here can be expressed in the diagonal form

$$(\alpha^2 + \beta^2)^{-1/2}\, \mathrm{dg}\,(1, -1)$$

when the final transformation

$$W = \begin{bmatrix} \mathrm{i}\,|\alpha + \mathrm{i}\beta| & \alpha + \mathrm{i}\beta \\ \alpha - \mathrm{i}\beta & \mathrm{i}\,|\alpha + \mathrm{i}\beta| \end{bmatrix} V \tag{4.11.27}$$

is made. The V-system then has the Levinson form, and the condition involving (4.11.20) is the relevant dichotomy condition L in § 1.3. Hence there are solutions $V_k(x)$ $(k = 1, 2)$ such that

$$V_k(x) = \{e_k + o(1)\} \exp\{(-1)^{k-1}\tau(x)\},$$

where $\tau(x)$ is defined by (4.11.23). The transformation back to (4.11.14) via (4.11.4) and (4.11.16) gives solutions y_1 and y_2 such that

$$\begin{bmatrix} y_1 \\ y_1' \end{bmatrix} = \begin{bmatrix} \mathrm{i}\,|\alpha+\mathrm{i}\beta|\,\mathrm{e}^{\mathrm{i}x} + (\alpha-\mathrm{i}\beta)\mathrm{e}^{-\mathrm{i}x} + o(1) \\ -|\alpha+\mathrm{i}\beta|\,\mathrm{e}^{\mathrm{i}x} - \mathrm{i}(\alpha-\mathrm{i}\beta)\mathrm{e}^{-\mathrm{i}x} + o(1) \end{bmatrix} \exp \tau(x)$$

$$\begin{bmatrix} y_2 \\ y_2' \end{bmatrix} = \begin{bmatrix} (\alpha+\mathrm{i}\beta)\mathrm{e}^{\mathrm{i}x} + \mathrm{i}\,|\alpha+\mathrm{i}\beta|\,\mathrm{e}^{-\mathrm{i}x} + o(1) \\ \mathrm{i}(\alpha+\mathrm{i}\beta)\mathrm{e}^{\mathrm{i}x} - |\alpha+\mathrm{i}\beta|\,\mathrm{e}^{-\mathrm{i}x} + o(1) \end{bmatrix} \exp\{-\tau(x)\}.$$

Then, to obtain (4.11.21) and (4.11.22), we adjust y_1 by the constant multiple

$$\tfrac{1}{2}\,|\alpha+\mathrm{i}\beta|^{-1} \exp\{\tfrac{1}{2}\mathrm{i}(\phi - \tfrac{1}{2}\pi)\}$$

and y_2 by

$$\tfrac{1}{2}\,|\alpha+\mathrm{i}\beta|^{-1} \exp\{-\tfrac{1}{2}\mathrm{i}(\phi + \tfrac{1}{2}\pi)\}. \quad \square$$

The formulae (4.11.21) generalize (4.3.3) because, in the case

$$r(x) = \xi(x)p(\lambda x) \qquad (\lambda = 2/N)$$

considered previously, (4.11.18) is satisfied by $\alpha + \mathrm{i}\beta = c_N$, and then (4.11.21)–(4.11.24) reduce to (4.3.3) and (4.3.4). Also, (4.11.17) corresponds to the condition $c_0 = 0$ in (4.3.2), and (4.11.19) corresponds to the case $M = 1$ of (4.1.17).

Both the statement and the proof of Theorem 4.11.3 extend readily to the nth-order equation in (4.1.12) and (4.5.1), and the results similarly generalize those in § 4.5 for the case $c_0 = 0$. The theorem is as follows.

Theorem 4.11.4 *Let $r(x)$ be real-valued. Let J and K be a given pair of unequal integers in $[1, n]$ and suppose that*

(i) the infinite integrals

$$q_{jk}(x) = -\int_x^\infty r(t) \exp\{\mathrm{i}(a_k - a_j)t\}\, \mathrm{d}t \tag{4.11.28}$$

converge for all pairs (j, k) different from (J, K) and (K, J);

(ii) there are constants α and β such that the integral

$$q(x) = -\int_x^\infty r(t)[\alpha \sin\{(a_J - a_K)t\} + \beta \cos\{(a_J - a_K)t\}]\, \mathrm{d}t \tag{4.11.29}$$

converges;

(iii) $rq_{jk} \in L(a, \infty)$, $\quad rq \in L(a, \infty)$;

(iv) for $a \leqslant t \leqslant x < \infty$, there is a constant κ such that

$$\int_t^x r(s)[\alpha \cos\{(a_J - a_K)s\} - \beta \sin\{(a_J - a_K)s\}]\, \mathrm{d}s$$

is either bounded below by κ or bounded above by κ.

Then (4.1.12) *has solutions* $y_k(x)$ $(1 \leqslant k \leqslant n)$ *such that* (4.11.13) *holds for* $k \neq J$ *and* $k \neq K$, *while*

$$\left.\begin{aligned} y_J(x) &= \{\delta e^{i(a_J x + \psi)} - d_K e^{i(a_K x - \psi)} + o(1)\} \exp \tau(x) \\ y_K(x) &= \{d_J e^{i(a_J x + \psi)} - \delta e^{i(a_K x - \psi)} + o(1)\} \exp \{-\tau(x)\}, \end{aligned}\right\} \tag{4.11.30}$$

where

$$\delta = (-d_J d_K)^{1/2},$$

$$\tau(x) = \delta(\alpha^2 + \beta^2)^{-1/2} \int_a^x r(t)[\alpha \cos\{(a_J - a_K)t\} - \beta \sin\{(a_J - a_K)t\}]\, dt,$$

and

$$\psi = \tfrac{1}{2} \arg(\alpha + i\beta) + \tfrac{1}{4}\pi.$$

Proof By (4.11.2), (4.11.9), and (4.11.10), the general (j, k) entry in the coefficient matrix $E^{-1}RE$ in (4.11.3) is

$$id_j r(x) \exp\{i(a_k - a_j)x\}.$$

In the (J, K) and (K, J) entries, we use (4.11.25) but with $2x$ replaced by $(a_J - a_K)x$. Then, corresponding to (4.11.26), we define

$$Q = D(iq_{jk})$$

except that, in place of iq_{JK} and iq_{KJ}, we write

$$(\alpha - i\beta)^{-1}q, \qquad -(\alpha + i\beta)^{-1}q$$

respectively. The transformed system (4.11.5) is then

$$W' = (r[\alpha \cos\{(a_J - a_K)x\} - \beta \sin\{(a_J - a_K)x\}]\Gamma + S)W,$$

where Γ is the constant matrix whose only non-zero entries are

$$id_J(\alpha - i\beta)^{-1}, \qquad id_K(\alpha + i\beta)^{-1}$$

in the (J, K) and (K, J) positions. Then we make a final transformation $W = CV$ similar to (4.11.27) to diagonalize Γ. The columns of C are e_k $(k \neq J,\ k \neq K)$; the Jth column has Jth component $i\delta\,|\alpha + i\beta|$, Kth component $-d_K(\alpha - i\beta)$, and other components zero; the Kth column has Jth component $d_J(\alpha + i\beta)$, Kth component $i\delta\,|\alpha + i\beta|$, and other components zero. The V-system then has the Levinson form in which the diagonal matrix has entries ia_k $(1 \leqslant k \leqslant n)$ except for $k = J$ and $k = K$, where the entries are

$$\pm\delta(\alpha^2 + \beta^2)^{-1/2}.$$

The proof now draws to an end in the same way as in Theorem 4.11.3. □

There is a significant extension of Theorem 4.11.4 to the situation where there are several exceptional values of J and K in (4.11.28). Suppose then that (4.11.28) converges for all (j, k) except (J_ν, K_ν) and (K_ν, J_ν), where $1 \leqslant \nu \leqslant \nu_0 \leqslant \frac{1}{2}n$ and each J_ν does not coincide with any other J_ν or K_ν. Correspondingly, in (4.11.29) and subsequently in the statement of the theorem, we replace q, α, β, δ, ψ, and τ by $q_\nu, \ldots, \tau_\nu$. The only difference in the proof is that the diagonalization of Γ now involves what is effectively the diagonalization of ν_0 separate 2×2 blocks each of which is of the type considered in the above proof. The conclusion is that (4.1.12) has $n - 2\nu_0$ solutions of the form (4.11.13) and $2\nu_0$ solutions of the form (4.11.30) (with $\tau = \tau_\nu$, etc).

4.12 SYSTEMS WITH $\Lambda_0 = 0$

When $\Lambda_0 = 0$ in (4.1.2), the parameter λ which was introduced in (4.1.7) no longer has a purpose, and the system can be written simply as

$$Y'(x) = \xi(x)P(x)Y(x), \qquad (4.12.1)$$

where $P(x)$ has period 2π. Although the question of resonance does not arise for (4.12.1) in the same way as before, the transformation (4.9.2) can still be used to obtain asymptotic results which have significance in a different direction. Lemma 4.9.1 now simplifies as follows.

Lemma 4.12.1 *Let ξ satisfy* (4.1.16) *and* (4.1.17). *Then there are matrices $P_m(x)$ with period 2π and mean-value zero such that the transformation*

$$Y = \{\exp(\xi P_1 + \xi^2 P_2 + \cdots + \xi^M P_M)\}Z \qquad (4.12.2)$$

takes (4.12.1) *into the form*

$$Z' = (\xi\Gamma_1 + \cdots + \xi^M \Gamma_M + S)Z, \qquad (4.12.3)$$

where the Γ_m are constant and $S \in L(a, \infty)$.

Proof In place of (4.9.5), we require that

$$\xi \mathrm{e}^{-\mathscr{P}} P \mathrm{e}^{\mathscr{P}} - \mathrm{e}^{-\mathscr{P}}(\mathrm{e}^{\mathscr{P}})' = \xi\Gamma_1 + \cdots + \xi^M \Gamma_M + O(\xi^{M+1}), \quad (4.12.4)$$

where

$$\mathscr{P} = \xi P_1 + \cdots + \xi^M P_M.$$

Starting with the terms in ξ on each side of (4.12.4), we have to arrange that

$$P - P_1' = \Gamma_1.$$

Hence we define

$$\Gamma_1 = (2\pi)^{-1}\int_{-\pi}^{\pi} P(t)\,\mathrm{d}t \qquad (4.12.5)$$

and then, on integration, the equation

$$P_1' = P - \Gamma_1 \tag{4.12.6}$$

gives a 2π-periodic P_1. The constant of integration can be chosen so that P_1 has mean-value zero. For a general power ξ^m, (4.12.4) gives

$$Q_m - P_m' = \Gamma_m,$$

where Q_m is a 2π-periodic matrix involving $P, P_1, \dots, P_{m-1}$. Hence we define

$$\Gamma_m = (2\pi)^{-1}\int_{-\pi}^{\pi} Q_m(t)\,\mathrm{d}t$$

and then we integrate $P_m' = Q_m - \Gamma_m$ to obtain the P_m and Γ_m inductively. Again, the mean-value of P_m can be made zero by choice of the constant of integration. This completes the proof of the lemma. □

As in the case of Lemma 4.7.1, the Γ_m are increasingly complicated to calculate in terms of P. However, the results for $m = 2$ and $m = 3$ can still be obtained easily and the details are similar to, but simpler than, those in Lemma 4.7.1 partly because (4.12.4) corresponds to $\Lambda_0 = 0$ in (4.7.4). The terms in ξ^2 in (4.12.4) give

$$\begin{aligned}\Gamma_2 &= PP_1 - P_1P + \tfrac{1}{2}(P_1P_1' - P_1'P_1) - P_2' \\ &= \tfrac{1}{2}(PP_1 - P_1P) + \tfrac{1}{2}(\Gamma_1P_1 - P_1\Gamma_1) - P_2',\end{aligned}$$

by (4.12.6). Since P_1 has mean-value zero, this gives

$$\Gamma_2 = (4\pi)^{-1}\int_{-\pi}^{\pi} (PP_1 - P_1P)(t)\,\mathrm{d}t. \tag{4.12.7}$$

Next, the terms in ξ^3 in (4.12.4) give

$$\begin{aligned}\Gamma_3 = P(P_2 + \tfrac{1}{2}P_1^2) &- P_1PP_1 - (P_2 - \tfrac{1}{2}P_1^2)P \\ &+ P_1\{P_2' + \tfrac{1}{2}(P_1^2)'\} + (P_2 - \tfrac{1}{2}P_1^2)P_1' \\ &- \{P_3' + \tfrac{1}{2}(P_1P_2 + P_2P_1)' + \tfrac{1}{6}(P_1^3)'\}.\end{aligned}$$

In the first and third terms on the right, we substitute for P from (4.12.6) to obtain

$$\begin{aligned}\Gamma_3 = \Gamma_1(P_2 + \tfrac{1}{2}P_1^2) &- P_1PP_1 - (P_2 - \tfrac{1}{2}P_1^2)\Gamma_1 \\ &- \{P_3' + \tfrac{1}{2}(P_2P_1 - P_1P_2)' + \tfrac{1}{3}(P_1^3)'\}.\end{aligned}$$

Since P_2 has mean-value zero, we obtain

$$\Gamma_3 = (4\pi)^{-1}\int_{-\pi}^{\pi} (\Gamma_1P_1^2 - 2P_1PP_1 + P_1^2\Gamma_1)(t)\,\mathrm{d}t. \tag{4.12.8}$$

Returning to (4.12.3), we can deduce the asymptotic form of the solutions of (4.12.1) as in the case of (4.9.11). Let Γ_L be the first non-zero Γ_m and assume again that (4.9.16) holds. Then, corresponding to (4.9.17), we obtain solutions $Y_k(x)$ of (4.12.1) such that

$$Y_k(x) = \{v_k + o(1)\} \exp\left(\int_a^x \xi^L(t)\mu_k(t)\,\mathrm{d}t\right), \qquad (4.12.9)$$

where the μ_k are the eigenvalues of (4.9.18).

We note that (4.9.14) is again a consequence of (4.12.4) and therefore, as in Lemma 4.9.2(i),

$$\operatorname{tr} \Gamma_m = 0 \qquad (m \geq 2).$$

Once again, as in Lemma 4.10.1, this property of the Γ_m has a useful implication for μ_k when $n = 2$. Suppose then that $n = 2$, P is real-valued, and $\operatorname{tr} \Gamma_1 = 0$. Then the Γ_m are all real, and (4.9.18) is a real 2×2 matrix with zero trace, leading to the conclusion that

$$\mu_k^2 \text{ is real} \qquad (4.12.10)$$

in these circumstances.

An example which is covered by (4.12.1) is provided by the system

$$\mathrm{d}V(u)/\mathrm{d}u = P\{g(u)\}V(u) \qquad (a \leq u < \infty), \qquad (4.12.11)$$

where $P(s)$ has period 2π in s and $g'(u) \to \infty$ as $u \to \infty$. Thus $P\{g(u)\}$ is rapidly oscillating in terms of u. The change of variable

$$x = g(u) \qquad (4.12.12)$$

transforms (4.12.11) into (4.12.1) with

$$Y(x) = V(u), \qquad \xi(x) = 1/g'(u). \qquad (4.12.13)$$

Then (4.1.16) is satisfied if, for example, g' is differentiable and monotonic and, with the assumption that (4.1.17) holds for some M, the theory leading to (4.12.9) is applicable.

Let us consider further the case where

$$P(s) = \begin{bmatrix} 0 & 1 \\ -c - p(s) & 0 \end{bmatrix} \qquad (4.12.14)$$

with a non-zero constant c and a 2π-periodic $p(s)$ whose mean-value is zero. Then (4.12.11) is equivalent to the second-order equation

$$\mathrm{d}^2y/\mathrm{d}u^2 + [c + p\{g(u)\}]y = 0 \qquad (4.12.15)$$

and (4.12.5) gives

$$\Gamma_1 = \begin{bmatrix} 0 & 1 \\ -c & 0 \end{bmatrix}. \qquad (4.12.16)$$

Further, (4.12.14) and (4.12.6) give

$$P_1(s)=\begin{bmatrix} 0 & 0 \\ -p_1(s) & 0 \end{bmatrix},$$

where $p_1'=p$ and p_1 has mean-value zero. Then, in turn, (4.12.7) gives $\Gamma_2=0$ and, since $P_1^2=0$, (4.12.8) gives

$$\Gamma_3=\begin{bmatrix} 0 & 0 \\ -\beta & 0 \end{bmatrix}, \qquad (4.12.17)$$

where

$$\beta=(2\pi)^{-1}\int_{-\pi}^{\pi} p_1^2(s)\,\mathrm{d}s. \qquad (4.12.18)$$

In (4.12.9) and (4.9.18), we have $L=1$ and

$$\mu_k^2+c\xi^2+\beta\xi^4+O(\xi^5)=0.$$

Hence

$$\mu_k=\pm\mathrm{i}c^{1/2}\xi\{1+\tfrac{1}{2}(\beta/c)\xi^2+O(\xi^3)\}.$$

Then, by (4.12.9) with $L=1$, the transformation (4.12.13) back to the original variables gives solutions $y_k(u)$ of (4.12.15) such that

$$y_k(u)\sim\exp\left\{\pm\mathrm{i}c^{1/2}\left(u+\tfrac{1}{2}(\beta/c)\int_a^u \{g'(t)\}^{-2}\,\mathrm{d}t\right)\right\} \qquad (4.12.19)$$

if $\{g'(u)\}^{-3}\in L(a,\infty)$. Thus, in the case where

$$g(u)=u^{\alpha}, \qquad (4.12.20)$$

(4.12.19) holds when $\alpha>\frac{4}{3}$. We also note that $\operatorname{tr}\Gamma_1=0$ in (4.12.16), and therefore (4.12.10) applies when c and p are real in (4.12.14).

When $p(s)$ has only a finite Fourier series, it is possible to calculate the Γ_m in (4.12.3) beyond Γ_3, although still with some labour. In the case of the equation

$$\mathrm{d}^2y/\mathrm{d}u^2+\{c+\cos g(u)\}y=0, \qquad (4.12.21)$$

for example, we find that

$$\Gamma_4=0, \qquad \Gamma_5=\begin{bmatrix} 0 & 0 \\ -2c & 0 \end{bmatrix}.$$

The asymptotic formula (4.12.19) is then extended to

$$y_k(u)\sim\exp\left[\pm\mathrm{i}c^{1/2}\left\{u+\tfrac{1}{2}(\beta/c)\int_a^u \{g'(t)\}^{-2}\,\mathrm{d}t\right.\right.$$
$$\left.\left.+\left(1-\frac{1}{32c^2}\right)\int_a^u \{g'(t)\}^{-4}\,\mathrm{d}t\right\}\right] \qquad (4.12.22)$$

if $\{g'(u)\}^{-5} \in L(a, \infty)$. This last condition extends the range of α in (4.12.20) to $\alpha > \frac{6}{5}$. Staying with the example (4.12.20), we note that as α decreases to unity further terms appear in (4.12.22), arising from higher powers of ξ. However, when α takes the limiting value unity, the y_k have a different character since then (4.12.21) becomes the Mathieu equation and, for certain ranges of c, the instability domains are applicable.

4.13 SLOWLY DECAYING OSCILLATORY COEFFICIENT

A further application of the system (4.12.1) is provided by the second-order equation

$$y''(x) + \xi(x)p(x)y(x) = 0 \tag{4.13.1}$$

in which, once again, $\xi(x)$ satisfies (4.1.16) and (4.1.17), while $p(x)$ has period 2π and mean-value zero. The condition $M \geqslant 1$ in (4.1.17) implies that the coefficient of $y(x)$ is slowly decaying to zero, and this situation complements the one in § 2.7 where the coefficient $q(x)$ decayed sufficiently rapidly for the solutions of $y'' - qy = 0$ to resemble those of $y'' = 0$ in some sense. Here, as may be expected, the situation is less simple. However, we begin as in the proof of Theorem 2.7.5 by writing (4.13.1) as the system

$$Y_0' = \begin{bmatrix} 0 & 1 \\ -\xi p & 0 \end{bmatrix} Y_0,$$

where Y_0 has components y and y', and then making the transformation

$$Y_0 = \begin{bmatrix} 1 & 0 \\ \phi & 1 \end{bmatrix} Y_1, \tag{4.13.2}$$

where

$$\phi = \int_x^\infty \xi(t)p(t)\,dt. \tag{4.13.3}$$

Then we obtain the system

$$Y_1' = \begin{bmatrix} \phi & 1 \\ -\phi^2 & -\phi \end{bmatrix} Y_1 \tag{4.13.4}$$

as in (2.7.37). Now, in order to apply the theory of § 4.12, we need additional information on the nature of ϕ and therefore additional conditions on ξ. In the following theorem, we give these conditions and we also require the 2π-periodic functions p_1 and p_2 defined by

$$p_1' = p, \qquad p_2' = p_1, \tag{4.13.5}$$

where both have mean-value zero.

Theorem 4.13.1 *Let $\xi(x)$ satisfy* (4.1.16) *and* (4.1.17), *and let $\xi(x)$ be real-valued and positive. In addition, let $\xi'(x)$ be locally absolutely continuous in $[a, \infty)$, and let*

$$\xi'(x) = o\{\xi^2(x)\} \qquad (x \to \infty); \tag{4.13.6}$$

$$\{\xi'(x)/\xi^2(x)\}' \in L(a, \infty); \tag{4.13.7}$$

$$\int_x^\infty |\xi''(t)|\, \mathrm{d}t \in L(a, \infty). \tag{4.13.8}$$

Finally, let $p(x)$ be real-valued with period 2π and mean-value zero. Then (4.13.1) *has solutions y_k $(k = 1, 2)$ such that*

$$y_k \sim \xi^{-1/2} \exp\left((-1)^{k-1} \mathrm{i}c \int_a^x \xi(t)\chi(t)\, \mathrm{d}t\right), \tag{4.13.9}$$

$$y_k' = \xi^{1/2}\{(-1)^{k-1}\,\mathrm{i}c - p_1 + o(1)\} \times \exp\left((-1)^{k-1}\mathrm{i}c \int_a^x \xi(t)\chi(t)\, \mathrm{d}t\right), \tag{4.13.10}$$

where

$$c = \left((2\pi)^{-1} \int_{-\pi}^{\pi} p_1^2(t)\, \mathrm{d}t\right)^{1/2} \tag{4.13.11}$$

and

$$\chi = (1 - \tfrac{1}{4}c^{-2}\xi'^2/\xi^4)^{1/2}\{1 + O(\xi)\}. \tag{4.13.12}$$

Proof We begin by noting a consequence of (4.13.6) and (4.13.8). Let

$$f(x) = \int_x^\infty |\xi''(t)|\, \mathrm{d}t.$$

Then f is non-negative, decreasing, and $L(a, \infty)$. It follows that

$$\tfrac{1}{2}xf(x) \leqslant \int_{\frac{1}{2}x}^x f(t)\, \mathrm{d}t = o(1),$$

and hence

$$\int_x^\infty |\xi''(t)|\, \mathrm{d}t = o(x^{-1}). \tag{4.13.13}$$

Further, by (4.13.6), we have $\xi'/\xi^2 = o(1)$ and integration gives

$$1/\xi(x) = o(x).$$

Then (4.13.13) gives the consequence we require, which is

$$\int_x^\infty |\xi''(t)|\, \mathrm{d}t = o\{\xi(x)\}. \tag{4.13.14}$$

Returning now to (4.13.3) and integrating by parts twice, we have

$$\phi = -\xi p_1 + \phi_1, \tag{4.13.15}$$

where

$$\phi_1 = \xi' p_2 + \int_x^\infty \xi''(t) p_2(t)\, dt \tag{4.13.16}$$

with p_1 and p_2 as in (4.13.5). Then also

$$\phi^2 = \xi^2 p_1^2 + \xi\phi_2,$$

where

$$\phi_2 = -2p_1\phi_1 + \xi^{-1}\phi_1^2.$$

It follows from (4.1.16) and (4.13.8) that $\phi_1 \in L(a, \infty)$, and then from (4.13.6) and (4.13.12) that $\phi_2 \in L(a, \infty)$. When these expressions for ϕ and ϕ^2 are substituted into (4.13.4), the transformation

$$Y_1 = \begin{bmatrix} \xi^{-1/2} & 0 \\ 0 & \xi^{1/2} \end{bmatrix} Y \tag{4.13.17}$$

takes (4.13.4) into

$$Y' = \{\xi P + (\xi'/\xi)C + S\}Y, \tag{4.13.18}$$

where

$$P = \begin{bmatrix} -p_1 & 1 \\ -p_1^2 & p_1 \end{bmatrix}, \tag{4.13.19}$$

$$C = \tfrac{1}{2}\,\mathrm{dg}\,(1, -1), \tag{4.13.20}$$

and $S \in L(a, \infty)$. Despite the appearance of the C and S terms in (4.13.18), we can proceed with the transformation (4.12.2), where the P_m are defined in terms of P just as before. Then (4.13.18) becomes

$$Z' = \{\xi\Gamma_1 + \cdots + \xi^M\Gamma_M + (\xi'/\xi)C + S\}Z, \tag{4.13.21}$$

where we have a new $S \in L(a, \infty)$, and

$$\Gamma_1 = \begin{bmatrix} 0 & 1 \\ -c^2 & 0 \end{bmatrix}, \tag{4.13.22}$$

by (4.12.5), (4.13.19), and (4.13.11). In order to be in a position to apply Theorem 1.6.1, we express Γ_1 in its diagonal form, which is

$$T^{-1}\Gamma_1 T = \mathrm{i}\,\mathrm{dg}\,(c, -c), \tag{4.13.23}$$

where

$$T = \begin{bmatrix} 1 & 1 \\ \mathrm{i}c & -\mathrm{i}c \end{bmatrix} = [v_1\ v_2],$$

say. The transformation $Z = TW$ gives a W-system to which Theorem 1.6.1 is applicable and, transforming back immediately to (4.13.21), we obtain solutions

$$Z_k(x) = \{v_k + o(1)\} \exp\left(\int_a^x \mu_k(t)\, dt\right) \tag{4.13.24}$$

($k = 1, 2$), where the μ_k are the eigenvalues of the matrix

$$\xi\Gamma_1 + \cdots + \xi^M \Gamma_M + (\xi'/\xi)C. \tag{4.13.25}$$

We note that (4.12.10) continues to hold since C has zero trace, and it follows that the necessary dichotomy condition is satisfied. Also, after a short calculation, it follows from (4.13.20), (4.13.22), and (4.13.25) that

$$\mu_k = (-1)^{k-1} \mathrm{i} c \xi \chi, \tag{4.13.26}$$

where χ satisfies (4.13.12). Finally, we transform back to (4.13.1) via (4.12.2), (4.13.17), and (4.13.2), exercising some care however. The complete transformation is

$$\begin{aligned}\begin{bmatrix} y_k \\ y_k' \end{bmatrix} &= \begin{bmatrix} 1 & 0 \\ \phi & 1 \end{bmatrix} \begin{bmatrix} \xi^{-1/2} & 0 \\ 0 & \xi^{1/2} \end{bmatrix} \{I + o(1)\} Z_k \\ &= \begin{bmatrix} \xi^{-1/2} & 0 \\ \phi\xi^{-1/2} & \xi^{1/2} \end{bmatrix} \{I + o(1)\} Z_k. \end{aligned} \tag{4.13.27}$$

By (4.13.15), (4.13.16), (4.13.6), and (4.13.14), we have

$$\phi\xi^{-1/2} = -\xi^{1/2} p_1 + o(\xi^{1/2}).$$

Then (4.13.9) and (4.13.10) follow from (4.13.24) and (4.13.27), and the proof is complete. □

When

$$\xi(x) = x^{-\alpha}, \tag{4.13.28}$$

we have $\xi'(x)/\xi^2(x) = -\alpha x^{\alpha-1}$, and the conditions in the theorem are satisfied if $0 < \alpha < 1$. If, on the other hand, $\alpha > 1$, the example (2.7.45) shows that the asymptotic analysis of (4.13.1) is covered by Theorem 2.7.5. There remains the case $\alpha = 1$, and this case can also be covered by the methods of this section if we replace (4.13.6) by the condition

$$\xi'(x) = -\kappa \xi^2(x)\{1 + o(1)\}, \tag{4.13.29}$$

where κ is a non-zero constant. Clearly (4.13.28) satisfies (4.13.29) when $\alpha = 1$. Conversely, on dividing (4.13.29) by ξ^2 and integrating, we obtain

$$\xi(x) \sim (\kappa x)^{-1} \qquad (x \to \infty). \tag{4.13.30}$$

The condition (4.13.29) represents a situation of Euler type, such as we first met in § 2.6, where the terms involving C and Γ_1 in (4.13.21) are now of comparable size. In the following theorem, we give the result which deals with this remaining case.

Theorem 4.13.2 *Let the conditions of Theorem* 4.13.1 *hold except that* (4.13.6) *is replaced by* (4.13.29). *Let*

$$\kappa^2 \neq 4c^2. \tag{4.13.31}$$

Then (4.13.1) *has solutions* y_k $(k = 1, 2)$ *such that*

$$y_k \sim \xi^{-1/2} \exp\left((-1)^{k-1}\int_a^x \xi(t)\psi(t)\,dt\right),$$

$$y'_k = \xi^{1/2}\{\tfrac{1}{2}\kappa + (-1)^{k-1}(\tfrac{1}{4}\kappa^2 - c^2)^{1/2} - p_1 + o(1)\}$$
$$\times \exp\left((-1)^{k-1}\int_a^x \xi(t)\psi(t)\,dt\right),$$

where

$$\psi = (\tfrac{1}{4}\xi'^2/\xi^4 - c^2)^{1/2}\{1 + O(\xi)\}. \tag{4.13.32}$$

Proof The proof of Theorem 4.13.1 continues to hold down to (4.13.22). By (4.13.7) and (4.13.29), we can write

$$\xi'/\xi^2 = -\kappa + \eta,$$

when $\eta = o(1)$ and $\eta' \in L(a, \infty)$. Further, by (4.13.30), we have $M = 1$ in (4.13.21), and then the Z-system becomes

$$Z' = \{\xi(\Gamma_1 - \kappa C) + \eta\xi C + S\}Z.$$

By (4.13.20) and (4.13.22), the eigenvalues and eigenvectors of $\Gamma_1 - \kappa C$ are

$$(\tfrac{1}{4}\kappa^2 - c^2)^{1/2}, \qquad -(\tfrac{1}{4}\kappa^2 - c^2)^{1/2},$$

$$u_1 = \begin{bmatrix} 1 \\ \tfrac{1}{2}\kappa + (\tfrac{1}{4}\kappa^2 - c^2)^{1/2} \end{bmatrix}, \qquad u_2 = \begin{bmatrix} 1 \\ \tfrac{1}{2}\kappa - (\tfrac{1}{4}\kappa^2 - c^2)^{1/2} \end{bmatrix},$$

where we use (4.13.31). Then, by Theorem 1.6.1 once again, we obtain (4.13.24) but with u_k in place of v_k and, corresponding to (4.13.26),

$$\mu_k = (-1)^{k-1}\xi\psi,$$

where ψ satisfies (4.13.32). The theorem now follows from (4.13.27) as before. □

We note that Theorem 4.13.2 separates into oscillatory and non-oscillatory types according as $\kappa^2 < 4c^2$ or $\kappa^2 > 4c^2$. As an example, we consider the equation

$$y'' + (bx^{-1}\sin x)y = 0, \tag{4.13.33}$$

where b is a constant. Here we have

$$\xi(x) = x^{-1}, \qquad p(x) = b\sin x.$$

Hence (4.13.29) and (4.13.11) give

$$\kappa = 1, \qquad c = |b|/\sqrt{2}.$$

Then, choosing real-valued solutions provided by Theorem 4.13.2, we find that (4.13.33) has re-defined solutions y_1 and y_2 such that

(i) if $|b| > \sqrt{2}$ and $\gamma = (\frac{1}{2}b^2 - \frac{1}{4})^{1/2}$,

$$y_1(x) = x^{1/2}\{\cos(\gamma \log x) + o(1)\},$$
$$y_2(x) = x^{1/2}\{\sin(\gamma \log x) + o(1)\},$$
$$y_1'(x) = x^{-1/2}\{(\tfrac{1}{2} + b\cos x)\cos(\gamma \log x) - \gamma \sin(\gamma \log x) + o(1)\},$$
$$y_2'(x) = x^{-1/2}\{(\tfrac{1}{2} + b\cos x)\sin(\gamma \log x) + \gamma \cos(\gamma \log x) + o(1)\};$$

(ii) if $0 < |b| < \sqrt{2}$ and $\gamma = (\frac{1}{4} - \frac{1}{2}b^2)^{1/2}$,

$$y_1(x) \sim x^{\gamma+\frac{1}{2}}, \qquad y_2(x) \sim x^{-\gamma+\frac{1}{2}},$$
$$y_1'(x) = x^{\gamma-\frac{1}{2}}\{\tfrac{1}{2} + \gamma + b\cos x + o(1)\},$$
$$y_2'(x) = x^{-\gamma-\frac{1}{2}}\{\tfrac{1}{2} - \gamma + b\cos x + o(1)\}.$$

NOTES AND REFERENCES

4.1 The first systematic analysis of resonance and non-resonance was conducted by Atkinson (1954) for the equation (4.1.8) using methods based on the Prüfer transformation. The knowledge that resonance can occur in (4.1.8) goes back to Perron (1930). Other early references are given in the notes on § 4.3.

Systems of the type (4.1.1), with $\Lambda(x) = \mathrm{i}\Lambda_0$ and small $R(x)$, were first identified by Harris and Lutz (1975, § 3) in connection with resonance and non-resonance; see also Samokhin and Fomin (1981) for the case $M = 1$ of (4.1.17). The first extended discussion of perturbations of the type (4.1.3) is due to Cassell (1982). Cassell's scheme is obtained from (4.1.2) by making the initial transformation $Y = (\exp \mathrm{i}\Lambda_0 x)Z$ which, in our account, appears later in (4.9.10).

The equation (4.1.12) and the set σ in (4.1.18) were first formulated by Eastham and El-Sharif (1986). However, the set σ has a longer history in the related theory of parametric resonance, where the $o(1)$ function $\xi(x)$ is replaced by a small parameter ϵ. A comprehensive account of this theory is given by Yakubovich and Starzhinskii (1975, Chapters 5 and 6) and, in this context, the numbers in σ are known as critical frequencies (Yakubovich and Starzhinskii 1975, Chapter 5, § 1); see also Atkinson (1968), where a connection between the two theories is established.

4.2 Although Theorem 4.2.1 can be deduced from the earlier work of Atkinson (1954, § 2.1), Harris and Lutz (1975), Becker (1979), and

Cassell (1982, § 6), it seems not to have been stated explicitly until recently, by Maksudov and Dilbazov (1982, 1983) and Eastham and El-Sharif (1986). The proof can be extended readily to cover non-periodic $p(s)$ which satisfy certain boundedness conditions instead, and this extended result is given by Maksudov and Dilbazov (1982, 1983).

An asymptotic formula such as (4.2.3) is described by Atkinson (1954, Chapter III) as exhibiting phase resonance. Other examples of phase resonance appear in (4.10.30) and (4.12.19).

4.3 The exceptional nature of $\lambda = 2$ in relation to (4.3.5) has long been known in terms of either stability or asymptotic analysis; see Perron (1930), Ascoli (1950), Barbuti (1952*a,b*), Bellman (1953, pp. 136–7), Cesari (1971, § 5.4), Prodi (1951), and Wintner (1946*a*, 1947*a*, 1956). The possibility that there may be an infinity of resonant values was indicated for example by Atkinson (1954, p. 359) and Skriganov (1977) but the first explicit proof appears to be by Eastham and El-Sharif (1986) in terms of (4.3.8) and (4.3.10).

The formulae (4.3.7) are due to Atkinson (1954, Theorem 4.9). Later proofs using the techniques discussed in this book are given by Harris and Lutz (1975, Theorem 2.2), Becker (1979), Cassell (1982, § 7), and Pinto (1985, Theorem 3).

The first application of the asymptotic formulae (4.3.3) to the question of embedded eigenvalues was made by Everitt (1972). For a general survey of this area, we refer to Eastham and Kalf (1982) and, of the specific papers relating to the values (4.3.18), we mention Eastham (1976) and Atkinson and Everitt (1978).

4.4 The construction in Example 4.4.1 is due to Wintner (1946*a*, 1947*a*) in the case $q_0(x) = 1$. The first example of the construction based on (4.4.6) is due to Weidmann (1967, Anhang zu § 4). The general formulae in Example 4.4.2 are due to Eastham and Thompson (1973). We refer again to Eastham and Kalf (1982, Chapter 4) for further details of both Examples 4.4.1 and 4.4.2.

4.5 The results concerning (4.5.1) are due to Eastham and El-Sharif (1986), and we refer to this paper for the details of the proof. The equation (4.5.9) was introduced into resonance theory by Becker (1980).

4.6 The first systematic use of a transformation involving a power series in ξ was developed by Cassell (1981 and, more especially, 1982). Cassell's series was introduced directly and not in the exponential form (4.6.4). The crucial fact about (4.6.4) is the *DU* property of the P_m given in Lemma 4.6.2.

The theory in §§ 4.6–4.8 is due to Eastham and McLeod (1988). As in the method of Cassell (1982, § 4), the series in (4.6.4) can be extended to an infinite series since the convergence of the procedure can be established (Eastham and McLeod 1988, § 6). The condition (4.1.17) is not then required and the method would cover $\xi(x)=(\log x)^{-1}$ for example. A different extension of the theory, in which $P(s)$ is almost periodic, has been pointed out by Cassell (private communication).

4.9 The power series procedure is again closely related to that of Cassell (1982) but, once more, the emphasis here is on the *DU* property. Theorem 4.9.1 and part (i) of Lemma 4.9.2 are due to Cassell (1982, § 3). Theorem 4.9.2 is new here.

4.10 The asymptotic formulae in Example 4.10.1 are due to Atkinson (1954), who made intricate use of the Prüfer transformation. Later proofs using the techniques in this book are given by Harris and Lutz (1975, § 4, 1977, § 4) (in the case $\xi(x)=x^{-1/2}$) and by Cassell (1982, § 7), who established $\lambda=\pm 2$ and $\lambda=\pm 1$ as the only resonant values; see also Skriganov (1977), where the factor $\sqrt{5}/24$ is again identified. The remaining theory in § 4.10, including Examples 4.10.2 and 4.10.3, is new here.

In a simple case like (4.10.26), the asymptotic analysis can be taken further by calculating some of the Γ_m in (4.9.4) beyond Γ_2. M. B. Sadrieh shows that the odd Γ_m are zero and, for example,

$$\Gamma_4=\tfrac{1}{64}\mathrm{i}(-2\lambda^4+35\lambda^2-60)(\lambda^2-4)^{-3}(\lambda^2-1)^{-1}\,\mathrm{dg}\,(1,-1).$$

In the earlier papers of Kelman and Madsen (1968) and Hartman (1968), results of this kind are obtained by means of the Prüfer transformation, but with restrictions on the values taken by λ.

4.11 Theorem 4.11.1 is in essence due to Harris and Lutz (1975, § 3). Theorem 4.11.2 was first proved for second-order equations by Atkinson (1954, Chapter II) using the Prüfer transformation, and it improved the earlier results of Prodi (1951) (see also Bellman (1953, p. 136) and Coppel (1965, pp. 123–4)), which established only the boundedness of solutions. More recent proofs using the methods of this book are due to Wintner (1956), Harris and Lutz (1975, Theorem 2.1), Becker (1979, Example 1), and Read (1980, Corollary 3.4). For the even-order equation (4.5.9), the theorem is established by Becker (1980, Theorem 3).

Several similar but slightly specialized versions of Theorem 4.11.3 have appeared in the literature; see Coppel (1965, pp. 125–8), Harris and Lutz (1975, Theorem 2.2), Becker (1979, Example 2), and Pinto (1985, Theorem 3). The method of proof can be iterated to give a generalization corresponding to the case $M=2$ of (4.1.17), but the detailed conditions on $r(x)$ become cumbersome; see Harris and Lutz (1975, Theorem 3.2) and Pinto (1985, § 3). Theorem 4.11.4 itself is new here but, in the case of the even-order equation (4.5.9) with $a_J=-a_K$, the extended version reduces to the result of Becker (1980, Theorem 2). We refer also to Becker (1979, Theorem 2, 1982) for further discussion and refinements of the asymptotic formulae.

After a change of variable (in effect), Theorems 4.11.3 and 4.11.4 have applications to certain specialized questions concerning the calculation of deficiency indices, and we refer to Atkinson (1976) and Becker (1981*b*) for these results.

4.12 The equation (4.12.21) was introduced by Atkinson (1954, § 3.5), who established (4.12.19) in this case by means of the Prüfer transformation. Otherwise, the results concerning (4.12.11) are new here.

4.13 Theorems 4.13.1 and 4.13.2, and the ideas involved in their proofs, are due to Cassell (1981, 1982). We also refer to these papers for several detailed examples and refinements. The first asymptotic analysis of (4.13.1) is due to Mahoney (1972) who considered $\xi(x)=x^{-\alpha}$ (where $0<\alpha\leq 1$) and used a complicated integral equation technique. Subsequently, Atkinson *et al.* (1977) dealt with $0<\alpha<1$ using the idea of iterating a transformation of the type mentioned in this book in (2.5.23).

The earlier papers of Willett (1969) and Wong (1968, 1969) investigate the oscillatory nature of equations of the type (4.13.1). In terms of oscillation, they identified the critical value $|b|=1/\sqrt{2}$ for the equation (4.13.33).

Bibliography

V. S. Adamchik and A. D. Lizarev (1986). Asymptotic expansions of solutions of differential equations of hyper-Bessel type. *Dokl. Adak. Nauk BSSR* **30,** 393–6, 476. (*Mathematical Reviews* **87j**: 34107)

N. I. Akhiezer and I. M. Glazman (1981). *Theory of linear operators in Hilbert space* (2 vols, 3rd edn). Pitman, London.

A. S. A. Alhammadi (1988). Asymptotic theory for third-order differential equations, *Mathematika,* **35,** 225–32.

L. I. Anikeeva (1976). The asymptotic behaviour of solutions of the equation $y^{(4)} - a(x^{\alpha}y')' + bx^{\beta}y = \lambda y$ as $x \to \infty$, *Vest. Mosk. Univ. Ser. I Mat. Meh.* **31**(6), 44–52.

L. I. Anikeeva (1977). On the deficiency index of a high order differential operator, *Uspehi Mat. Nauk* **32**(1), 179–80.

G. Ascoli (1941). Sulla forma asintotica degli integrali dell'equazione differenziale $y'' + A(x)y = 0$ in un caso notevole di stabilità, *Univ. Nac. Tucuman Revista A* **2,** 131–40 (*Mathematical Reviews* **4** (1943), 43).

G. Ascoli (1947). Sopra un caso di stabilità per l'equazione $y'' + A(x)y = 0$, *Ann. Mat. Pura Appl.* **26,** 199–206.

G. Ascoli (1950). Osservazioni sopra alcune questioni di stabilità II, *Rend. Accad. Lincei* **9,** 210–13.

G. Ascoli (1953*a*). Sul comportamento asintotico delle soluzioni dell'equazione $y'' - (1 + f)y = 0$, *Boll. Un. Mat. Ital.* **8,** 115–23.

G. Ascoli (1953*b*). Sul comportamento asintotico degli integrali dell'equazione $y'' = (1 + f(t))y$ in un caso notevole, *Riv. Mat. Univ. Parma* **4,** 11–29.

F. V. Atkinson (1954). The asymptotic solution of second-order differential equations, *Ann. Mat. Pura Appl.* **37,** 347–78.

F. V. Atkinson (1956/8). Asymptotic formulae for linear oscillations, *Proc. Glasgow Math. Assoc.* **3,** 105–11.

F. V. Atkinson (1964). *Discrete and continuous boundary value problems.* Academic Press, New York and London.

F. V. Atkinson (1968). On asymptotically periodic linear systems, *J. Math. Anal. Appl.* **24,** 646–53.

F. V. Atkinson (1976). Asymptotic integration and the L^2-classification of squares of second-order differential operators, *Quaestiones Mathematicae* **1,** 155–80.

F. V. Atkinson, M. S. P. Eastham, and J. B. McLeod (1977). The limit-point, limit-circle nature of rapidly oscillating potentials, *Proc. Roy. Soc. Edinburgh* **76A,** 183–96.

F. V. Atkinson and W. N. Everitt (1978). Bounds for the point spectrum for a Sturm–Liouville equation, *Proc. Roy. Soc. Edinburgh* **80A,** 57–66.

U. Barbuti (1952*a*). Sulla stabilità delle soluzioni per la equazione $x'' + B(t)x = 0$, *Rend. Accad. Lincei* **12,** 170–75.

U. Barbuti (1952*b*). Sopra un caso di 'risonanza' per la equazione $x'' + A(t)x = 0$, *Boll. Un. Mat. Ital., Ser. 3* **7,** 154–9.

U. Barbuti (1954). Su alcuni teoremi di stabilità, *Ann. Scuola Norm. Super. Pisa* **8,** 81–91.

R. I. Becker (1979). Asymptotic expansions of second order linear differential equations having conditionally integrable coefficients, *J. London Math. Soc., Ser. 2* **20,** 472–84.

R. I. Becker (1980). Asymptotic expansions of formally self-adjoint differential equations with conditionally integrable coefficients, *Q. J. Math. (Oxford), Ser. 2* **31,** 49–64.

R. I. Becker (1981*a*). Limit-circle criteria for 2*n*th order differential operators, *Proc. Edin. Math. Soc.* **24,** 59–72.

R. I. Becker (1981*b*). Limit-circle criteria for fourth-order differential operators with an oscillatory coefficient, *Proc. Edin. Math. Soc.* **24,** 105–17.

R. I. Becker (1982). Asymptotic expansions of formally self-adjoint differential equations with conditionally integrable coefficients II, *Q. J. Math. (Oxford), Ser. 2* **33,** 11–25.

R. Bellman (1947). The boundedness of solutions of linear differential equations, *Duke Math. J.* **14,** 83–97.

R. Bellman (1950). On the asymptotic behaviour of solutions of $u'' - (1 + f(t))u = 0$, *Ann. Mat. Pura Appl.* **31,** 83–91.

R. Bellman (1953). *Stability theory of differential equations.* McGraw-Hill, New York.

M. Ben–Artzi (1980). On the absolute continuity of Schrödinger operators with spherically symmetric, long-range potentials II, *J. Diff. Equations* **38,** 51–60.

M. Ben–Artzi (1983). An application of asymptotic techniques to certain problems of spectral and scattering theory of Stark-like Hamiltonians, *Trans. Am. Math. Soc.* **278,** 817–39.

M. Ben–Artzi and A. Devinatz (1979). Spectral and scattering theory for the adiabatic oscillator and related potentials, *J. Math. Phys.* **20,** 594–607.

G. D. Birkhoff (1909). Singular points of ordinary linear differential equations, *Trans. Am. Math. Soc.* **10,** 436–70.

M. Bôcher (1900). On regular singular points of linear differential equations of the second order whose coefficients are not necessarily analytic, *Trans. Am. Math. Soc.* **1,** 40–52.

B. L. J. Braaksma (1971). Asymptotic analysis of a differential equation of Turrittin, *SIAM J. Math. Anal.* **2,** 1–16.

B. L. J. Braaksma (1984). Asymptotics and deficiency indices for certain pairs of differential operators, *Proc. Roy. Soc. Edinburgh* **97A,** 43–58.

J. S. Cassell (1981). The asymptotic behaviour of a class of linear oscillators, *Q. J. Math. (Oxford), Ser. 2* **32,** 287–302.

J. S. Cassell (1982). The asymptotic integration of some oscillatory differential equations, *Q. J. Math. (Oxford), Ser. 2* **33,** 281–96.

J. S. Cassell (1985). An extension of the Liouville–Green asymptotic formula for oscillatory second-order differential equations, *Proc. Roy. Soc. Edin.* **100A,** 181–90.

J. S. Cassell (1986). Generalised Liouville–Green asymptotic approximations for second-order differential equations, *Proc. Roy. Soc. Edin.* **103A,** 229–39.

J. S. Cassell (1988). Liouville–Green asymptotic theory for second-order equations with complex-valued coefficients, *Q. J. Math. (Oxford), Ser. 2* **39,** 135–49.

L. Cesari (1940). Un nuovo criterio di stabilità per le soluzioni delle equazioni differenziali lineari, *Ann. Sc. Norm. Sup. Pisa* **9,** 163–86.

L. Cesari (1971). *Asymptotic behaviour and stability problems in ordinary differential equations* (3rd edn). Springer-Verlag, Berlin, Heidelberg, New York.

E. A. Coddington and N. Levinson (1955). *Theory of ordinary differential equations.* McGraw-Hill, New York.

W. A. Coppel (1965). *Stability and asymptotic behaviour of differential equations.* Heath, Boston.

E. T. Copson (1935). *An introduction to the theory of functions of a complex variable.* Clarendon Press, Oxford.

A. Devinatz (1965). The asymptotic nature of the solutions of certain systems of differential equations, *Pacific J. Math.* **15,** 75–83.

A. Devinatz (1966). The deficiency index of ordinary self-adjoint differential operators, *Pacific J. Math.* **16,** 243–57.

A. Devinatz (1971). An asymptotic theorem for systems of linear differential equations, *Trans. Am. Math. Soc.* **160,** 353–63.

A. Devinatz (1972*a*). The deficiency index of a certain class of ordinary self-adjoint differential operators, *Adv. Math.* **8,** 434–73.

A. Devinatz (1972*b*). The deficiency index of certain fourth-order ordinary self-adjoint differential operators, *Q. J. Math. (Oxford), Ser. 2* **23,** 267–86.

A. Devinatz (1973). The deficiency index problem for ordinary self-adjoint differential operators, *Bull. Am. Math. Soc.* **79,** 1109–27.

A. Devinatz and J. L. Kaplan (1972). Asymptotic estimates for solutions of linear systems of ordinary differential equations having multiple characteristic roots, *Indiana Univ. Math. J.* **22,** 355–66.

O. Dunkel (1902/3). Regular singular points of a system of homogeneous linear differential equations of the first order, *Proc. Am. Acad. Arts Sci.* **38,** 341–70.

M. S. P. Eastham (1976). On the absence of square-integrable solutions of the Sturm–Liouville equation, *Proc. Dundee Conf. Ordinary and Partial Differential Equations, Lecture Notes in Mathematics* 564, Springer-Verlag, Berlin, Heidelberg, New York.

M. S. P. Eastham (1979). The limit-2*n* case of symmetric differential operators of order 2*n*, *Proc. London Math. Soc., Ser. 2* **38,** 272–94.

M. S. P. Eastham (1981*a*). The deficiency index of a second-order differential system, *J. London Math. Soc., Ser. 2* **23,** 311–20.

M. S. P. Eastham (1981*b*). Asymptotic theory and deficiency indices for differential equations of odd order, *Proc. Roy. Soc. Edin.* **90A,** 263–79.

M. S. P. Eastham (1982*a*). The deficiency index of even-order differential equations, *J. London Math. Soc., Ser. 2* **26,** 113–6.

M. S. P. Eastham (1982*b*). Asymptotic theory for a critical class of fourth-order differential equations, *Proc. Roy. Soc. London* **383A,** 465–76.

M. S. P. Eastham (1982*c*). On the number of solutions of right-definite problems with a convergent Dirichlet integral, *Proc. Roy. Soc. Edin.* **91A,** 347–60.

M. S. P. Eastham (1983*a*). The Liouville–Green asymptotic theory for second-order differential equations: a new approach and some extensions. *Proc. Dundee Conf. Ordinary Differential Equations and Operators, Lecture Notes in Mathematics* 1032. Springer-Verlag, Berlin, Heidelberg, New York.

M. S. P. Eastham (1983*b*). Asymptotic formulae of Liouville–Green type for higher-order differential equations, *J. London Math. Soc., Ser. 2* **28,** 507–18.

M. S. P. Eastham (1984). On the eigenvectors for a class of matrices arising from quasi-derivatives, *Proc. Roy. Soc. Edin.* **97A,** 73–8.

M. S. P. Eastham (1985). The asymptotic solution of linear differential systems, *Mathematika* **32,** 131–8.

M. S. P. Eastham (1986). A repeated transformation in the asymptotic solution of linear differential systems, *Proc. Roy. Soc. Edin.* **102A,** 173–88.

M. S. P. Eastham (1987). Repeated diagonalization and extended Liouville–Green asymptotic formulae, *J. London Math. Soc., Ser. 2* **36,** 115–25.

M. S. P. Eastham (1988). The asymptotic solution of higher-order differential equations with small final coefficient, *Portugaliae Mathematica* **45,** in press.

M. S. P. Eastham and N. S. A. El-Sharif (1986). Resonant perturbations of harmonic oscillation, *J. London Math. Soc., Ser. 2* **34,** 291–9.

M. S. P. Eastham and C. G. M. Grundniewicz (1981*a*). Asymptotic theory and deficiency indices for fourth and higher order self-adjoint equations: A simplified approach. *Proc. Dundee Conf. Ordinary and Partial Differential Equations, Lecture Notes in Mathematics* 846. Springer-Verlag, Berlin, Heidelberg, New York.

M. S. P. Eastham and C. G. M. Grundniewicz (1981*b*). Asymptotic theory and deficiency indices for higher-order self-adjoint differential equations, *J. London Math. Soc., Ser. 2* **24,** 255–71.

M. S. P. Eastham and H. Kalf (1982). *Schrödinger-type operators with continuous spectra.* Research Notes in Mathematics 65 (Pitman, London, 1982).

M. S. P. Eastham and J. B. McLeod (1988). Non-resonance for linear differential systems, *Proc. Roy. Soc. Edin.* **110A,** in press.

M. S. P. Eastham and M. L. Thompson (1973). On the limit-point, limit-circle classification of second-order ordinary differential equations, *Q. J. Math. (Oxford), Ser. 2* **24,** 531–5.

D. E. Edmunds and W. D. Evans (1987). *Spectral theory and differential operators.* Clarendon Press, Oxford.

W. N. Everitt (1959). Integrable-square solutions of ordinary differential equations, *Q. J. Math. (Oxford)* **10,** 145–55.

W. N. Everitt (1962). Integrable-square solutions of ordinary differential equations II. *Q. J. Math. (Oxford)* **13,** 217–20.

W. N. Everitt (1963). Integrable-square solutions of ordinary differential equations III, *Q. J. Math. (Oxford)* **14,** 170–80.

W. N. Everitt (1972*a*). Integrable-square, analytic solutions of odd-order, formally symmetric, ordinary differential equations, *Proc. London Math. Soc., Ser. 3* **25,** 156–82.

W. N. Everitt (1972*b*). On the spectrum of a second-order linear differential equation with a p-integrable coefficient. *Applicable Analysis* **2,** 143–60.

W. N. Everitt (1974). Integral inequalities and the Liouville transformation. *Proc. Dundee Conf. Ordinary and Partial Differential Equations, Lecture Notes in Mathematics* 415. Springer-Verlag, Berlin, Heidelberg, New York.

W. N. Everitt (1982). On the transformation theory of ordinary second-order linear symmetric differential expressions, *Czechoslovak Math. J.* **32,** 275–306.

W. N. Everitt and A. Zettl (1979). Generalised symmetric ordinary differential expressions I: the general theory, *Nieuw Arch. Wisk.* **27,** 363–97.

S. Faedo (1946). Il teorema di Fuchs per le equazione differenziali lineari a coefficienti non analitici e proprietà asintotiche delle soluzioni, *Ann. Mat. Pura Appl.* **25,** 111–33.

S. Faedo (1947). Proprietà asintotiche delle soluzioni dei sistemi differenziali lineari omogenei, *Ann. Mat. Pura Appl.* **26,** 207–15.

M. V. Fedorjuk (1965). The asymptotic behaviour of solutions of ordinary linear differential equations of order *n*, *Soviet Math. Dokl.* **6,** 1528–30.

M. V. Fedorjuk (1966*a*). Asymptotic properties of the solutions of ordinary n-th order linear differential equations, *Differential Equations* **2,** 492–507.

M. V. Fedorjuk (1966*b*). Asymptotic methods in the theory of one-dimensional singular differential operators, *Trans. Moscow Math. Soc.* **15,** 333–86 in the *Am. Math. Soc.,* translation.

M. V. Fedorjuk (1966*c*). Asymptotic behaviour of eigenvalues and eigenfunctions of one-dimensional singular differential operators, *Soviet Math. Dokl.* **7,** 929–32.

G. Fubini (1937). Studi asintotici per alcuni equazione differenziale, *Rend. Reale Accad. Lincei* **26,** 253–9.

A. Ghizzetti (1949). Un teorema sul comportamento asintotico degli integrali delle equazioni differenziali lineari omogenee, *Rend. Mat. Appl.* **8,** 28–42.

R. C. Gilbert (1978). Asymptotic formulas for solutions of a singular linear ordinary differential equation, *Proc. Roy. Soc. Edin.* **81A,** 57–70.

R. C. Gilbert (1979). A class of symmetric ordinary differential operators whose deficiency numbers differ by an integer, *Proc. Roy. Soc. Edin.* **82A,** 117–34.

H. Gingold (1983). On the location of zeros of oscillatory solutions, *Trans. Am. Math. Soc.* **279,** 471–96.

H. Gingold (1985*a*). On the location of zeroes of oscillatory solutions of $y^{(n)} = c(x)y$, *Pacific J. Math.* **119,** 317–36.

H. Gingold (1985*b*). Almost diagonal systems in asymptotic integration, *Proc. Edin. Math. Soc.* **28,** 143–58.

I. M. Glazman (1949). The deficiency indices of differential operators, *Dokl. Akad. Nauk SSSR* **64,** 151–4.

G. Green (1837). On the motion of waves in a variable canal of small depth and width, *Trans. Camb. Phil. Soc.* **6,** 457–62. Also in: *Mathematical papers of the late George Green* (Chelsea, New York, 1970), pp. 223–30.

C. G. M. Grundniewicz (1980*a*). Asymptotic analysis and spectral theory for higher-order differential equations. Unpublished Ph.D. thesis. London University.

C. G. M. Grudniewicz (1980*b*). Asymptotic methods for fourth-order differential equations, *Proc. Roy. Soc. Edin.* **87A,** 53–63.

T. G. Hallam (1970). Asymptotic integration of second order differential equations with integrable coefficients, *SIAM J. Appl. Math.* **19,** 430–9.

T. G. Hallam (1971). Asymptotic integration of a nonhomogeneous differential equation with integrable coefficients, *Czechoslovak Math. J.* **96,** 661–71.

B. J. Harris (1978). A systematic method of estimating gaps in the essential spectrum of self-adjoint operators, *J. London Math. Soc., Ser. 2* **18,** 115–32.

B. J. Harris (1982). Bounds for the point spectra of Sturm–Liouville equations, *J. London Math. Soc., Ser. 2* **25,** 145–61.

B. J. Harris (1983). On the zeros of solutions of differential equations, *J. London Math. Soc., Ser. 2* **27,** 447–64.

B. J. Harris (1984). Limit-circle criteria for second-order differential expressions, *Q. J. Math. (Oxford), Ser. 2* **35,** 415–27.

W. A. Harris and D. A. Lutz (1974). On the asymptotic integration of linear differential systems, *J. Math. Anal. Appl.* **48,** 1–16.

W. A. Harris and D. A. Lutz (1975). Asymptotic integration of adiabatic oscillators, *J. Math. Anal. Appl.* **51,** 76–93.

W. A. Harris and D. A. Lutz (1977). A unified theory of asymptotic integration, *J. Math. Anal. Appl.* **57,** 571–86.

P. Hartman (1948). Unrestricted solution fields of almost separable differential equations, *Trans. Am. Math. Soc.* **63,** 560–80.

P. Hartman (1968). On a class of perturbations of the harmonic oscillator, *Proc. Am. Math. Soc.* **19,** 533–40.

P. Hartman (1982). *Ordinary differential equations* (2nd edn). Birkhauser, Boston.

P. Hartman and A. Wintner (1953). On nonoscillatory linear differential equations, *Am. J. Math.* **75,** 717–30.

P. Hartman and A. Wintner (1955). Asymptotic integrations of linear differential equations, *Am. J. Math.* **77,** 48–86 and 932.

O. Haupt (1942). Über das asymptotische Verhalten der Losungen gewisser linearer gewöhnlicher Differentialgleichungen, *Math. Z.* **48,** 289–92.

E. Hille (1948). Nonoscillation theorems, *Trans. Am. Math. Soc.* **64,** 234–52.

D. B. Hinton (1968). Asymptotic behaviour of solutions of $(ry^{(m)})^{(k)} + qy = 0$, *J. Diff. Equations* **4,** 590–6.

D. B. Hinton (1972). Asymptotic formulae for linear equations, *Glasgow Math. J.* **13,** 147–52.

D. B. Hinton (1984). Asymptotic behaviour of solutions of disconjugate differential equations. *Proc. Alabama Conf. Differential Equations, North-Holland Mathematics Studies* 92. Amsterdam.

J. Horn (1897). Ueber eine Classe linearer Differentialgleichungen, *Math. Ann.* **50,** 525–56.

P. F. Hsieh and Y. Sibuya (1966). On the asymptotic integration of second-order linear ordinary differential equations with polynomial coefficients, *J. Math. Anal. Appl.* **16,** 84–103.

H. Jeffreys (1924). On certain approximate solutions of linear differential equations of the second order, *Proc. London Math. Soc.* **23,** 428–36.

R. M. Kauffman, T. T. Read, and A. Zettl. (1977). *The deficiency index problem for powers of ordinary differential expressions, Lecture Notes in Mathematics* 621. Springer-Verlag, Berlin, Heidelberg, New York.

R. B. Kelman and N. K. Madsen (1968). Stable motions of the linear adiabatic oscillator, *J. Math. Anal. Appl.* **21,** 458–65.

A. Kneser (1896/9). Untersuchung und asymptotische Darstellung der Integrale gewisser Differentialgleichungen bei grossen reellen Werthen des Arguments, *J. reine angew. Math.* **116,** 178–212; **117,** 72–103; **120,** 267–75.

A. Kneser (1897). Einige Sätze über die asymptotische Darstellung von Integralen linearer Differentialgleichungen, *Math. Ann.* **49,** 383–99.

V. I. Kogan and F. S. Rofe-Beketov (1975). On the question of deficiency indices of differential operators with complex coefficients, *Proc. Roy. Soc. Edin.* **72A,** 281–98.

V. I. Kogan and F. S. Rofe-Beketov (1975). On the question of deficiency indices of differential operators with complex coefficients, *Proc. Roy. Soc. Edin.* **72A,** 281–98.

N. Levinson (1946). The asymptotic behaviour of a system of linear differential equations, *Am. J. Math.* **68,** 1–6.

N. Levinson (1948). The asymptotic nature of solutions of linear systems of differential equations, *Duke Math. J.* **15,** 111–26.

J. Liouville (1837). Sur le développment des fonctions ou parties de fonctions en séries . . . , *J. Math. Pures Appl.* **2,** 16–35.

V. Longani (1982). Asymptotic theory and deficiency indices for systems of differential equations. Unpublished Ph.D. thesis. London University.

J. Lutzen (1984). Sturm and Liouville's work on ordinary linear differential equations. The emergence of Sturm–Liouville theory, *Arc. Hist. Exact Sci.* **29,** 309–76.

J. A. M. McHugh (1971). An historical survey of ordinary linear differential equations with a large parameter and turning points, *Arc. Hist. Exact Sci.* **7,** 277–324.

J. J. Mahoney (1972). Asymptotic results for the solutions of a certain differential equation, *J. Australian Math. Soc.* **13,** 147–58.

F. G. Maksudov (1963). The spectrum of singular non-self-adjoint differential operators of order $2n$, *Soviet Math. Doklady* **4,** 1742–5.

F. G. Maksudov and K. A. Dilbazov (1982). An asymptotic formula for solutions of the one-dimensional Schrödinger equation for large values of the argument, *Izv. Akad. Nauk Azerbaidzhan SSSR Ser. Fiz.-Techn. Mat. Nauk* **3,** 8–11. (*Mathematical Reviews* **85a,** 34078).

F. G. Maksudov and K. A. Dilbazov (1983). Some asymptotic formulas for solutions of the one-dimensional Schrödinger equation as $t \to \infty$, *Izv. Akad. Nauk Azerbaidzhan SSSR Ser. Fiz.-Techn. Mat. Nauk* **4,** 19–25. (*Mathematical Reviews* **85k,** 34132)

R. Medina and M. Pinto (1985). Conditionally integrable perturbations of linear differential systems, preprint, Facultad de Ciencias, Universidad de Chile.

J. Miklo (1986). On an asymptotic behaviour of solutions of the differential equation of the fourth-order, *Math. Slovaca* **36,** 69–83. (*Mathematical Reviews* **87j,** 34097).

M. A. Naimark (1968). *Linear differential operators,* Part 2. Harrap, London.

M. Naito (1984). Asymptotic behaviour of solutions of second order differential equations with integrable coefficients, *Trans. Am. Math. Soc.* **282,** 577–88.

F. A. Neimark (1962). On the deficiency index of linear differential operators, *Uspehi Mat. Nauk* **17,** 157–63.

F. W. J. Olver (1961). Error bounds for the Liouville-Green (or WKB) approximation. *Proc. Camb. Phil. Soc.* **57,** 790–810.

F. W. J. Olver (1974). *Asymptotics and special functions.* Academic Press, New York.

S. A. Orlov (1953). On the deficiency index of linear differential operators, *Dokl. Akad. Nauk SSSR* **92,** 483–6.

R. B. Paris (1980). On the asymptotic expansions of solutions of an n-th order linear differential equation, *Proc. Roy. Soc. Edin.* **85A,** 15–57.

R. B. Paris and A. D. Wood (1980). The asymptotic expansion of solutions of the differential equation . . . for large z, *Phil. Trans. Roy. Soc. London* **293A,** 511–33.

R. B. Paris and A. D. Wood (1981). On the L^2 nature of solutions of nth order symmetric differential equations and McLeod's conjecture, *Proc. Roy. Soc. Edin.* **90A,** 209–36.

R. B. Paris and A. D. Wood (1982). On a fourth order symmetric differential operator, *Q. J. Math. (Oxford), Ser. 2* **33,** 97–113.

R. B. Paris and A. D. Wood (1984). Deficiency indices of an odd order differential operator, *Proc. Roy. Soc. Edin.* **97A,** 223–31.

R. B. Paris and A. D. Wood (1985). On the asymptotic expansions of an n-th order linear differential equation with power coefficients, *Proc. Roy. Irish Acad.* **85A,** 201–20.

R. B. Paris and A. D. Wood (1986). *Asymptotics of high order differential equations, Pitman Research Notes in Mathematics* 129. Longman, London.

M. C. Pease (1965). *Methods of linear algebra.* Academic Press, New York.

O. Perron (1913). Über lineare Differentialgleichungen, bei denen die unabhängig Variable reel ist, *J. reine angew. Math.* **142,** 254–70; **143,** 29–50.

O. Perron (1929). Über Stabilität und asymptotisches Verhalten der Integrale von Differentialgleichungenssystemen, *Math. Z.* **29,** 129–60.

O. Perron (1930). Über ein vermeintliches Stabilitätskriterium, *Göttinger Nach.* 28–9.

O. Perron (1960). Über lineare Differentialgleichungen mit reeler unabhänigig Variabler, *Bayer. Akad. Wiss. Math.-Nat. Kl. S.-B.* 1–10.

G. W. Pfeiffer (1972*a*). Asymptotic solutions of $y''' + qy' + ry = 0$, *J. Diff. Equations* **11,** 145–55.

G. W. Pfeiffer (1972*b*). Deficiency indices of a third-order equation, *J. Diff. Equations* **11,** 474–90.

M. Pinto (1985). Asymptotic integration of second-order linear differential systems, *J. Math. Anal. Appl.* **111,** 388–406.

H. Poincaré (1885). Sur les équations lineares aux differentielles ordinaires et aux différences finies, *Am. J. Math.* **7,** 203–58.

G. Prodi (1951). Nuovo criteri di stabilità per l'equazione $y''(x) + A(x)y(x) = 0$, *Rend. Accad. Lincei* **10,** 447–51.

M. Ráb (1958). Über lineare Perturbationen eines Systems von linearen Differentialgleichungen, *Czechoslovak Math. J.* **8,** 222–9.

M. Ráb (1966*a*). Les développements asymptotiques des solutions de l'équation $(py')' + qy = 0$, *Arch. Math.* (*Brno*) **2,** 1–17.

M. Ráb (1966*b*). Note sur les formules asymptotiques pour les solutions d'un système d'équations differentielles linéaires, *Czechoslovak Math. J.* **16,** 127–9.

M. Ráb (1969). Asymptotic formulas for the solutions of linear differential equations $y' = \{A + B(x)\}y$, *Časopis Pěst. Math.* **94,** 78–83, 107.

M. Ráb (1972). Asymptotic expansions of solutions of the equation $(p(x)y')' - q(x)y = 0$ with complex-valued coefficients, *Arch. Math.* (*Brno*) **8,** 1–15.

I. M. Rapoport (1951). On the asymptotic behaviour of solutions of linear differential equations, *Dokl. Akad. Nauk SSSR* **78,** 1097–100.

I. M. Rapoport (1954). *On some asymptotic methods in the theory of differential equations.* Kiev.

C. H. Rasmussen (1979). Oscillation and asymptotic behaviour of systems of ordinary linear differential equations, *Trans. Am. Math. Soc.* **256,** 1–48.

T. T. Read (1979). Higher-order differential equations with small solutions, *J. London Math. Soc., Ser. 2* **19,** 107–22.

T. T. Read (1980). Second-order differential equations with small solutions, *J. Math. Anal. Appl.* **77,** 165–74.

Yu. A. Samokhin and V. N. Fomin (1981). Asymptotic integration of a system of differential equations with oscillatory decreasing coefficients, *Problems in the modern theory of periodic motions No.* 5, 45–50, *Bibl. Trudov Izhevsk Mat. Sem. Izhevsk*, (*Mathematical Reviews* **86b,** 34098).

A. Schlissel (1977). The development of asymptotic solutions of linear ordinary differential equations 1817–1920, *Arch. Hist. Exact Sci.* **16,** 307–78.

W. E. Schreve (1971). Asymptotic behaviour in a second-order linear matrix differential equation, *J. Diff. Equations* **9,** 13–24.

R. A. Shirikyan (1967). Asymptotic methods in the theory of one dimensional singular differential operators of odd order, *Differential Equations* **3,** 1010–17.

J. Šimša (1982). The second order differential equation with an oscillatory coefficient, *Arch. Math.* (*Brno*) **18,** 95–100.

J. Šimša (1984). Asymptotic integration of perturbed linear differential equations under conditions involving ordinary convergence, *SIAM J. Math. Anal.* **15,** 116–23.

J. Šimša (1985). The condition of ordinary integral convergence in the asymptotic theory of linear differential equations with almost constant coefficients, *SIAM J. Math. Anal.* **16,** 757–69.

J. Šimša (1987). Asymptotic integration of a second-order ordinary differential equation, *Proc. Am. Math. Soc.* **101,** 96–100.

M. M. Skriganov (1977). The eigenvalues of the Schrödinger operator situated on the continuous spectrum, *J. Soviet Math.* **8,** 464–7.

D. R. Smith (1986). Liouville–Green approximations via the Riccati transformation, *J. Math. Anal. Appl.* **116,** 147–65.

W. Sternberg (1919). Über die asymptotische Integration von Differentialgleichungen, *Math. Ann.* **81,** 118–86.

Ya. T. Sultanaev (1984). On the deficiency indices and the spectrum of a non-semibounded Sturm-Liouville operator, *Soviet Math. Dokl.* **29,** 652–3.

M. Švec (1958). Sur le comportement asymptotique des intégrales de l'équation différentielle $y^{(4)} + Q(x)y = 0$, *Czechoslovak Math. J.* **8,** 230–45.

M. Švec (1963). Asymptotische Darstellung der Lösungen der Differentialgleichung $y^{(n)} + Q(x)y = 0$, $n = 3, 4$, *Czechoslovak Math. J.* **12,** 572–81.

J. G. Taylor (1978). Error bounds for the Liouville–Green approximation to initial-value problems, *Z. Angew. Math. Mech.* **58,** 529–37.

J. G. Taylor (1982). Improved error bounds for the Liouville–Green (or WKB) approximation, *J. Math. Anal. Appl.* **85,** 79–89.

E. C. Titchmarsh (1962). *Eigenfunction expansions,* Part 1 (2nd edn). Clarendon Press, Oxford.

E. C. Tomastik (1984). Asymptotic behaviour of systems of linear ordinary differential equations, *Proc. Am. Math. Soc.* **90,** 381–90.

W. F. Trench (1976). Asymptotic integration of linear differential equations subject to integral smallness conditions involving ordinary convergence, *SIAM J. Math. Anal.* **7,** 213–21.

W. F. Trench (1984). Asymptotic integration of linear differential equations subject to mild integral conditions, *SIAM J. Math. Anal.* **15,** 932–42.

W. F. Trench (1986). Linear perturbations of a nonoscillatory second-order equation, *Proc. Am. Math. Soc.* **97,** 423–28.

K. Unsworth (1973). Asymptotic expansions and deficiency indices associated with third-order self-adjoint differential operators, *Q. J. Math.* (*Oxford*), *Ser. 2* **24,** 177–88.

H. Väliaho (1986). An elementary approach to the Jordan form of a matrix, *Am. Math. Monthly* **93,** 711–14.

P. W. Walker (1971*a*). Asymptotica of the solutions to $((ry'')' - py')' + qy = \sigma y$, *J. Diff. Equations* **9,** 108–32.

P. W. Walker (1971*b*). Deficiency indices of fourth-order singular differential operators, *J. Diff. Equations* **9,** 133–40.

P. W. Walker (1971*c*). Asymptotics for a class of non-analytic second-order differential equations, *SIAM J. Math. Anal.* **2,** 328–39.

P. W. Walker (1972*a*). Asymptotics for a class of fourth-order differential equations, *J. Diff. Equations* **11,** 321–34.

P. W. Walker (1972*b*). Asymptotics for a class of weighted eigenvalue problems, *Pacific J. Math.* **40,** 501–10.

W. Wasow (1985). *Linear turning point theory.* Springer-Verlag, Berlin, Heidelberg, New York.

J. Weidmann (1967). Zur Spektraltheorie von Sturm–Liouville Operatoren, *Math. Zeit.* **98,** 268–302.

H. Weyl (1910). Über gewöhnliche Differentialgleichungen mit Singularitäten und die zugehörigen Entwicklungen willkürlicher Funktionen, *Math. Ann.* **68,** 220–69.

H. Weyl (1946). Comment on the preceding paper, *Am. J. Math.* **68,** 7–12.

D. Willett (1969). On the oscillatory behaviour of the solutions of second order linear differential equations, *Ann. Polon. Math.* **21,** 175–94.

A. Wintner (1946*a*). The adiabatic linear oscillator, *Am. J. Math.* **68,** 385–97.

A. Wintner (1946*b*). An Abelian lemma concerning asymptotic equilibria, *Am. J. Math.* **68,** 451–4.

A. Wintner (1947*a*). Asymptotic integrations of the adiabatic oscillator, *Am. J. Math.* **69,** 251–72.

A. Wintner (1947*b*). On the normalization of characteristic differentials in continuous spectra, *Phys. Rev.* **72,** 516–7.

A. Wintner (1948). Asymptotic integrations of the adiabatic oscillator in its hyperbolic range, *Duke Math. J.* **15,** 55–67.

A. Wintner (1949*a*). On almost free linear motions, *Am. J. Math.* **71,** 595–602.

A. Wintner (1949*b*). On linear asymptotic equilibria, *Am. J. Math.* **71,** 853–8.

A. Wintner (1954). On a theorem of Bôcher in the theory of ordinary linear differential equations, *Amer. J. Math.* **76,** 183–90.

A. Wintner (1956). Addenda to the paper on Bôcher's Theorem, *Am. J. Math.* **78,** 895–7.

J. S. W. Wong (1968). Second order linear oscillation with integrable coefficients, *Bull. Am. Math. Soc.* **74,** 909–11.

J. S. W. Wong (1969). Oscillation and nonoscillation of solutions of second order linear differential equations with integrable coefficients, *Trans. Am. Math. Soc.* **44,** 197–215.

A. D. Wood (1971). Deficiency indices of some fourth-order differential operators, *J. London Math. Soc., Ser. 2* **3,** 96–100.

H. G. Wood and J. B. Morton (1980). Onsager's pancake approximation for the fluid dynamics of a gas centrifuge, *J. Fluid Mech.* **101,** 1–31.

V. A. Yakubovich and V. M. Starzhinskii (1975). *Linear differential equations with periodic coefficients* (2 vols). Wiley, New York.

Subject index

adjoint operator 152
almost periodic 228
amplitude factor 167, 174, 179, 210
asymptotically constant system 3, 34, 52
asymptotically diagonal system 3, 52, 55, 166

constant coefficient system 1

deficiency indices 152, 155, 164, 229
diagonal form 4, 22, 34, 56, 91, 105
diagonal system 2
dichotomy condition 9, 10, 52

embedded eigenvalues 176, 227
error terms 102, 147
Euler type 74, 93, 125, 131, 134, 138, 224
explicit estimates 15

fourth-order equation 126, 143, 163
fundamental matrix 6

generalized hypergeometric equation 138, 164, 165
generalized hypergeometric function 104, 163

Hamiltonian system 138, 164
Hartman–Wintner theorem 17, 84
higher-order equation 36, 105, 168, 180, 215

Jordan matrix 6, 42, 56, 96

L-diagonal theory 53
Levinson form 3
Levinson theorem 8, 52

Liouville–Green asymptotic formulae 58, 90, 102
Liouville–Green transformation 68, 101, 102
Liouville–Jacobi formula 7, 52, 201
lower triangular form 188

Mathieu equation 221

non-absolute convergence 46, 82

odd-order equation 41, 134, 154, 160, 163, 164

parametric resonance 226
phase resonance 227
Prüfer transformation 101, 226, 228

quasi-derivatives 38, 53, 113, 123, 153

rapidly varying 68
repeated diagonalization 60, 101
resonance 167, 174, 187, 203, 226
resonance set 188, 199

shearing transformation 162
simplifying condition 26, 28, 92, 111, 137
skew-hermitian 185, 200, 203
Sturm–Liouville equation 55, 90, 123, 176

third-order equation 163

variation of parameters 7
von Neumann formula 152, 164

WKBJ theory 101